P9-CCW-641

Optical Mineralogy

Mineral Descriptions

Volume 2

Optical Mineralogy

Mineral Descriptions

Ernest G. Ehlers
Professor Emeritus
Ohio State University

BLACKWELL SCIENTIFIC PUBLICATIONS
Palo Alto Oxford London Edinburgh Boston Melbourne

Editorial Offices

667 Lytton Avenue
Palo Alto, California 94301

Osney Mead
Oxford, OX2 0EL, UK

8 John Street
London WC 1N 2ES, UK

23 Ainslie Place
Edinburgh, EH3 6AJ, UK

52 Beacon Street
Boston, Massachusetts 02108

107 Barry Street
Carlton, Victoria 3053, Australia

Distributors

USA and Canada
Blackwell Scientific Publications
P.O. Box 50009
Palo Alto, California 94303
(415) 965-4081

Australia
Blackwell Scientific Publications (Australia) Pty Ltd
107 Barry Street, Carlton
Victoria 3053

United Kingdom
Blackwell Scientific Publications
Osney Mead
Oxford OX2 0EL
011 44 865-240201

Sponsoring Editor: John H. Staples
Production: Ruth Cottrell Inc.
Text and Cover Design: Gary Head
Manuscript Editor: Andrew Alden
Illustrations: Robin Mouat
Compositor: G&S Typesetters

Library of Congress Cataloging-in-Publication Data

Ehlers, Ernest G.
Optical mineralogy.
Includes index.
Contents: v. 1. Theory and technique—v. 2. Mineral descriptions.
1. Optical mineralogy. I. Title.
QE369.06E37 1987 549′.125 86-14793
ISBN 0-86542-324-5 (v. 2)

British Library Cataloguing in Publication Data

Ehlers, Ernest G.
Optical mineralogy.
Vol. 2: Mineral descriptions
1. Optical mineralogy 2. Microscope and microscopy
I. Title
549′.125 QE369.06
ISBN 0-86542-324-5

For Diane

Contents

Preface

Optical Mineralogy provides in two volumes the necessary information to identify the common rock-forming minerals in both thin section and grain mounts by use of the petrographic (polarizing) microscope.

Volume 1 describes optical theory and techniques. A background in crystallography and mineralogy is assumed, and the order of presentation is arranged to be compatible with a standard undergraduate course in optical crystallography.

Volume 2 presents descriptions of more than 150 common rock-forming minerals. The mineral descriptions often include both colored and black-and-white photomicrographs of thin sections. A departure from standard procedure is the alphabetic arrangement of mineral groups and species in both the descriptions and color plates. In addition to the standard descriptions, information is furnished on indices of refraction, extinction angles, and interference figure orientations of fragments lying on principal cleavage surfaces.

The mineral descriptions are prefaced by a short review chapter on optical techniques as well as a variety of determinative tables and charts. The charts and tables list a large variety of features in order to decrease the amount of time involved in cross-checking.

I wish to thank all of those individuals who aided in the development of this book—the many mineralogists who determined the optical properties of the rock-forming minerals, and the individuals who compiled the fundamental data, such as Larsen and Berman (1934), Winchell and Winchell (1951), Deer, Howie, and Zussman (1962, 1963, 1978, 1982), Troeger (1979), Phillips and Griffen (1981), and Fleischer, Wilcox, and Matzko (1984). Valuable suggestions on organization and content were supplied by Charles Shultz and Richard Dietrich.

I particularly wish to thank Jan Hinsch and Irma Knaak of E. Leitz, Inc. and Charles Shultz of Slippery Rock University, who generously supplied a large number of photomicrographs. Charles Corbato of the Ohio State University aided in some of the crystal drawings. Endless amounts of word processing were performed by my wife, Diane. I am extremely grateful to all. Errors of omission and commission are (unfortunately) my own.

Ernest G. Ehlers

PART 1

Optical Techniques in Brief

Let us begin by committing ourselves to the truth—to see it like it is, and tell it like it is—to find the truth, to speak the truth, and to tell the truth.

RICHARD NIXON, ACCEPTING THE PRESIDENTIAL NOMINATION IN 1968

It is presumed that the reader has previously acquired a background in the theory and techniques of optical mineralogy, either from Volume 1 or from a comparable text. This chapter, therefore provides only a brief review of standard determinative techniques without any attempt to discuss the theoretical background.

THIN SECTIONS AND GRAIN MOUNTS

Most work done with the petrographic microscope involves *thin sections.* Thin sections are slices of rock that have been sawed, ground, and polished to a uniform thickness (commonly 0.03 mm). The slice is glued to a glass slide; a thin cover slip is commonly glued on top of the rock slice. The mounting medium is most commonly epoxy (n = 1.540–1.555), but may be Canada balsam (n = 1.534) or Lakeside 70 (n = 1.54) in older sections.

The major advantage of the thin-section approach is that all of the minerals of the rock can be viewed with retention of the fabric. Each mineral maintains its original habit, relative orientation, and grain size. Easily noted features are mineral associations and incompatibilities, alteration products associated with their hosts, reaction rims, compositional zoning, inclusions, and exsolution lamellae. The general character of the rock, the crystallization sequence, and partial melting relationships can often be deduced from observation of rock texture. In addition, most of the common rock-forming minerals are easily recognized, or can be identified after determination of a few optical parameters. The major disadvantage of the thin-section approach is that it does not normally permit precise optical determination of the indices of refraction; consequently, only approximate rather than specific mineral compositions can normally be determined.

Grain mounts do not permit an examination of the whole rock and its fabric. A single mineral is separated from the rock, ground, pulverized, and sieved (commonly to a size between about 200 and 300 mesh or 50 to 75 μm). The grains are examined during immersion in a liquid whose index of refraction is known. The mount is made by sprinkling a few grains (perhaps 20–30) on a glass slide, covering them with a cover slip, and then immersing them in a drop or two of liquid. The immersion liquid is commonly applied to the edge of the cover slip with a dropping rod, and is drawn to the mineral fragments by capillarity. A set of immersion liquids, ranging in index of refraction from 1.44 to 1.72 (in steps of 0.01), is sufficient for most purposes, but sets of liquids with smaller intervals (0.005) are common. A series of mounts in various immersion liquids permits precise determination of optical parameters, particularly the indices of refraction. An experienced microscopist is capable of precisely identifying almost any of the nonopaque minerals within a half-hour or less.

ISOTROPIC VERSUS ANISOTROPIC BEHAVIOR

Substances that are either noncrystalline (such as volcanic glass) or have isometric symmetry display optically isotropic behavior. A ray of monochromatic light travels at the same velocity when transmitted in any direction through an isotropic material. Furthermore, the polarization characteristics of the transmitted light remain unchanged: a beam of incident plane-polarized light remains plane-polarized during transmission through an isotropic substance.

Anisotropic substances are composed of crystalline materials that have tetragonal, hexagonal, orthorhombic, monoclinic, or triclinic symmetry. A ray of monochromatic light transmitted through an anisotropic material has a variety of velocities that are dependent upon the direction of transmission. Furthermore, the type of polarization of an incident beam of light may change during and after transmission through an anisotropic substance. Thus, an incident beam of plane-polarized light may become doubly polarized during transmission through an anisotropic substance, and circularly polarized upon emergence.

Isotropic and anisotropic materials are easily distinguished by use of the polarizing devices in the petrographic microscope. All light transmitted through a substance on the stage of the microscope has passed through the lower polarizer (lower nicol), which normally converts nonpolarized light into light that is plane-polarized in an east-west (EW) or left-right direction relative to the microscopist (who is facing a direction that is taken as north). The upper polarizer, when inserted into the tube of the microscope, permits transmission of light that is vibrating in the north-south (NS) direction, or has a component in that direction. If an isotropic material (such as glass, liquid, air, or an isometric crystalline material) is viewed between crossed polarizers, no light reaches the observer; the substance appears black and is said to be "in extinction." Extinction is maintained during rotation of the microscope stage. Interference figures cannot be obtained.

Anisotropic materials, when viewed in most orientations between crossed polarizers, normally display interference colors (see the Michel-Lévy interference color chart in Plate 1). Rotating the stage causes the substance to reach extinction at 90° intervals. In either one or two special orientations, anisotropic substances may remain in constant extinction during rotation of the stage, but the anisotropic character can be verified by use of interference figures.

ESTIMATION OF INDICES OF REFRACTION

Light of any wavelength traveling through a vacuum has a constant velocity of 2.997925×10^{10} cm/s. During transmission through a liquid or solid the velocity is decreased—as a function of the substance through which the light is transmitted, the direction of transmission, and the initial wavelength or wavelengths of the incident light. The index of refraction (n) of the substance provides a measure of this change of velocity (V), as

$$n = \frac{V_{\text{air}}}{V_{\text{substance}}}$$

The index (or indices) of refraction of a substance is normally determined not by a direct measurement of the velocity of light transmitted through the substance, but rather by a consideration of refraction (bending) of a light ray as it enters or leaves the substance. This relationship is described by Snell's law (Fig. 1-1), which states that

$$n_1 \sin i = n_2 \sin r$$

A ray of light in substance 1 has an index of refraction n_1. The ray is incident to an interface with substance 2. The angle of incidence i is the angle between the incident ray and a perpen-

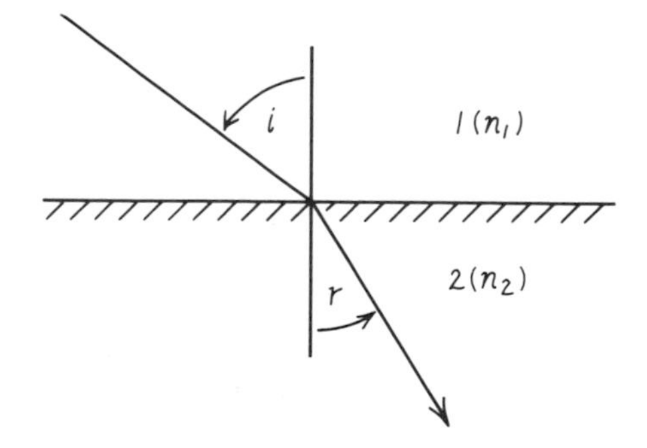

Figure 1-1. A ray of light in material 1 (index of refraction n_1) with an angle of incidence i, is refracted into material 2 (index of refraction n_2) with an angle of refraction r.

dicular to the interface. In substance 2, whose index of refraction is n_2, the ray may continue without deviation or may be refracted into a different direction. The angle of refraction r in substance 2 is the angle between the refracted ray and a perpendicular to the interface. By consideration of angles i and r, and one of the indices of refraction, the other index of refraction is easily calculated. If i and r are identical, refraction does not occur, and velocities in the two materials are identical. If r is less than i, the velocity in substance 2 is less than that in 1. If r is greater than i, the velocity in substance 2 is greater than in 1.

Comparative values of indices of refraction between adjacent substances are obtained by use of the *Becke test* (Fig. 1-2). This test is performed with uncrossed nicols as follows. A mineral grain is brought into focus; the amount of light is then decreased by partial closure of the substage diaphragm. The grain is brought slightly out of focus by increasing the distance between the stage and the lowermost lens (the objective) of the microscope. As this is done, a thin band of white light leaves the interface between the grain and the adjacent substance. The band of light, called the Becke line, moves into the higher index material (due to refraction at the interface). Upon a return to focus, the Becke line returns to the interface. In the event that the dispersion curves of the grain and adjacent material overlap, two colored Becke lines are seen, moving in opposite directions from the interface. If monochromatic, rather than the usual polychromatic light is used, the Becke line is not present when the grain and the adjacent substance have the same index of refraction. When working with grain mounts, the index is determined by making a number of mounts with different immersion liquids until a match is achieved between the index of the grain and the surrounding liquid. A match is indicated with polychromatic light when the two Becke lines are of equal intensity.

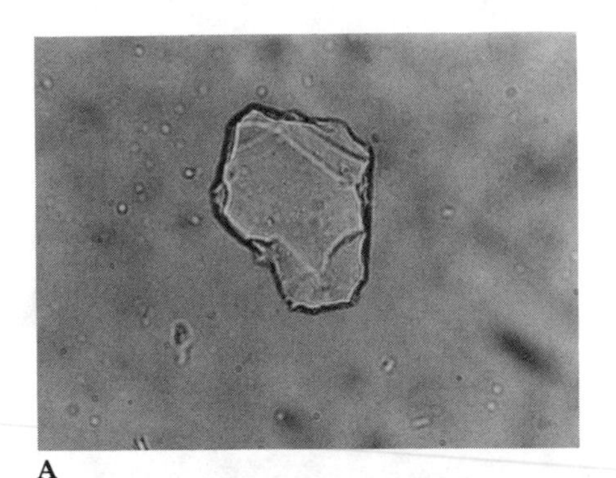

A

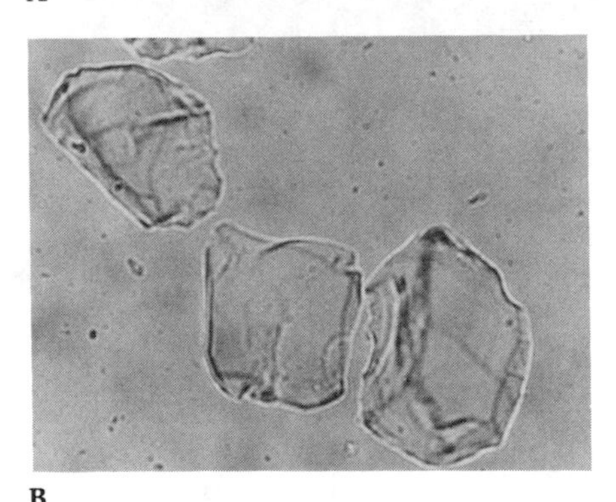

B

Figure 1-2. The Becke line test. Part A shows mineral grains that have a higher index of refraction than the surrounding immersion liquid. Part B shows mineral grains that have a lower index of refraction than the surrounding immersion liquid.

A second measure of the index difference between adjacent materials is provided by *relief*. A grain whose index differs considerably from an adjacent substance has a high relief: light rays crossing the interface are strongly refracted, giving the image of a shadowed grain possessing significant thickness. Relief is decreased when adjacent materials have similar indices of refraction. With a monochromatic light source and identical indices of adjacent materials, the interface becomes invisible.

In an isotropic material, the estimation of index of refraction can be done without any consideration of grain orientation. With an anisotropic substance, however, where transmitted light is commonly doubly polarized and more than one index of refraction is present, orientation must be considered.

In order to estimate the indices of refraction of an anisotropic material, first insert the upper polarizer. The grain of interest will normally show interference colors. Rotate the stage until the grain becomes extinct. Uncross the upper polarizer and estimate the index of refraction. The vibration plane of the index being estimated is parallel to the lower polarizer (normally EW). The vibration plane of a second index of refraction is brought into position for estimation by reinserting the upper polarizer, rotating the grain to the next extinction position (at 90°), and then removing the upper polarizer. If there is a significant index difference between the two indices of refraction, this can be noted by a distinct change of relief during stage rotation.

Minerals are listed according to index of refraction in the tables and charts of Part 2.

BIREFRINGENCE, OPTIC SIGN, AND INTERFERENCE COLORS

The birefringence of a substance, as measured at a particular wavelength of light, is the numerical difference between the maximum and minimum indices of refraction possessed by that substance. Thus, isotropic materials, with a single index of refraction, have zero birefringence, whereas all anisotropic materials are characterized by a specific birefringence.

Anisotropic materials are subdivided into two categories: uniaxial (with tetragonal or hexagonal symmetry) or biaxial (with orthorhombic, monoclinic, or triclinic symmetry). *Uniaxial materials* possess two principal indices of refraction, called ω and ε. The ω index is measured in a vibration plane that is normal to the *c* axis of the substance, whereas the ε index is measured in vibration planes that are parallel to the *c* axis. Intermediate vibration planes yield intermediate indices, called ε' (that is, "epsilon-prime"). If ε exceeds ω, the material is classified as uniaxial positive (+), and if ω exceeds ε the material is classified as uniaxial negative (−). The spatial relationships of ε, ε', and ω can be represented by the optical indicatrix—a three-dimensional geometric representation of the indices of refraction of a substance—the radii of which are proportional in length to the refractive indices, in their direction of vibration, for all waves originating at a point source (Fig. 1-3). The birefringence is the numerical difference between ε and ω.

Biaxial materials possess three principal indices of refraction, called α, β, and γ, with the fixed relationship $\alpha < \beta < \gamma$. Intermediate indices are α' (between α and β) and γ' (between β and

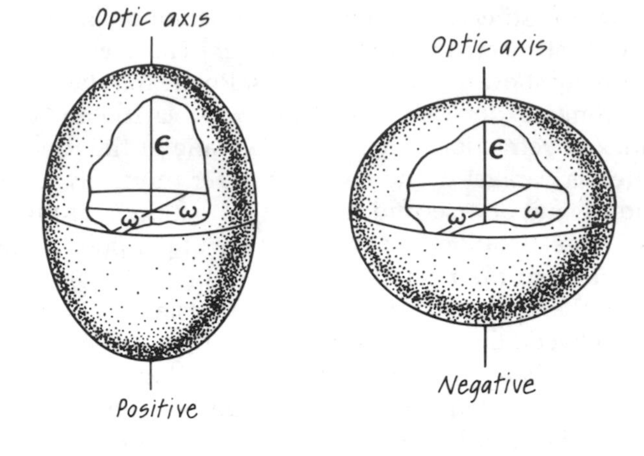

Figure 1-3. The uniaxial indicatrix for optically positive and negative minerals.

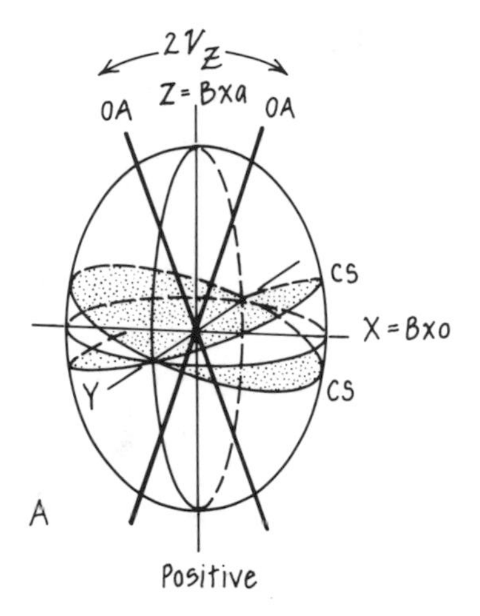

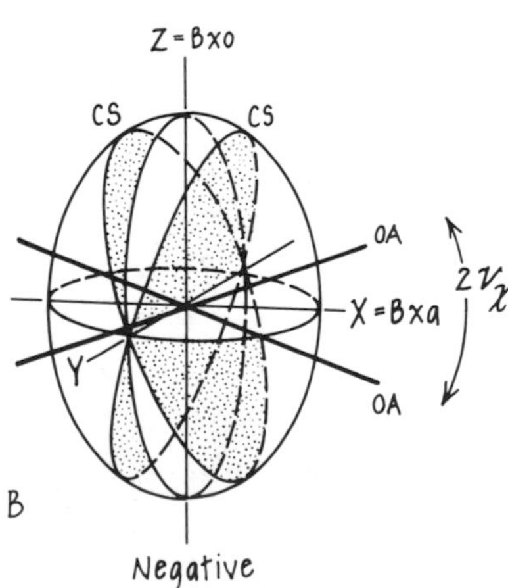

Figure 1-4. The biaxial indicatrix for optically positive minerals (A) and optically negative minerals (B).

γ). If β is closer in value to α than to γ, the substance is classified as biaxial positive (+). If β is closer in value to γ than α, the substance is generally classified as biaxial negative (−).

The biaxial indicatrix (Fig. 1-4) has the form of a triaxial ellipsoid. The directions along which the three principal indices, α, β, and γ, are located are indicated as X, Y, and Z, respectively. A characteristic property of a triaxial ellipsoid is the presence of two circular sections that intersect along the principal intermediate radius, Y. The angle between the two circular sections varies as a function of the value of the index β as compared to α and γ. If the value of β is closer to α than to γ, the circular sections are similarly inclined closer to α than to γ (Fig. 1-4A). Directions normal to the circular sections are called optic axes. There are two optic axes (hence the term "biaxial"). The smaller angle between the optic axes is called the *optic angle* or $2V$. In this case, the (acute) optic angle is bisected by Z, and Z is referred to as the acute bisectrix (Bxa). The obtuse angle between the optic axes is bisected by X, and X is referred to as the obtuse bisectrix (Bxo). If the value of β is closer to γ than to α (Fig. 1-4B), this relationship is reversed, and Bxa is X and Bxo is Z. The birefringence of a biaxial material is the numerical difference between the maximum and minimum indices (γ and α).

The *interference colors* of a mineral viewed between crossed nicols are a consequence of the vibrational constraints imposed by the atomic arrangements of anisotropic materials. Light transmission, in the form of electronic orbital disturbances, is confined to specific directions. As a result, the plane-polarized light from the lower polarizer is converted into doubly polarized (mutually perpendicular) light while transmitted (in most orientations) through anisotropic crystalline materials. The two vibrations travel at different velocities, with the consequence that upon emergence, one vibration has advanced in position relative to the other. The amount of (positional) retardation of the slower vibration to the faster is indicated by the level of the interference colors that are observed.

Notice in the Michel–Lévy interference color chart (Plate 1) that the array of colors is subdivided into orders. Higher order colors are indicative of greater amounts of retardation. In contrast, an absence of retardation is indicated by blackness when a substance is viewed between crossed nicols.

The observed retardation is influenced by three main factors. (1) A high birefringence permits a large velocity difference between the two transmitted vibrations, which results in greater retardation and higher level interference colors. (2) Thicker grains lengthen the transmitted path within the grain, with a greater opportunity for increased retardation. (3) Grain orientations that favor the maximum and minimum difference in vibrational velocities result in greater retardation than other orientations.

Minerals are listed according to birefringence on the Michel–Lévy chart and shown in Charts 1, 2, 3, and 4 in Part 2. Note that a few minerals display *anomalous interference colors* (because of selective absorption of particular wavelengths or crossing of dispersion curves). In these cases, the interference colors will not exactly match those of the Michel–Lévy chart. These minerals are listed in Part 2.

THE USE OF ACCESSORIES IN DETERMINING VIBRATION DIRECTIONS

Commonly used accessories are the first-order red plate, the mica plate (quarter-wave plate), and the quartz wedge. These accessories consist of a cut and polished clear crystalline material (Fig. 1-5) that can be inserted in a diagonally oriented slot located just above the objective lens. The privileged vibration directions of each accessory are parallel to the edges, with the higher index (slower) vibration direction (indicated with an arrow labeled γ) generally parallel to the short edge, and the lower index (faster) direction parallel to the longer edge. It is the purpose of accessory places to furnish information on the nature of the vibration directions within anisotropic grains or interference figures.

The *quartz wedge* is a wedge-shaped slice of quartz; observations are made *during* insertion of the wedge. When the wedge

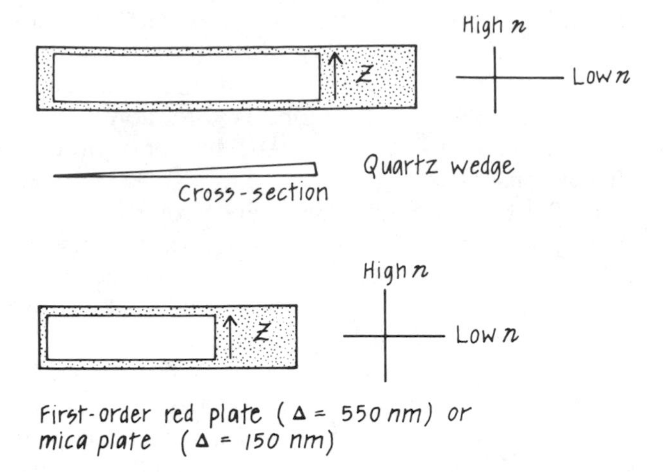

Figure 1-5. Accessories used to characterize vibration directions and determine optic sign.

is inserted (thin end first) between crossed nicols, the effect seen is a rise of interference colors, matching those shown in the Michel–Lévy chart. Alternatively, the first-order red plate is of uniform thickness, and produces a first-order red color across the field, whereas the uniformly thick mica plate yields a gray color.

When using an accessory to determine the nature of the vibration directions of a mineral grain, the first step is to orient the grain so that its vibration directions are located in a diagonal (45°) position, parallel to those of the accessory. This is accomplished by crossing the nicols, and then rotating the stage 45° from an extinction position. Insertion of the accessory causes the combined interference colors of the grain and the accessory to either add to or subtract from each other. An addition of interference colors indicates that the vibration directions of the accessory and the grain are similarly oriented, with the high and low index directions of one parallel to those of the other. A subtraction of colors indicates an opposite relationship.

The quartz wedge is commonly used with grains that show a number of different interference colors. Here, an additive relationship is confirmed as an outward movement of color bands, and a subtractive relationship as an inward movement (Fig. 1-6).

The first-order red and mica plates are commonly used with grains that have a uniform interference color, as is often the case with thin sections. As an example, consider a grain of quartz (uniaxial positive) in thin section having a uniform first-order white interference color. Perhaps the orientation of the ω vibration direction is needed, in order to compare with the indices of an adjacent feldspar. Cross the nicols and rotate the stage until the quartz grain is 45° from extinction. Insert the first-order red plate. If the reaction is additive, the combined color rises above

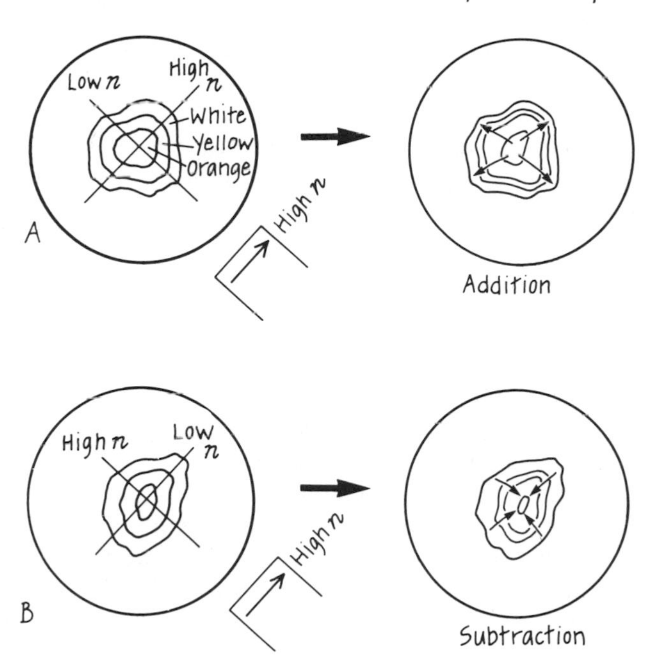

Figure 1-6. The use of the quartz wedge in determining the nature of the vibration directions of a diagonally oriented mineral grain. In part A the vibration directions of the grain and the accessory are similarly oriented (high parallels high, low parallels low). The combined retardations are additive, as seen by an outward flow of colors during insertion of the quartz wedge. In part B the vibration directions of the grain and the accessory are oppositely oriented, and the color bands flow inward during insertion.

that of the plate (white plus red yields blue), and the ω vibration direction is established as being oriented NW–SE. If the reaction is subtractive, the combined color is less than that of the plate (red minus white yields yellow), and the ω vibration direction is established as being NE–SW.

DETERMINATION OF BIREFRINGENCE

The determination of birefringence with the aid of grain mounts consists of making a series of mounts with appropriate immersion liquids, until matches have been achieved between the calibrated liquids and the maximum and minimum indices of refraction of the mineral. The process is facilitated with the use of interference figures, which reveal the optical orientation of the fragment.

The determination of birefringence of a mineral in a thin section is based upon two factors—orientation and thickness. Observe the interference colors of a number of grains (perhaps 10–15) of the mineral whose birefringence is to be determined, and note the highest level color. Next verify the thickness of the thin section; 0.03 mm is standard. Using Plate 1, find the appropriate thickness on the indicated scale; proceed from that point parallel to the long side of the chart, until the maximum interference color of the observed grains is reached. Then move outward from this point along (or almost parallel to) one of the radially oriented lines to the edge of the chart, where the birefringence is indicated. Obtaining an optic normal interference figure from the grain that yielded the maximum interference color will confirm that this grain displays the true birefringence.

COLOR AND PLEOCHROISM

Color is observed by viewing a mineral with uncrossed nicols. Note that certain minerals that appear colorless under the microscope may possess a color in the hand specimen. Colored minerals are listed in Part 2 of this book.

With certain anisotropic minerals, different colors or different intensities of color are associated with particular vibration directions. This phenomenon is known as *pleochroism*. Pleochroism is noted by observing minerals with uncrossed nicols during rotation of the stage. If a mineral changes color or intensity of color it is pleochroic; such minerals are listed in Part 2.

A proper description of pleochroism relates the observed color to the vibration directions of particular indices. Thus, the terms "Abs. $\omega > \varepsilon$" or "Abs. $O > E$" mean that the light absorption is greater (and the mineral has a darker color) in the vibration plane in which ω (the Ordinary vibration) is measured, rather than the direction in which ε (the Extraordinary vibration) is measured. Where the colors are different, a mineral might be described as "X = yellow, Y = light green, and Z = dark green." Such relationships are determined with the aid of interference figures and accessories.

INTERFERENCE FIGURES

Any anisotropic fragment or mineral grain in thin section, if of sufficient size, should produce an interference figure. Interference figures are obtained with the following procedure:

1. Find a grain of moderate size, and position it at the cross-hair intersection.
2. Insert the high-power objective and rotate the stage. If the grain does not remain at the cross-hair intersection, center the objective.
3. Open the diaphragm in the substage.
4. Insert the condensing (converging) lens (in the substage).

5. Insert the upper polarizer.
6. Insert the Bertrand-Amici lens in the upper end of the microscope tube. An interference figure should be visible. If the figure is diffuse, remove the Bertrand-Amici lens and the ocular (field lens). A smaller and sharper figure can be observed either by looking directly down the microscope tube or by looking through a pinhole lens (which is inserted in place of the ocular).

Clear and distinct interference figures are generally obtained from grains that are of large size, have no immediately adjacent neighbors, are unaltered, unzoned, and untwinned.

UNIAXIAL INTERFERENCE FIGURES

Light rays transmitted within uniaxial materials parallel to the *c* axis exhibit isotropic behavior, that is, the polarization characteristics of the ray remain unchanged. Rays traveling in any other direction within uniaxial materials may change their polarization characteristics (plane-polarized light, for example, might be converted into doubly polarized light). Because of the unique behavior of light transmitted along the *c* axis direction of uniaxial crystalline material, this direction has been called the *optic axis.* As there is only one such direction in crystalline materials of tetragonal and hexagonal symmetry, such materials are referred to as uniaxial. Materials of orthorhombic, monoclinic, or triclinic symmetry have two such directions and are classified as biaxial.

The type of interference figure produced by uniaxial crystalline materials is a function of the orientation of the optic axis relative to the microscope stage.

Optic Axis Figure

A thin section cut normal to the *c* axis of a uniaxial material produces a uniaxial optic axis (O.A.) figure; the microscopist's line of sight corresponds to the *c* axis. A uniaxial grain in this orientation is easily recognized, as it exhibits isotropic behavior when observed with crossed nicols; the grain remains in extinction during rotation of the stage, and permits an estimation of only the ω index of refraction.

The uniaxial optic axis figure (Fig. 1-7) is characterized by a black cross, which may be surrounded by concentric color bands. The arms of the cross are called *isogyres.* The intersection point

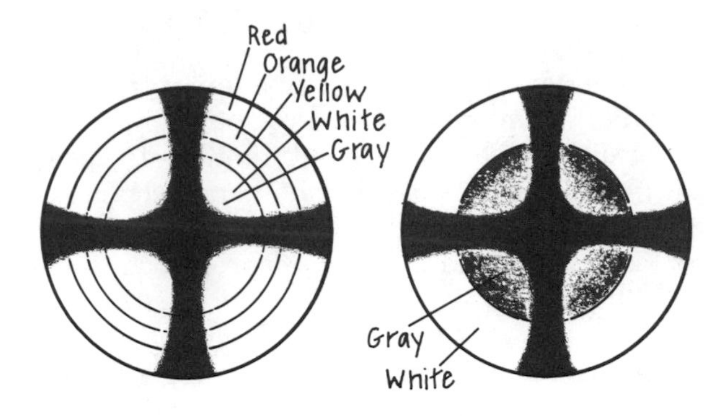

Figure 1-7. Uniaxial optic axis interference figures showing different numbers of isochromatic curves.

of the isogyres marks the position of the optic axis (here vertical), which is called a *melatope*. The isogyres divide the interference figure into four quadrants that can be distinguished directionally as NE, NW, SE, and SW. The four quadrants may be gray or white, or may contain color bands (*isochromes* or *isochromatic curves*) that are concentrically arranged about the melatope. The isochromes display the same sequence of interference colors as shown in the Michel–Lévy chart, with the lowest color adjacent to the melatope, and increasing outwardly. The number of isochromes present in a particular interference figure is variable, with greater numbers favored by increasing grain thickness and higher birefringence. In some few cases, the colors of the isochromes are partially masked by intense coloration of the mineral.

Rotation of the stage does not change the orientation or position of any part of the uniaxial optic axis figure.

Inclined Optic Axis Figure

This type of figure is obtained when the optic axis is neither perpendicular nor parallel to the stage of the microscope. The mineral grain, when viewed with crossed nicols, exhibits interference colors in most orientations; the interference colors are, however, not at the maximum level (consistent with grain thickness and birefringence), as vibration directions parallel to the stage are ε' and ω, rather than ε and ω.

This interference figure differs from the uniaxial optic axis interference figure in that the melatope does not coincide with the cross-hair intersection (Fig. 1-8). The melatope, marking the position of the optic axis, may be within or outside of the field of view, according to the angle of inclination of the optic axis. The isogyres are oriented NS and EW, and the isochromes are concentrically arranged about the melatope.

Figure 1-8. A uniaxial inclined optic axis figure.

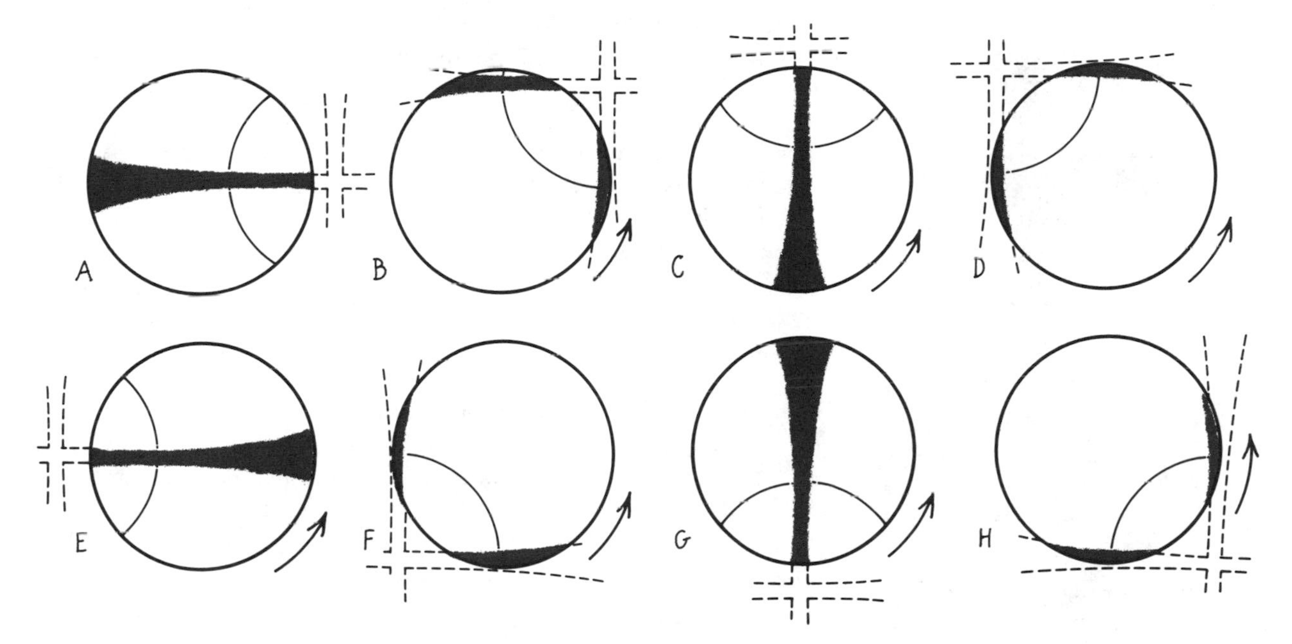

Figure 1-9. A uniaxial inclined optic axis interference figure during counterclockwise rotation of the stage. The isogyres maintain their NS and EW orientation, and the melatope (out of the field of view) moves in a circle about the center of the field.

Rotation of the microscope stage causes the melatope to move about the center of the field of view (Fig. 1-9). The isochromes and isogyres retain their positions relative to the melatope, and do not rotate.

Estimation of indices requires the mineral grain to be oriented first. If ω is to be estimated, rotate the stage until the melatope lies along the NS cross hair. The index ω is estimated with uncrossed nicols. If an estimation of ε' is wanted, rotate the microscope stage until the melatope lies along the EW cross hair. The index ε' is estimated with uncrossed nicols. The procedure assumes that the permissible vibration direction of the lower polarizer is EW, as is standard with all modern instruments.

Optic Normal Figure

The optic normal (O.N.) interference figure, also called a *flash figure,* is produced from grains whose *c* axis lies parallel to the microscope stage. The grain is viewed normal to the optic axis. Such grains are recognized fairly easily, as they possess the maximum level of interference color consistent with their thickness; the vibration directions ε and ω are parallel to the microscope stage.

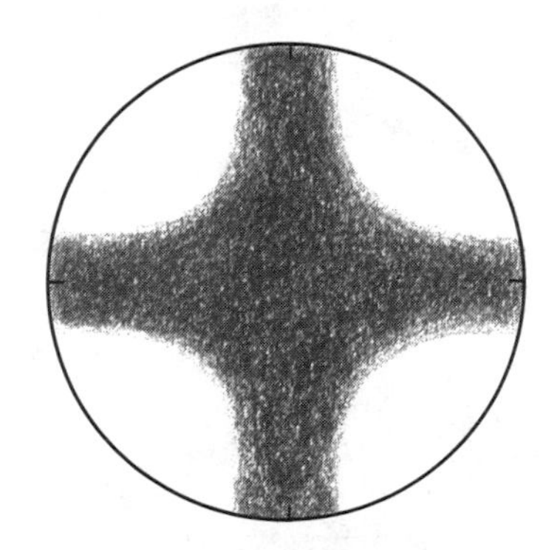

Figure 1-10. Appearance of the optic normal interference figure when the ε and ω vibration directions are parallel to the NS and EW directions.

The interference figure (Fig. 1-10) changes as a result of rotation of the microscope stage. When the ε or ω vibration directions are parallel to the NS or EW, the figure appears as an extremely diffuse black cross whose center corresponds to the cross-hair intersection. Rotation of the stage by a few degrees in

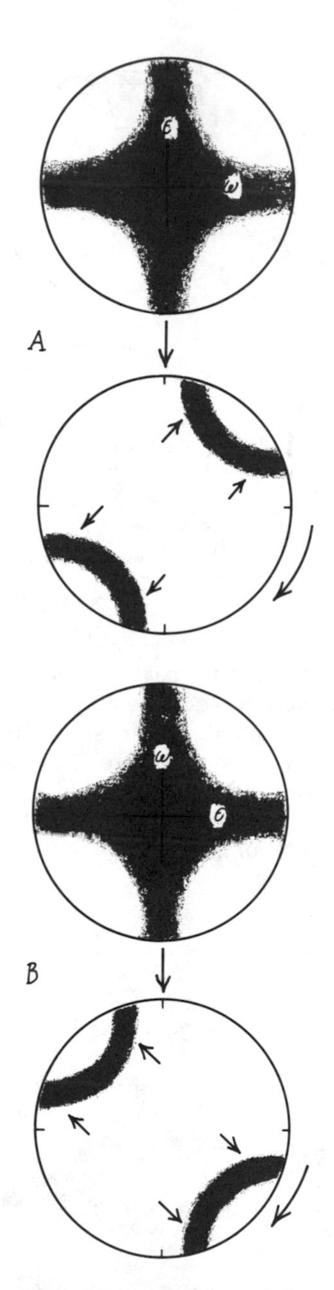

Figure 1-11. Determination of the ε and ω vibration directions in a uniaxial optic normal figure. (A) The ε vibration direction, originally NS, is rotated into the NE and SW quadrants. (B) The ε vibration direction, originally EW, is rotated into the NW and SE quadrants.

either direction causes the cross to break into two portions that rapidly leave the field in a diagonal direction; a stage rotation of 8° or less is sufficient to cause the isogyres to leave the field of view.

The isogyres leave the field in the quadrants within which the optic axis (the ε vibration direction) is located (Fig. 1-11). Hence, if during the cross position, the optic axis had been oriented NS, a clockwise stage rotation would cause the isogyres to leave the field in the NE and SW quadrants. Alternatively, if ε had been oriented EW in the cross position, a clockwise stage rotation would cause the isogyres to leave the field in the NW and SE quadrants.

The isochromes of an optic normal figure are generally quite diffuse and difficult to interpret. However, in those cases where the colors are distinct, those tending toward the lower orders are found in the two quadrants within which the isogyres have left the field of view.

DETERMINATION OF OPTIC SIGN OF UNIAXIAL MATERIALS

The determination of optic sign is generally carried out with the aid of interference figures and accessories (Fig. 1-12). Although any interference figure can theoretically provide a sign determination, the most easily interpreted is the optic axis type (see color photos 1 and 2 in the color insert). Consequently, grains with the lowest level interference colors should be examined first.

The sign reaction with an optic normal interference figure is often difficult to interpret. The best procedure is to use the figure to establish the grain orientation, and then work with the grain itself. With an optic normal figure, rotate the stage until the isogyres leave the field. The quadrants in which the isogyres have left the field locate the ε vibration direction. Remove the Bertrand-Amici and condenser lenses and view the grain between crossed polars. After rotating the grain into a diagonal orientation, use of the appropriate accessory will indicate whether the ε vibration direction corresponds to a higher or lower index of refraction than ω, thus establishing the optic sign.

BIAXIAL INTERFERENCE FIGURES

The interference colors shown by biaxial mineral grains are dependent upon birefringence, thickness, and grain orientation.

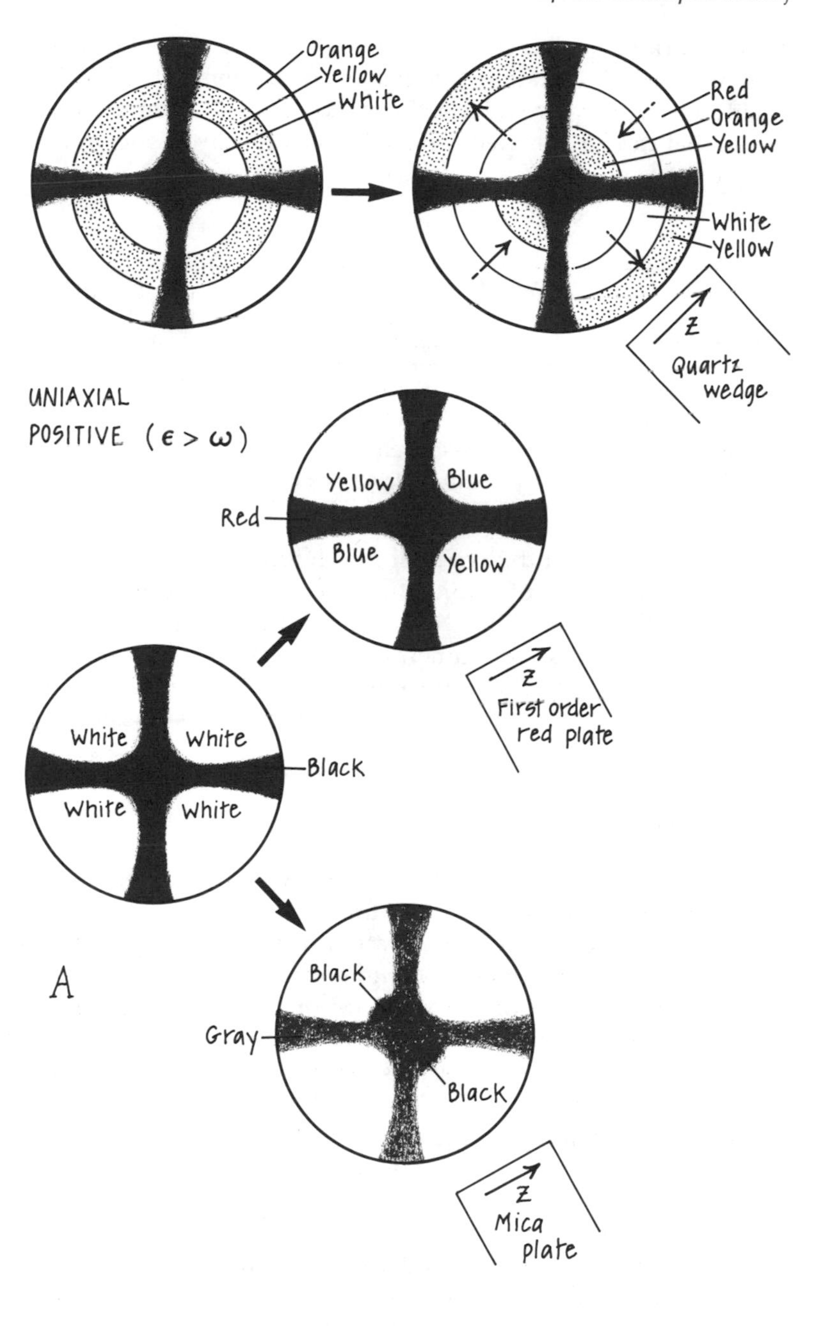

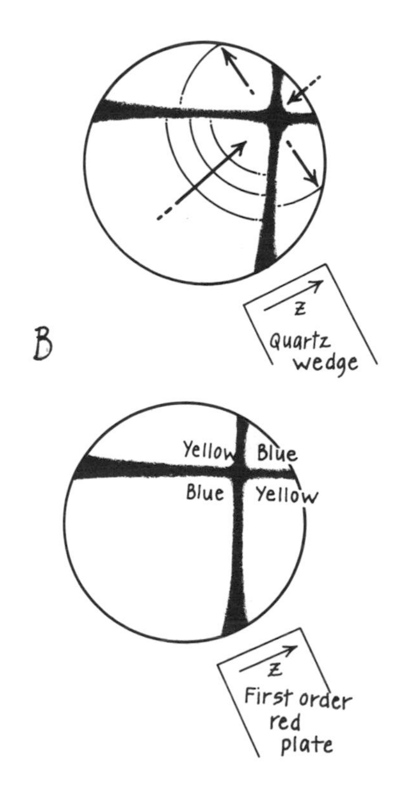

Figure 1-12. Sign reactions for uniaxial *positive* minerals. (A) Optic axis figures. The effect of quartz wedge insertion is to cause isochromes to move inward in the NE and SW quadrants, and move outward in the NW and SE quadrants. The first-order red plate and the mica plate cause an increase in interference colors in the NE and SW quadrants, and a decrease in the NW and SE quadrants. (B) Inclined uniaxial optic axis figures, showing the effect of insertion of the quartz wedge and the first-order red plate. The sign reactions are reversed for uniaxial negative figures.

When working with a single mineral in thin section, however, only the grain orientation is variable; consequently the level of interference colors provides a clue as to the type of interference figure that can be obtained. This relationship can be shown in a simple list:

Figure Type	Level of Interference Color	Vibration Directions Parallel to the Stage
Optic normal (O.N.)	Maximum	Bxa, Bxo
Obtuse bisectrix (Bxo)	Relatively high	Bxa, O.N.(β)
Acute bisectrix (Bxa)	Relatively low	Bxo, O.N.(β)
Optic axis (O.A.)	No interference colors	O.N.(β)

Note that in general the types of interference figures that yield the greatest amount of information are either the optic axis or acute bisectrix types. The O.N., Bxa, and Bxo interference figures are classified as bisymmetric, inasmuch as two symmetry planes are present when the figures are viewed in diagonal orientation. On the same basis, the O.A. figure is classified as monosymmetric.

Optic Normal Figure

The biaxial optic normal (O.N.) figure (frequently called a flash figure) is indistinguishable from the uniaxial optic normal figure. When the Bxa and Bxo vibration directions are parallel to the NS and EW cross hairs, the figure consists of a very diffuse black cross. Rotation of the stage (8° or less) causes the black cross to break into two portions that leave the field of view in opposite quadrants. The isogyres leave the field in those quadrants that contain the Bxa vibration direction (Fig. 1-13). The indices corresponding to the Bxa and Bxo vibration directions can be estimated when either is oriented parallel to the lower polarizer.

An optic sign is difficult to obtain directly from the figure, as the interference colors observed are frequently diffuse. The best approach is to determine first the positions of the Bxa and Bxo vibrations by stage rotation. These vibration directions are then oriented diagonally; when viewed between crossed nicols with an appropriate accessory, the relative values of the Bxa and Bxo indices are obtained. If Bxa exceeds Bxo, then Bxa is γ, and the mineral is optically positive. If Bxa is less than Bxo, then Bxa is α, and the mineral is optically negative.

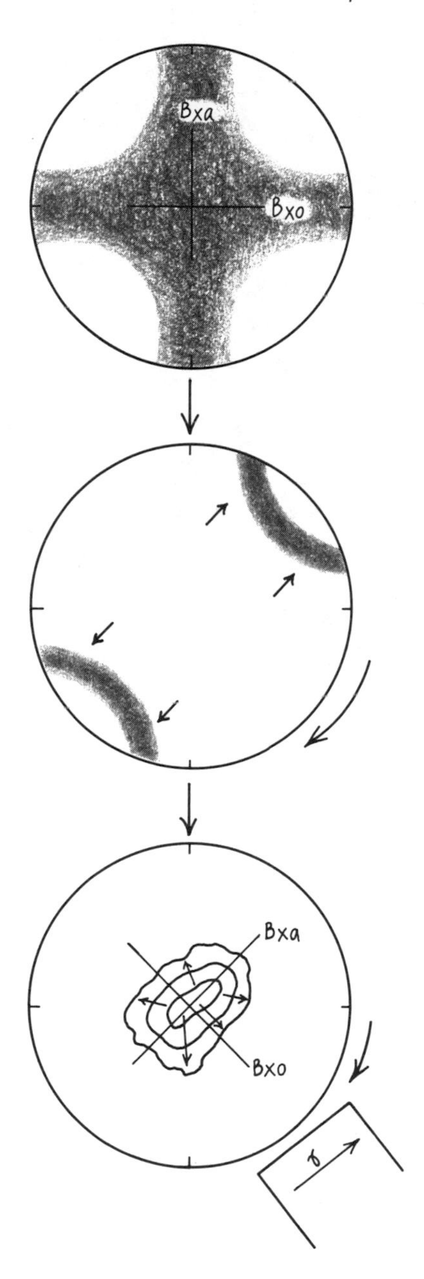

Figure 1-13. Sign determination for a *biaxial positive* optic normal figure. From the cross position, clockwise rotation of the stage makes the isogyres leave the field in the NE and SW quadrants; these quadrants must contain the Bxa vibration direction. Continued rotation orients the Bxa direction diagonally. Insertion of the quartz wedge when the grain is viewed between crossed nicols indicates that Bxa > Bxo; hence the mineral is established as being biaxial positive.

Obtuse Bisectrix Figure

The obtuse bisectrix (Bxo) figure changes in shape and orientation during stage rotation; isogyres enter the field of view from opposite quadrants, join to form a somewhat diffuse cross, sepa-

Figure 1-14. Sign determination for a *biaxial positive* obtuse bisectrix figure. From the cross position, clockwise rotation of the stage makes the isogyres leave the field in the NE and SW quadrants; these quadrants must contain the Bxa vibration direction. Continued rotation orients the Bxa direction diagonally. Insertion of the quartz wedge when the grain is viewed between crossed nicols indicates that Bxa > O.N.(β); hence the mineral is established as being biaxial positive.

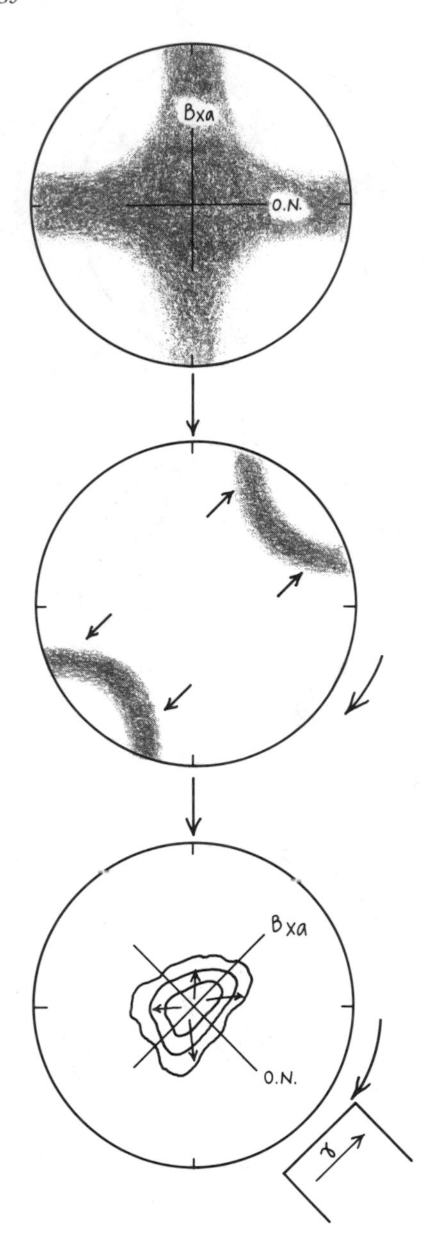

rate, and leave the field in the remaining opposite quadrants (Fig. 1-14). The sequence of isogyres entering, forming a cross, and leaving, frequently requires a stage rotation between 15° to 30°, the particular value being determined by the numerical aperture of the objective and the average index of the mineral (see Fig. 8-33 on p. 102, Volume 1, for precise values).

When the isogyres are in the cross position, the Bxa and O.N. vibration directions are parallel to the NS and EW cross hairs. Stage rotation causes the isogyres to leave the field of view in the quadrants that contain the Bxa vibration. To determine the sign, first rotate the Bxa and O.N. vibration directions into a diagonal orientation. Then, when viewing the grain between crossed nicols, determine with an appropriate accessory the relative index values of Bxa and O.N. If Bxa exceeds O.N., Bxa is γ and the mineral is optically positive. If Bxa is less than O.N., then Bxa is α and the mineral is optically negative.

Acute Bisectrix Figures

The acute bisectrix (Bxa) figure changes in shape and orientation during stage rotation as shown in Figure 1-15. During rotation the isogyres move together from opposite quadrants, form a cross, and then separate into the remaining quadrants. The extent of separation of the isogyres is determined by the $2V$, the β index of the mineral, and the numerical aperture of the objective. The isogyres will move to the edge of the field with a mineral of average β index (between 1.50 and 1.60) if the $2V$ is about 50° when using an objective having a numerical aperture of 0.65, or for a $2V$ of about 65° when using an objective having a numerical aperture of 0.85 (see Fig. 8-18 on p. 93, Volume 1, for specific values). For larger values of $2V$, the isogyres will leave the field of view during rotation, and for smaller values they will remain within the field of view during stage rotation. Estimates of $2V$ based upon the maximum isogyre separation can be made with the aid of Figure 1-16.

When the isogyres are in the cross position, the Bxo and O.N. vibration directions are parallel to the NS and EW cross hairs (see color photo 3). Stage rotation causes the isogyres to move

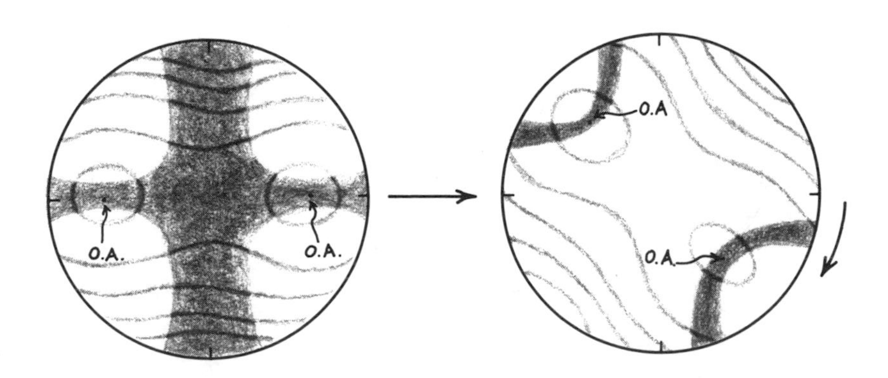

Figure 1-15. A biaxial Bxa interference figure. From the cross position, stage rotation causes the two isogyres to separate and move into opposite quadrants. Continued rotation repeats this pattern every 90° in alternate pairs of opposite quadrants.

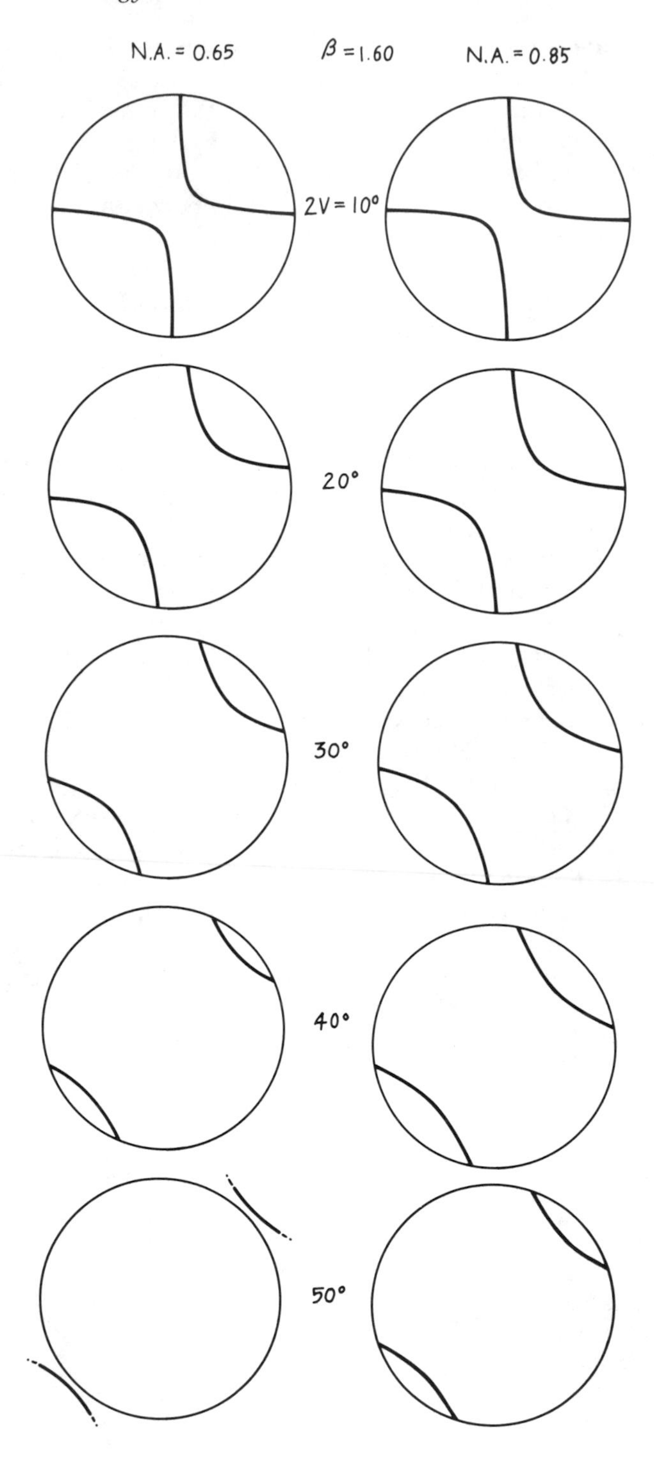

Figure 1-16. Isogyre positions for diagonally oriented Bxa figures. The grain has a β index of 1.60. The left column applies to an objective lens with a numerical aperture of 0.65, and the right column to an objective with a numerical aperture of 0.85.

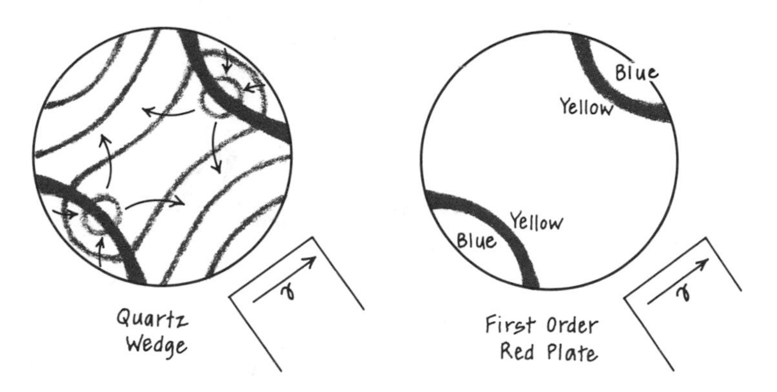

Figure 1-17. Sign reactions for *biaxial positive* acute bisectrix interference figures. (A) When isochromes are present, the quartz wedge is used. Isochromes move in from the NE and SW quadrants and leave in the NW and SE quadrants. (B) When no isochromes are present and the field is white or gray, the first-order red plate or mica plate produce higher level colors on the concave sides of the isogyres as compared to the convex sides. The reactions are reversed for biaxial negative minerals.

outward into opposite quadrants that contain the Bxo vibration direction.

The sign is easily determined with the proper accessory after the figure has been rotated into a diagonal position (see Fig. 1-17 and color photos 4 and 5). If the figure contains several isochromes, sign determination with the quartz wedge is preferred, whereas figures with few or no isochromes permit a sign determination with the first-order red plate or mica plate.

Optic Axis Figure

An optic axis (O.A.) figure is produced when one of the two optic axes is oriented perpendicular to the stage. In this orientation the grain displays no birefringence, and permits an estimation of only the β index of refraction.

To some extent the optic axis figure can be regarded as a half view of a Bxa figure (Fig. 1-18). When oriented diagonally, an isogyre crosses the center of the field. This isogyre intersects the melatope, which may be surrounded by concentrically arranged isochromatic curves. During rotation of the stage the isogyre changes in shape and orientation. When placed in diagonal orientation, the curvature of the isogyre provides an estimation of the $2V$ (Fig. 1-19). In this same orientation, the sign is determined by use of the proper accessory. An isogyre that shows no curvature in the diagonal orientation is indicative of a $2V$ of 90°, and the mineral is generally listed in both positive and negative categories. Biaxial minerals are plotted according to minimum values of $2V$ in Charts 5 and 6 in Part 2.

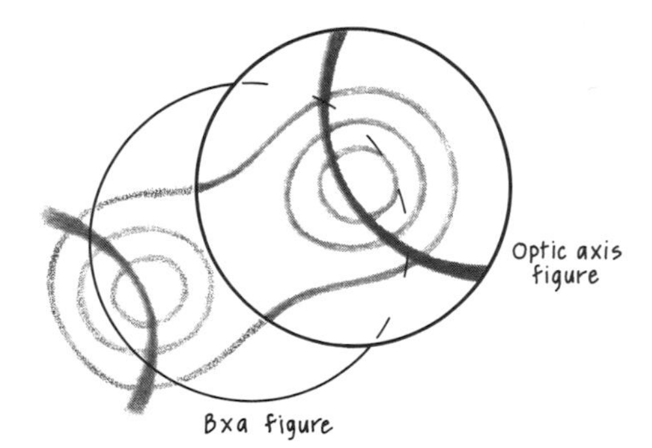

Figure 1-18. The biaxial optic axis interference figure, considered as being half of a Bxa figure.

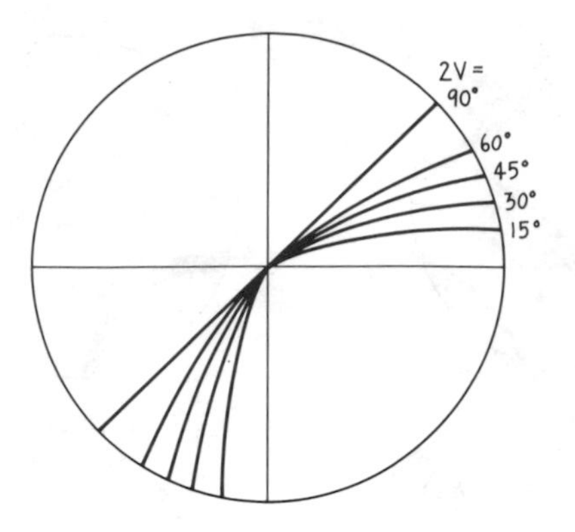

Figure 1-19. The curvature of isogyres at the optic axis, as a function of $2V$. A second isogyre may also be within the field of view for small values of $2V$.

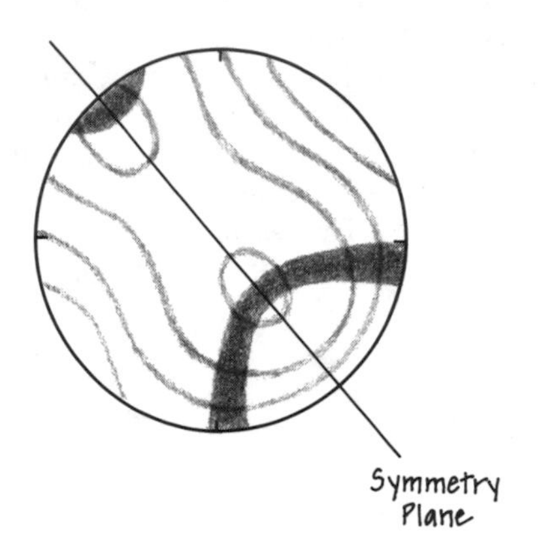

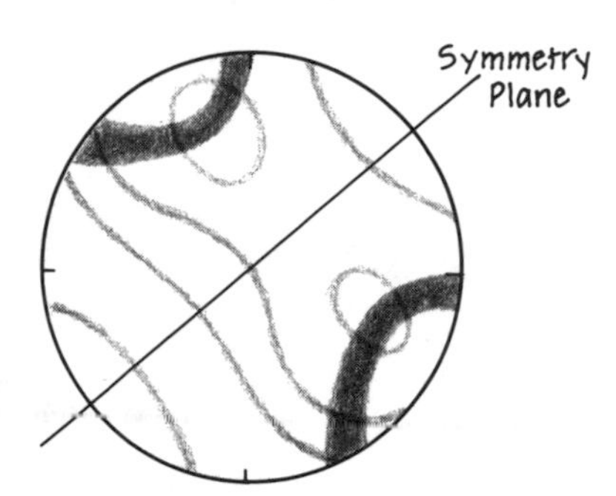

Figure 1-20. Monosymmetric Bxa figures contain only a single symmetry plane when in diagonal orientation. A single principal index can be estimated.

Monosymmetric and Nonsymmetric Figures

Minerals that are oriented with only a single principal vibration direction parallel to the stage yield monosymmetric interference figures; that is, only a single plane of symmetry is present when the figure is in diagonal orientation (Fig. 1-20). Figures of this type often permit estimation of the $2V$ and determination of sign, and will yield estimation of a single index of refraction.

Nonsymmetric figures (Fig. 1-21) are produced by minerals that have no principal vibration directions parallel to the stage. No principal indices can be estimated from such figures, but it is often possible to determine the sign and estimate the $2V$.

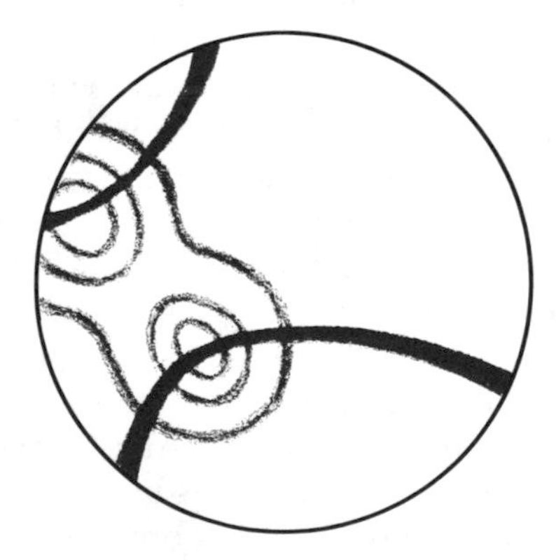

Figure 1-21. A nonsymmetric Bxa figure in diagonal orientation. No symmetry planes are present and no principal indices can be estimated.

Polarization Colors

Phase difference (nm)

Thickness 10^3 nm = μm: 10, 20, 30, 40

Birefringence Δn

Phase difference (nm): 100, 200, 300, 400, 500, 1λ, 600, 700, 800, 900, 1000, 2λ, 1100, 1200, 1300, 1400, 1500, 1600, 3λ, 1700, 1800, 1900, 2000, 2100, 4λ, 2200, 2300

Leucite 0.0005

0.002

α-Cristobalite 0.003
α-Tridymite 0.0035
0.004 Nepheline 0.004 Apatite Group 0.004
Orthoclase 0.005
Sanidine 0.0055

0.006 Gehlenite – akermanite series 0.006*
Anorthoclase 0.0065 Serpentine group 0.0065
Microcline 0.007

0.008
Corundum 0.0085
α-Quartz 0.009

0.010 Gypsum 0.010 Andalusite 0.010 Chlorite 0.010
Plagioclase series 0.0105
Chloritoid 0.011
Cordierite 0.0115*

0.012 Staurolite 0.012
Zoisite 0.0125*

Wollastonite 0.0135 Cancrinite 0.0135* Jadeite 0.0135

0.014
Brucite 0.015 Enstatite – orthoferrosilite series 0.015*

0.016
Kyanite 0.0165
Pumpellyite 0.017

0.018 Glaucophane – riebeckite series – 0.018*

0.020 Hornblende 0.020* Lawsonite 0.020
Tremolite – actinolite series 0.021 Sillimanite 0.021

0.022

0.024 Augite 0.024*

0.026 Anthophyllite 0.026 Pigeonite 0.026 Tourmaline group 0.026*
Clinozoisite – epidote series 0.0265*
Diopside – hedenbergite series 0.0275

0.028

0.030
Chondrodite 0.031*

0.032

0.034

0.036

0.038
Phlogopite 0.0385*

0.040

0.042
Muscovite 0.0425*

Forsterite – fayalite series 0.0435*
0.044 Anhydrite 0.044

Aegirine – aegirine-augite series 0.045*

0.046

Δn

0.300

0.200 Dolomite 0.179
Calcite 0.172
Aragonite 0.153

0.150 Sphene 0.146*

0.100

0.090

0.080

0.070 Stilpnomelane 0.070*

0.060

0.050 Talc 0.050
Oxyhornblende 0.0505*
Zircon 0.0535*
Pyrophyllite 0.0535*
Biotite 0.055*

* = The stated average birefringence varies more than ±0.005

Ernst Leitz Wetzlar GmbH Leitz

Plate 2
Interference figures (1–5) and minerals in thin section (6–75).

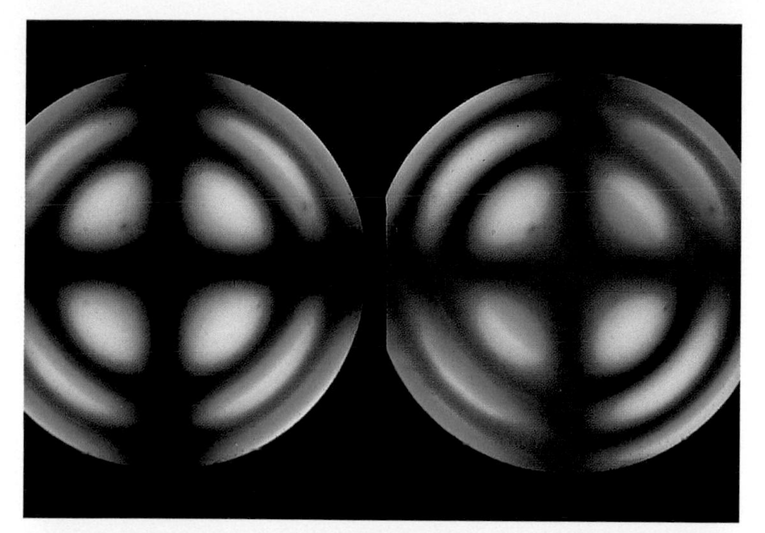

1. (Left) A uniaxial optic axis interference figure. (Right) With insertion of a first-order red plate, the optic sign is established as positive (+).

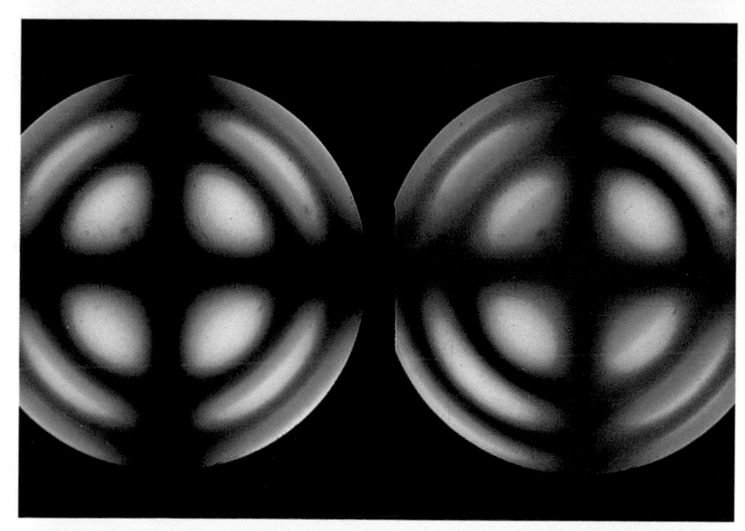

2. (Left) A uniaxial optic axis interference figure. (Right) With insertion of a first-order red plate, the optic sign is established as negative (−).

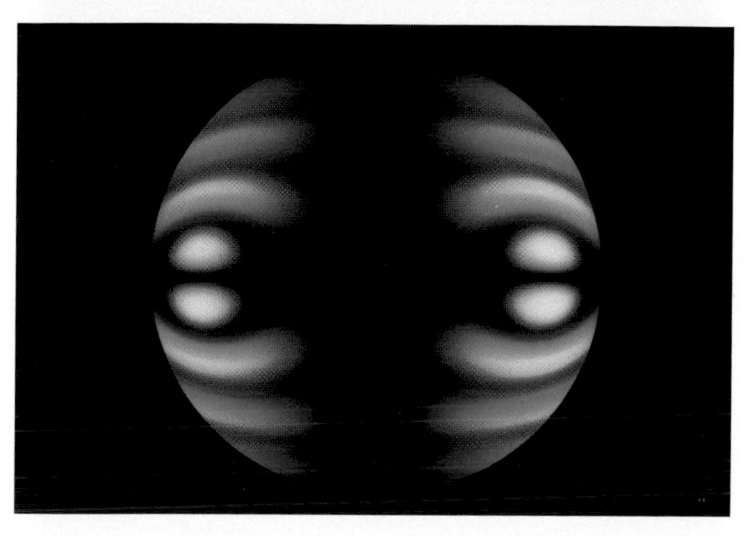

3. A biaxial acute bisectrix interference figure in the cross position. The melatopes are located along the EW isogyre. Notice that the NS isogyre is much broader than the EW isogyre.

4. (Left) A biaxial acute bisectrix interference figure in the diagonal position. (Right) With addition of a first-order red plate the colors have decreased between the melatopes, indicating that the optic sign is positive (+).

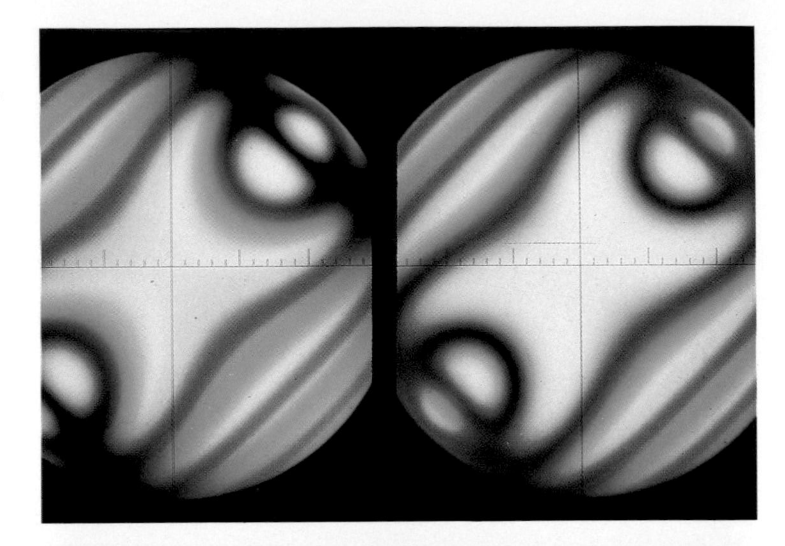

5. (Left) A biaxial acute bisectrix interference figure in the diagonal position. (Right) With addition of a first-order red plate the colors have increased between the melatopes, indicating that the optic sign is negative (−). (Figures 1 through 5 from J. Hinsch, E. Leitz, Inc.)

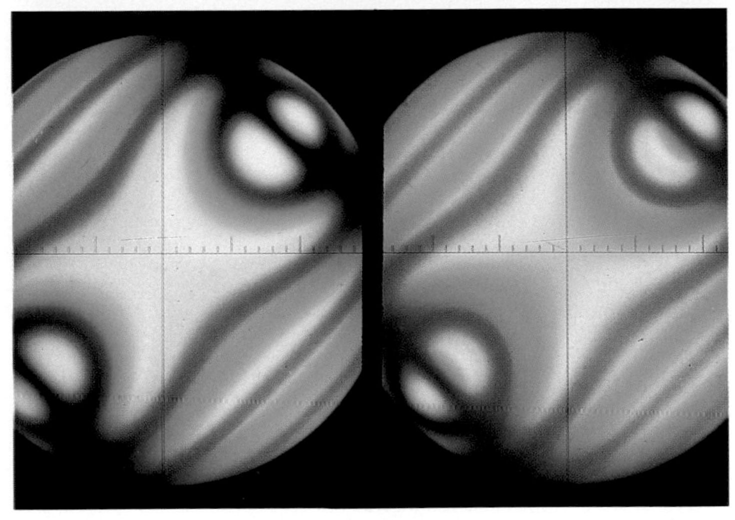

(A)

(B)

(C)

6. (A) Andalusite (A1) porphyroblasts in fine-grained quartz mica matrix. Crossed nicols. (B) The same, rotated so that the direction of elongation of andalusite (the *c* axis) is normal to the (EW) privileged vibration direction of the lower polarizer. The elongate andalusite grain is essentially colorless. Plane-polarized light. (C) The same, rotated so that the long direction is parallel to the lower polarizer. The grain has a subtle pink pleochroic color. Plane-polarized light. Photo length is 1.4 mm. (From J. Hinsch, E. Leitz, Inc.)

7. Andalusite (Al) (variety chiastolite) porphyroblasts in a fine-grained quartz-mica matrix. Crossed nicols. Photo length is 2.30 mm.

(A)

(B)

8. (A) Kyanite (Al) (center) in a muscovite-biotite schist. Crossed nicols. (B) The same, viewed in plane-polarized light. The kyanite has high relief, and typically shows one or two directions of cleavage. The surrounding muscovite is colorless, whereas the biotite is pleochroic in brown. Photo length is 3.50 mm. (From J. Hinsch, E. Leitz, Inc.)

(A)

(B)

9. (A) Sillimanite (A1) with minor biotite and quartz. The larger sillimanite grain (lower left) is oriented with its *c* axis at a small angle to the section, and displays upper first-order interference colors. Sillimanite grains whose *c* axes approach normality with the section (upper right) display low interference colors (gray), a prismatic cross-section, and pinacoidal cleavage; these grains yield a Bxa interference figure. Crossed nicols. (B) The same, viewed in plane-polarized light. Photo length is 1.4 mm. (From J. Hinsch, E. Leitz, Inc.)

10. Elongate grains of sillimanite (A1) (lower left), with quartz and garnet (isotropic). Crossed nicols. Photo length is 2.30 mm.

11. Acicular sillimanite (A1) (fibrolite variety) as individuals and parallel aggregates in quartz. Crossed nicols. Photo length is 0.82 mm.

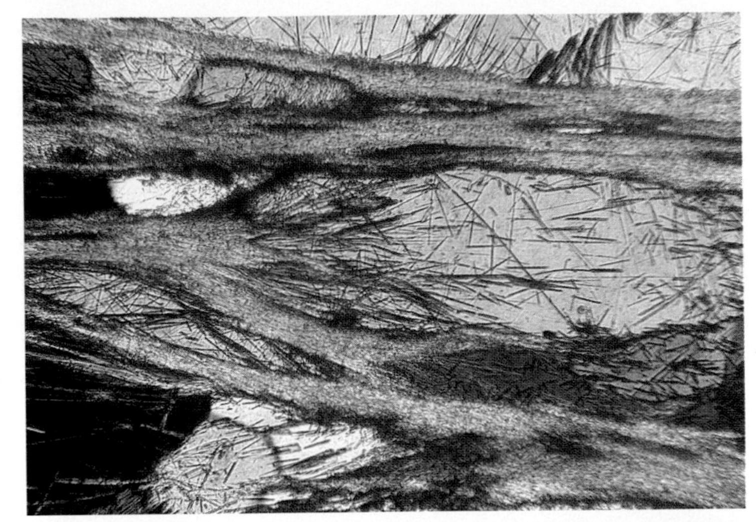

12. Hornblende (A2) grain viewed along the *c* axis, showing well-developed {110} cleavage. Crossed nicols. Photo length is 0.91 mm. (From C. Shultz, Slippery Rock University)

13. Elongate grains of tremolite (A2) in marble. Crossed nicols. Photo length is 2.30 mm.

14. Poikilitic tremolite (A2) in marble. Crossed nicols. Photo length is 2.30 mm.

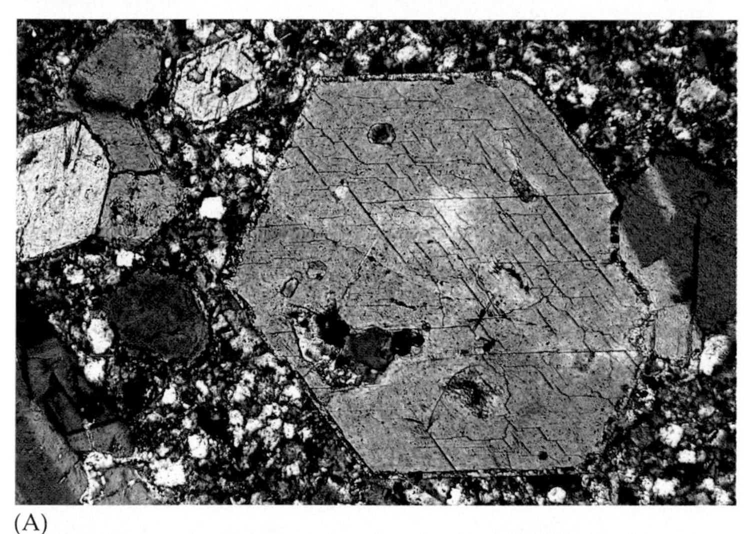

(A)

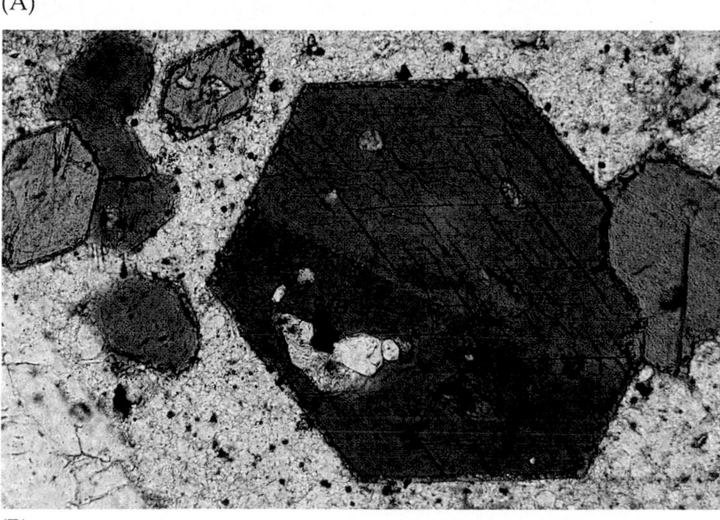

(B)

15. (A) Hornblende (A2) phenocrysts in plagioclase-rich matrix. The larger grain, oriented with its *c* axis approximately perpendicular to the section, displays {110} cleavage. Crossed nicols. (B) The same, viewed with plane-polarized light. Note the range in color of hornblende as a result of pleochroism. Photo length is 1.4 mm. (From J. Hinsch, E. Leitz, Inc.)

16. Hornblende (A2), shown here in an intrusive igneous rock, may display either no cleavage, or one or two cleavages as a function of orientation. The twinned grain at the lower left is plagioclase. Crossed nicols. Photo length is 2.30 mm.

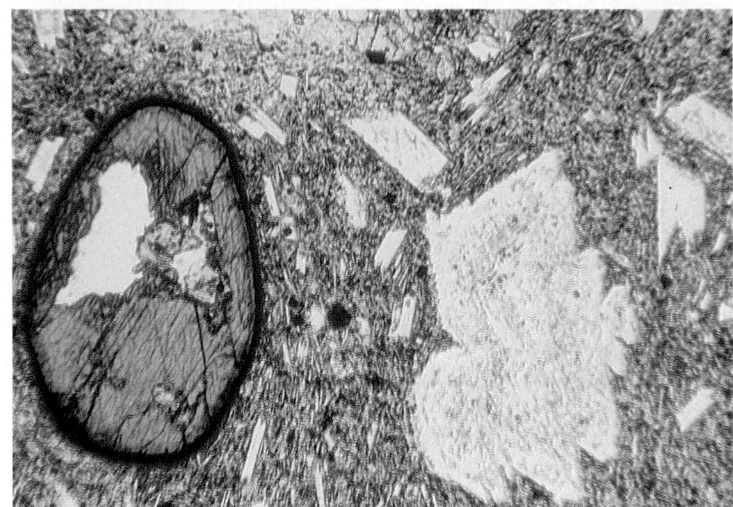

17. Oxyhornblende (A2) (left) and feldspar phenocrysts in an extrusive igneous rock. The black rim around the amphibole is composed mainly of opaque iron oxides, formed as a result of dehydration and oxidation during or after extrusion; a portion of the phenocryst has been plucked during sectioning, and permits estimation of relief against the cementing material. Plane-polarized light. Photo length is 2.30 mm.

(A)

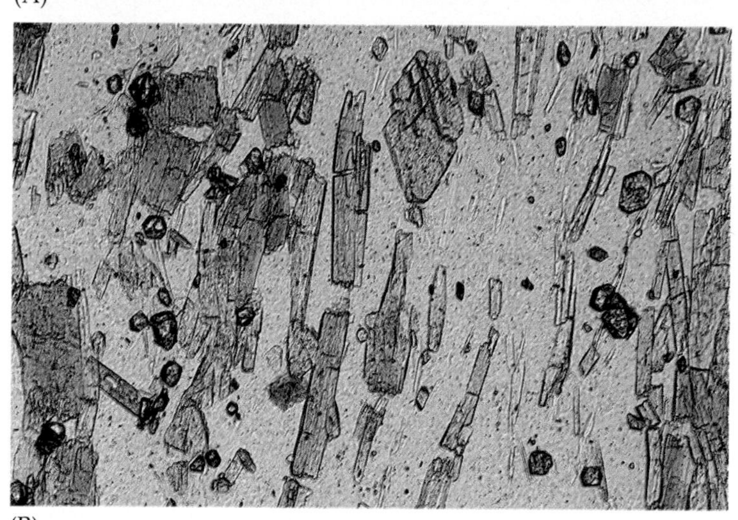
(B)

(C)

18. (A) A schist containing glaucophane (A2) (moderate birefringence), quartz (low birefringence), and garnet (isotropic). Crossed nicols. (B) The same, rotated so that the direction of elongation of glaucophane is roughly normal to the (EW) privileged vibration direction of the lower polarizer. Plane-polarized light. (C) The same, rotated so that the direction of elongation is roughly parallel to the lower polarizer. Plane-polarized light. Each photo length is 1.4 mm. (From J. Hinsch, E. Leitz, Inc.)

19. (A) Several small grains of apatite (A3) (low birefringence) within chlorite (blue-gray) and biotite (yellow-brown). Crossed nicols. (B) The same, viewed in plane-polarized light. Photo length is 1.4 mm. (From J. Hinsch, E. Leitz, Inc.)

(A)

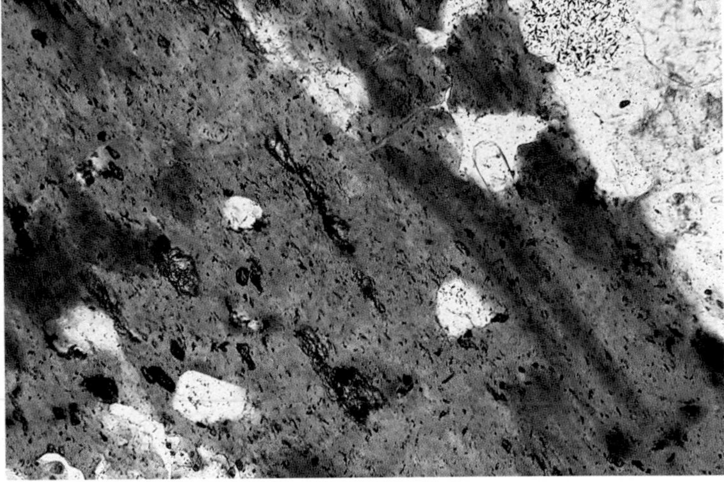

(B)

20. Interlocking grains of a carbonate mineral (C2) in a coarse grained marble. Both grains show rhombohedral cleavages whose angles of intersection are bisected by twin bands. Crossed nicols. Photo length is 2.96 mm. (From C. Shultz, Slippery Rock University)

(A)

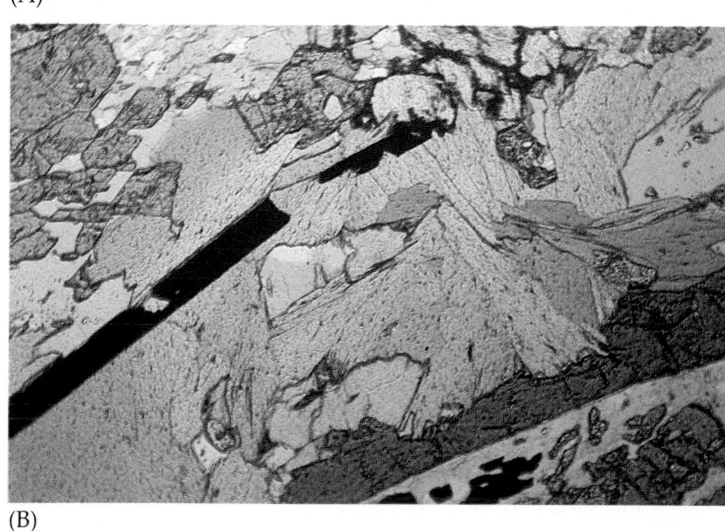

(B)

21. (A) Chlorite (C4) aggregate, with hornblende, epidote, calcite, and quartz. Crossed nicols. (B) The same, viewed in plane-polarized light. Light green chlorite is easily distinguished from dark green hornblende. Photo length is 2.45 mm.

22. (A) Chlorite (C4) with anomalous blue interference colors in a finer grained matrix composed mainly of quartz and biotite. Crossed nicols. (B) The same, viewed in plane-polarized light. The light green chlorite is easily distinguished from brown biotite and colorless quartz. Photo length is 2.09 mm. (From J. Hinsch, E. Leitz, Inc.)

(A)

(B)

23. Zircon within chlorite (C4) is surrounded by a pleochroic halo. Adjacent muscovite (top) is distinguished from chlorite by moderate interference colors. Plagioclase (right) has low birefringence and polysynthetic twinning. Crossed nicols. Photo length is 2.30 mm.

24. Twinned chloritoid (C5) porphyroblast. Surrounding minerals are biotite and quartz. Crossed nicols. Photo length is 2.22 mm.

25. Microcline (above) partially altered to fine-grained sericite (C6). Crossed nicols. Photo length is 0.91 mm. (From C. Shultz, Slippery Rock University)

26. Cordierite (C8) (low birefringence) and biotite (moderate birefringence). The cordierite displays several directions of polysynthetic twinning and pleochroic haloes about zircon inclusions. Crossed nicols. Photo length is 2.30 mm.

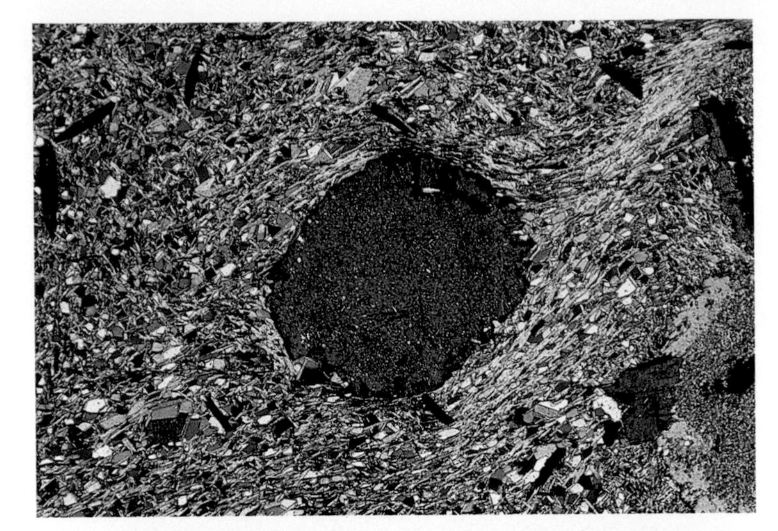

27. Cordierite (C8) poikiloblast in fine-grained mica schist. Crossed nicols. Photo length is 5.56 mm. (From J. Hinsch, E. Leitz, Inc.)

28. Elongate grains of zoisite (E1) showing anomalous blue interference colors. Crossed nicols. Photo length is 0.82 mm.

29. Replacement of plagioclase (F1) (low birefringence) by epidote (E1) (moderate birefringence). The replacement process (often referred to as saussuritization) is proved by the partial localization of epidote within the preexistent twin bands of plagioclase. Crossed nicols. Photo length is 0.82 mm.

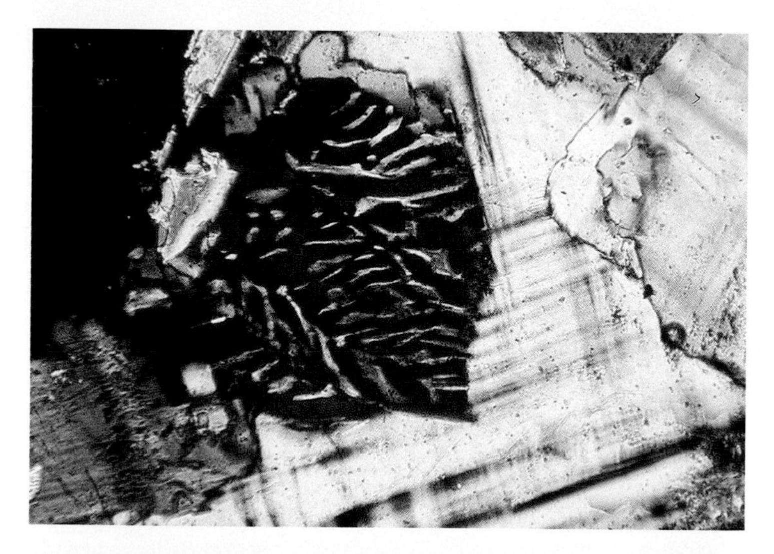

30. Microcline (F1) (right) is often characterized by two sets of polysynthetic twins. The wormy intergrowth (center), called myrmekite, is composed of quartz and sodic plagioclase. Crossed nicols. Photo length is 0.82 mm.

31. Sanidine (F1) phenocrysts (with simple twins) in trachyte. Crossed nicols. Photo length is 2.45 mm. (From J. Swope, Ohio State University)

32. Anorthoclase (F1) phenocryst (with typical fine-grained polysynthetic twinning) in siliceous extrusive rock. Crossed nicols. Photo length is 5.56 mm. (From J. Hinsch, E. Leitz, Inc.)

33. Microcline (F1) (left) typically has two sets of polysynthetic twins (albite and pericline), in contrast to plagioclase (right), which commonly has a single set of polysynthetic twins (albite). Crossed nicols. Photo length is 0.82 mm.

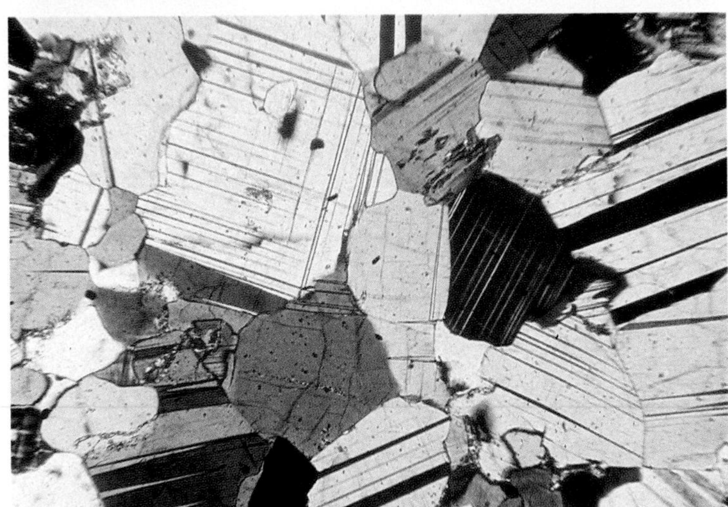

34. Plagioclase (F1), showing albite twinning and two directions of cleavage, in anorthosite. Crossed nicols. Photo length is 2.30 mm.

35. Plagioclase (F1) phenocryst showing oscillatory zoning. The matrix is mainly plagioclase microlites. Crossed nicols. Photo length is 0.82 mm.

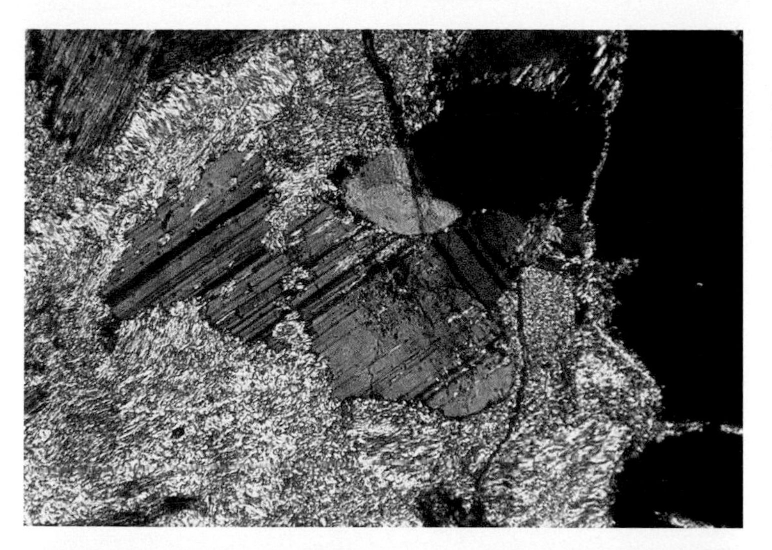

36. Plagioclase (F1) partially converted to fine-grained mica (sericite). Crossed nicols. Photo length is 0.82 mm.

37. Leucite (F2) phenocryst in phonolite. Leucite is characterized by its very low birefringence and complex polysynthetic twinning. Brightly colored grains are clinopyroxene. Crossed nicols. Photo length is 2.22 mm.

38. Anhedral nepheline (F2) in matrix composed mainly of feldspar. Nepheline, although uniaxial (−), is often mistaken for quartz or feldspar. Crossed nicols. Photo length is 2.30 mm.

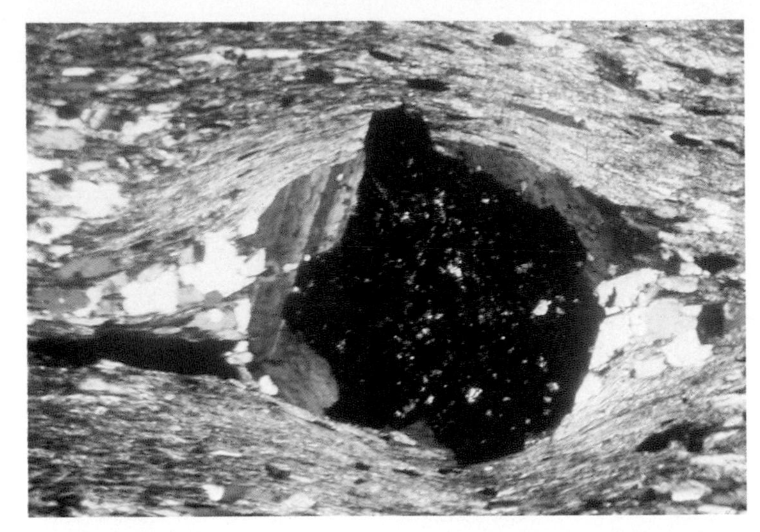

39. Garnet (G1) porphyroblast in quartz-mica schist. The garnet contains inclusions of quartz and is partially enclosed by chlorite (with anomalous olive-brown interference colors). Crossed nicols. Photo length is 2.30 mm.

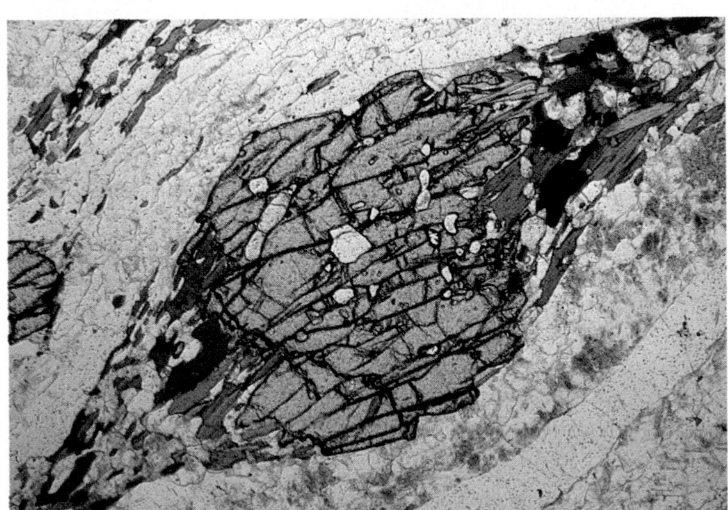

40. Garnet (G1) porphyroblast (high relief) with adjacent biotite, quartz, and apatite in fine-grained schist. Plane-polarized light. Photo length is 3.7 mm. (From C. Shultz, Slippery Rock University)

41. Deformed shards of pale brown glass (G2) and fragmented crystalline material in volcanic tuff. Plane-polarized light. Photo length is 2.30 mm.

42. Anhedral grains of chondrodite (H2) in a carbonate matrix. Chondrodite is generally distinguished from olivine by its lower birefringence and the presence of multiple twinning. Crossed nicols. Photo length is 2.30 mm.

43. Blocky melilite (M1) (white to gray) in clinopyroxene. Crossed nicols. Photo length is 0.55 mm.

44. (A) Biotite (M2) porphyroblasts within a quartz-mica schist. Crossed nicols. (B) The same, viewed in plane-polarized light, and rotated so that the cleavage in the larger porphyroblast is normal to the (EW) privileged vibration direction of the lower polarizer. Absorption is weak and the grain is light-colored. (C) The same, rotated so that the cleavage is parallel to the lower polarizer. Absorption is strong and the grain is dark-colored. All biotites show strong differential absorption, and the colors are typically green, brown, or reddish-brown. Photo length is 2.09 mm. (From J. Hinsch, E. Leitz, Inc.)

(A)

(B)

(C)

45. Muscovite (M2), biotite, and quartz (low birefringence), viewed with crossed nicols. The muscovite and biotite appear indistinguishable. However, when viewed in plane-polarized light the two are easily distinguished, as muscovite is colorless, and biotite is strongly colored and pleochroic. Photo length is 0.82 mm.

46. Mg-rich olivine (O1) in dunite. The lack of good cleavage and large $2V$ distinguish olivine from clinopyroxene. Crossed nicols. Photo length is 3.50 mm. (From J. Hinsch, E. Leitz, Inc.)

47. Embayed olivine (O1) phenocryst in a plagioclase-rich matrix. Olivine phenocrysts are commonly euhedral to subhedral, with {010}, {110}, {021}, and {001} forms predominant. Extinction is commonly parallel and symmetrical. Crossed nicols. Photo length is 1.4 mm. (From J. Hinsch, E. Leitz, Inc.)

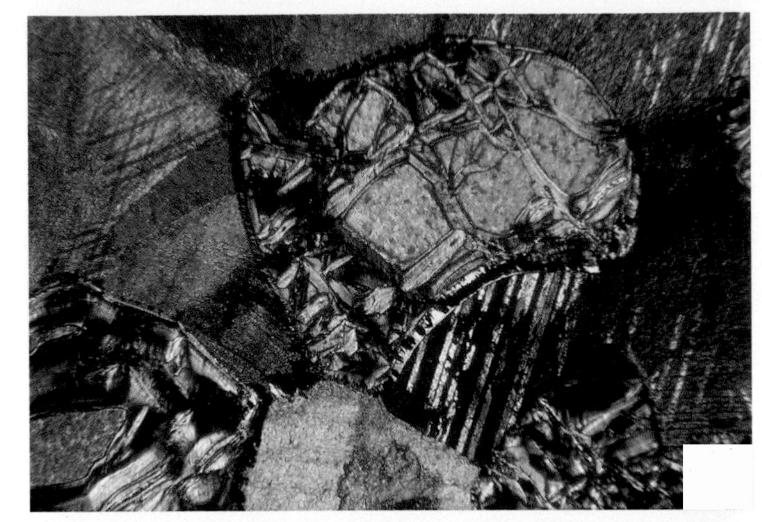

48. Olivine (O1) (blue-green and orange) in ophicalcite. The olivine has been partially altered to serpentine (gray to white). Other grains, showing twin bands, cleavage, and high order interference colors, are either calcite or dolomite. Photo length is 2.30 mm.

49. A single grain of hypersthene (P8) (gray) containing two generations of augite exsolution lamellae. Crossed nicols. Photo length is 0.32 mm.

50. Several grains of hypersthene (P8) (center) surrounded by plagioclase. Orthopyroxenes can be distinguished from clinopyroxenes on the basis of birefringence; the common orthopyroxenes never exceed first-order interference colors in standard thin section, whereas second-order interference colors are common in clinopyroxenes. Crossed nicols. Photo length is 3.50 mm. (From J. Hinsch, E. Leitz, Inc.)

(A)

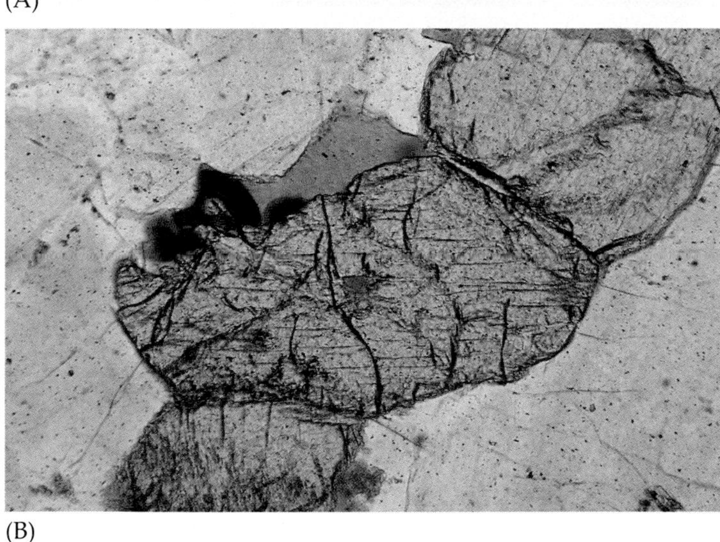

(B)

51. Hypersthene (P8) grains, viewed in plane-polarized light. (A) The *c* axis of the central crystal is normal to the (EW) privileged vibration direction of the lower polarizer. The grain has a subtle yellowish pink color. (B) The same, rotated 90°. The yellowish pink color of the central grain is absent. Photo length is 1.4 mm. (From J. Hinsch, E. Leitz, Inc.)

52. Hypersthene (P8) in a variety of orientations, with minor plagioclase. The good prismatic cleavage, {210}, is not visible in all grains. Crossed nicols. Photo length is 2.30 mm.

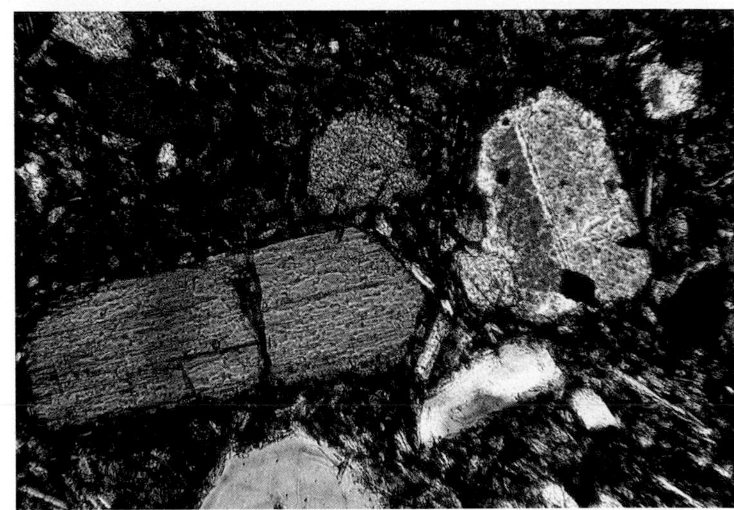

53. Hypersthene (P8) (lower left) and augite (upper right). The hypersthene has low birefringence and (typically) parallel extinction. The augite has moderate birefringence and (typically) inclined extinction. Crossed nicols. Photo length is 0.82 mm.

54. Compositionally zoned diopside (P8) phenocryst in altered alkali-rich basalt. Crossed nicols. Photo length is 3.7 mm. (From C. Shultz, Slippery Rock University)

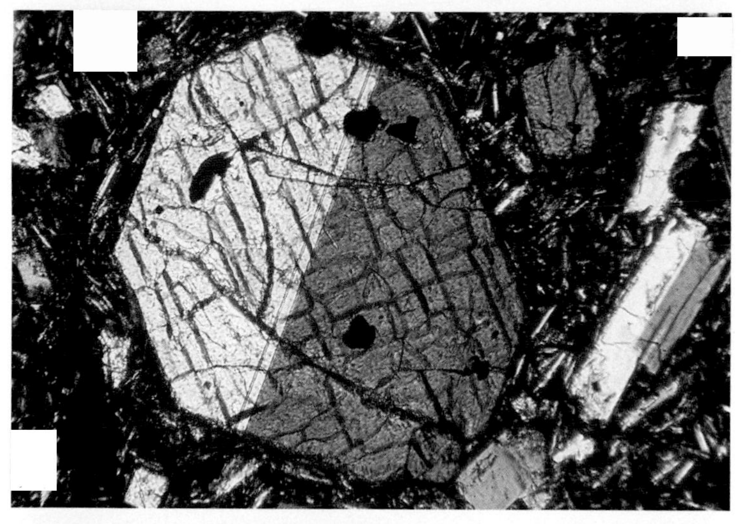

55. Augite (P8) phenocryst in plagioclase-containing lava. The augite is twinned and oriented with its *c* axis almost normal to the section. The almost perpendicular {110} cleavages and common eight-sided outline are visible. Crossed nicols. Photo length is 0.82 mm.

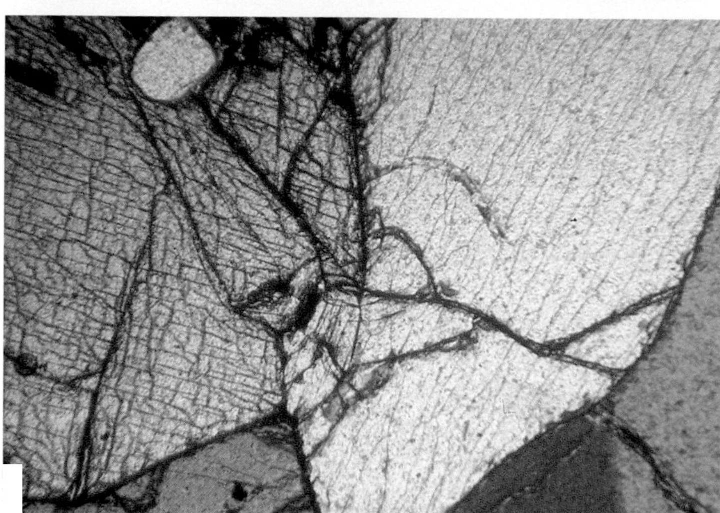

56. Augite (P8) (left) has conspicuous {110} cleavage, which distinguishes it from olivine (right). The olivine shows either no cleavage or poor {010} or {100} cleavage. Crossed nicols. Photo length is 2.30 mm.

57. The central clinopyroxene (P8) grain shows herringbone structure. This is caused by {100} simple twinning (left to right) combined with {001} polysynthetic twinning (almost vertical). The surrounding grains are clinopyroxene and plagioclase. Crossed nicols. Photo length is 0.91 mm. (From C. Shultz, Slippery Rock University)

58. Diabasic texture. Augite (P8) (blue) is surrounded by plagioclase laths. Crossed nicols. Photo length is 2.30 mm.

59. Aegirine-augite (P8) aggregate (center) is surrounded by perthitic feldspar and quartz. When viewed in plane-polarized light, the aegirine-augite is pleochroic in various shades of green. Crossed nicols. Photo length is 2.45 mm.

60. Serpentine (S3), consisting of an aggregate of fibers, plates, and semi-isotropic material. The larger grain (left) is a serpentine pseudomorph after olivine or pyroxene. Crossed nicols. Photo length is 2.45 mm.

61. Detrital rounded quartz (S4) grains (right) with chalcedony (larger fine-grained aggregates at left). Crossed nicols. Photo length is 2.30 mm.

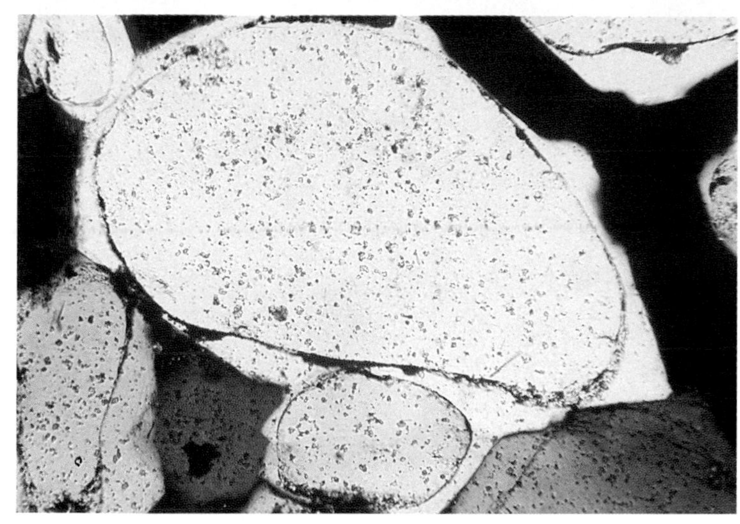

62. Detrital quartz (S4) grains with overgrowths. The detrital grains are outlined by tiny opaque inclusions. The overgrowths are in optical continuity with the detrital grains. Crossed nicols. Photo length is 2.30 mm.

63. (A) Anhedral sphene (S7) with biotite and feldspar. The yellow-brown color of the sphene almost masks the high-order interference colors. Crossed nicols. (B) The same, viewed in plane-polarized light. The biotite is now easily distinguished by its lower relief and greenish brown color. Photo length is 0.82 mm.

(A)

(B)

64. Euhedral sphene (S7) with feldspar and muscovite. The elongate rhombic cross-section and nonsymmetric extinction of sphene distinguish it from hexagonal carbonate minerals. Crossed nicols. Photo length is 0.82 mm.

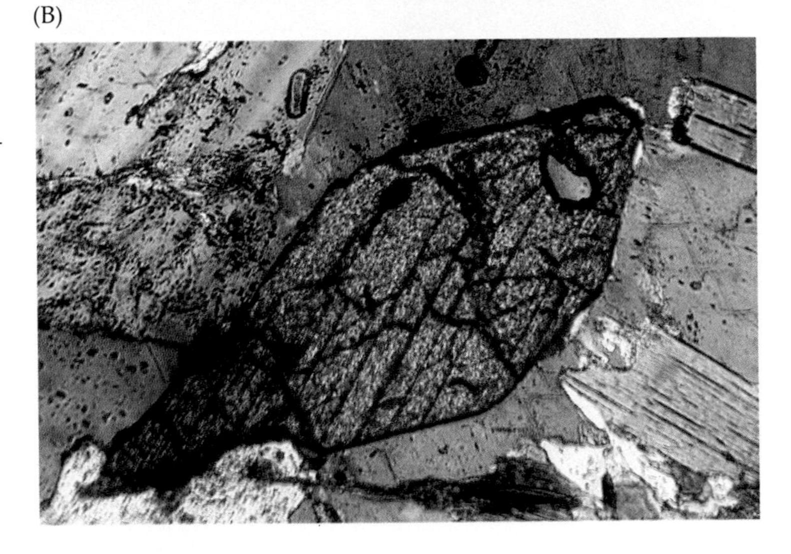

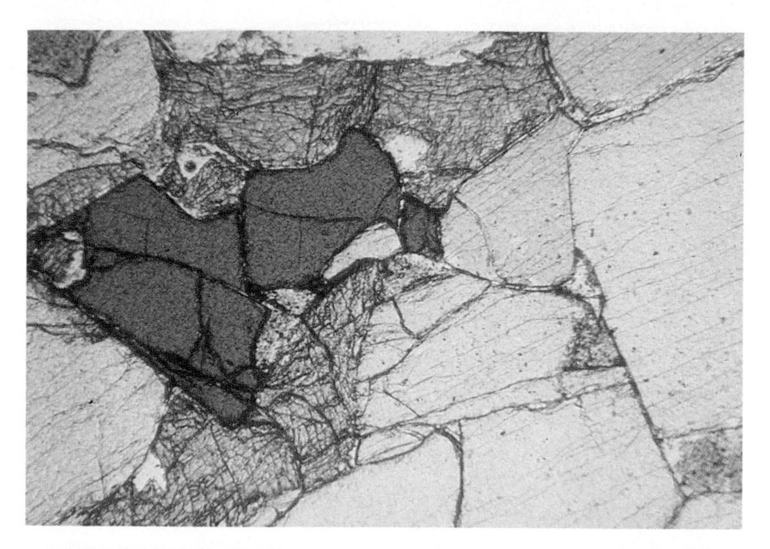

65. Spinel (S8) (brown) is isotropic. Adjacent clinopyroxene displays good {110} cleavage, whereas olivine (surrounding) does not. Plane-polarized light. Photo length is 2.30 mm.

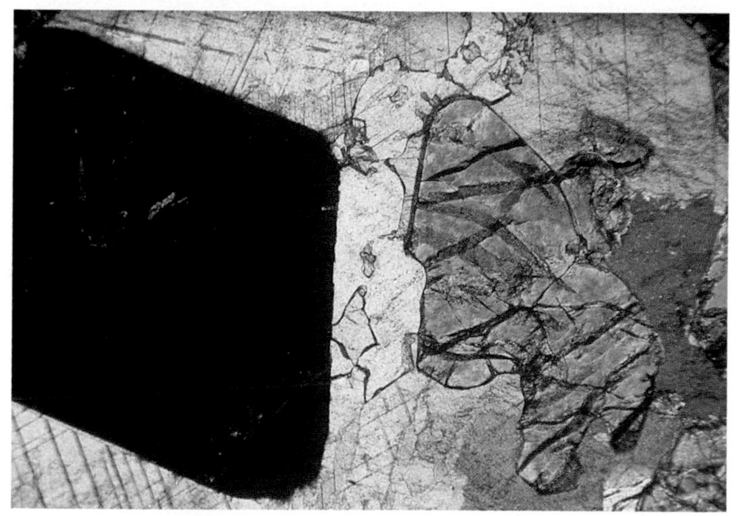

66. Spinel (S8) (left) and chondrodite in marble. Crossed nicols. Photo length is 2.22 mm.

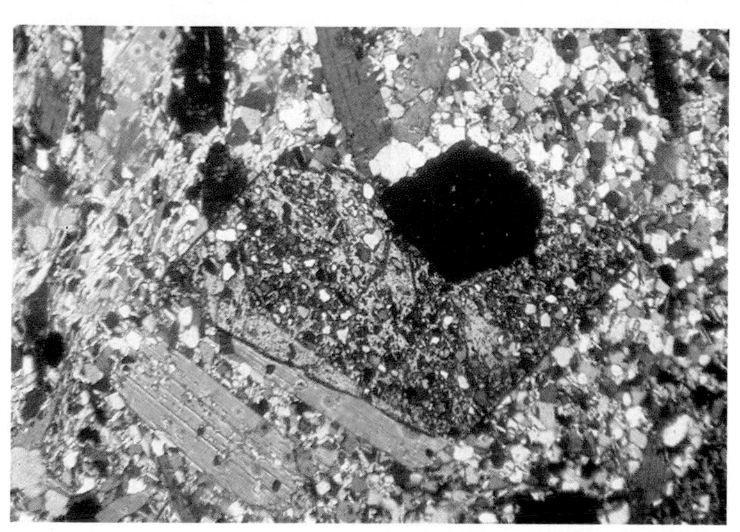

67. Poikilitic staurolite (S10) (center) in a fine-grained matrix of muscovite and quartz. Staurolite is usually identified easily on the basis of its relatively low interference colors, porphyroblastic habit, lack of good cleavage, and yellow pleochroism. Crossed nicols. Photo length is 2.30 mm.

68. Euhedral staurolite (S10) poikiloblast in a fine-grained matrix of muscovite and quartz. Crossed nicols. Photo length is 2.30 mm.

69. Staurolite (S10) porphyroblasts (gray) surrounded by biotite, quartz, and fibrolite (acicular sillimanite). Crossed nicols. Photo length is 2.30 mm.

70. Stilpnomelane (S11) (center) and quartz. Stilpnomelane is easily confused with biotite. Crossed nicols. Photo length is 2.30 mm.

71. Chlorite (C4) (gray) has been partially converted to a fine-grained talc (T1) aggregate. Talc is easily confused with muscovite and is commonly identified on the basis of paragenesis. Crossed nicols. Photo length is 0.82 mm.

(A)

(B)

72. (A) Tourmaline (T3) and quartz. The color of the larger tourmaline (left) masks the interference colors. The grain displays color zonation, and is oriented with its *c* axis at a high angle to the section. The smaller tourmaline grains (right) are oriented with their *c* axes at a small angle to the section. Crossed nicols. (B) The same, viewed in plane-polarized light. The extreme difference in absorption is seen among the differently oriented grains. Photo length is 2.45 mm.

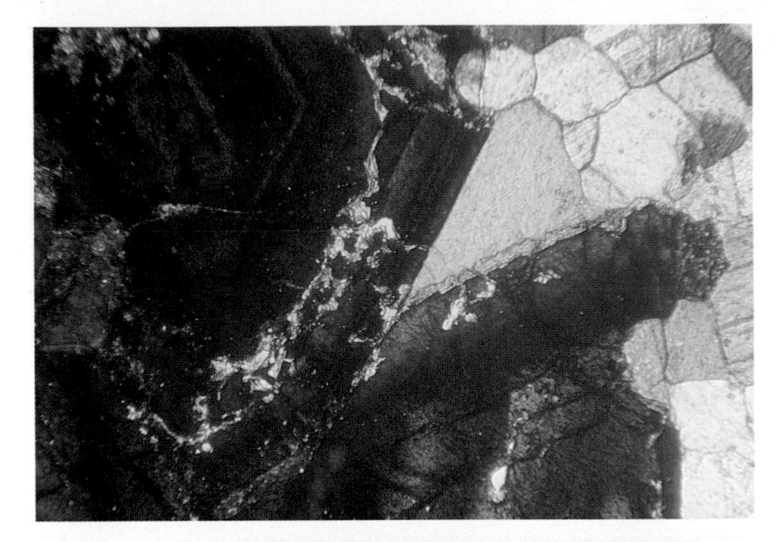

73. Vesuvianite (V1) (left) displays anomalous blue and brown interference colors. Grains at the right are a carbonate mineral. Crossed nicols. Photo length is 2.30 mm.

74. Wollastonite (W1) aggregate. Black areas are garnet. Crossed nicols. Photo length is 2.22 mm.

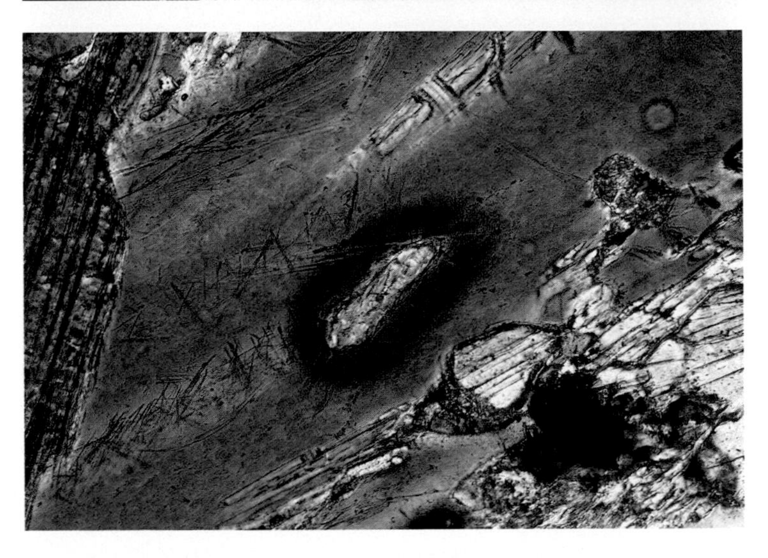

75. A grain of zircon (Z2) (center) included within biotite is surrounded by a pleochroic halo. Surrounding minerals are biotite (upper left) and sillimanite (lower right). Crossed nicols. Photo length is 0.87 mm. (From J. Hinsch, E. Leitz, Inc.)

PART 2

Identification of Unknown Minerals

Aristotle could have avoided the mistake of thinking that women have fewer teeth than men by the simple device of asking Mrs. Aristotle to open her mouth.

BERTRAND RUSSELL

Before examining an unknown material microscopically, it is good practice to obtain as much macroscopic information as possible beforehand. Knowledge of color, hardness, specific gravity, associated minerals, cleavage, and reactivity with various solvents will often simplify the microscopic identification.

The easiest (and most dangerous) method of mineral identification is to recognize the mineral, based upon having seen it before. While this is a common method of identification, it is *always* best to check a few of the diagnostic properties to be certain of the identification. For this reason, each of the mineral descriptions in Part 3 includes a section called "Look-Alikes," in which similar-appearing minerals are mentioned. When becoming familiar with a new mineral, it is wise to note the characteristics of the look-alikes.

The *opaque minerals* are easily recognized by examination of the thin section or grain mount with uncrossed nicols: being opaque, they appear black. Descriptions of the common opaque minerals are given in Part 3 under the name "Opaque Minerals." Identification of these minerals with the polarizing microscope is very questionable and is based upon color in reflected light, habit, associated minerals, cleavage, and alteration products. In general, firm identification requires use of a reflected-light microscope.

Isotropic materials are recognized by their property of remaining in extinction during stage rotation when viewed between crossed nicols. The isotropic character can be verified by lack of interference figures.

If the isotropic material is colored, refer to Table 1, which lists isotropic substances on the basis of color and increasing values of the index of refraction. Both colored and colorless isotropic materials are listed in Table 4 on the basis of index of refraction.

Identification of isotropic materials is based upon a combination of refractive index, color, habit, presence or absence of cleavage and twinning, and associated minerals.

Anistropic minerals are recognized by the frequent presence of interference colors when viewed between crossed nicols. The first step in identification is to determine whether the material is uniaxial or biaxial. This is accomplished with the use of *interference figures.* Those figures that yield the greatest amount of information have either one or two melatopes in the field of view. Figures of this type (optic axis or acute bisectrix) are obtained from grains that either remain in extinction during stage rotation (between crossed nicols) or possess a relatively low level of interference colors as compared to grains of similar thickness. The isogyre cross obtained with uniaxial materials remains as a cross during stage rotation, whereas the isogyre cross obtained with biaxial materials breaks into two portions during stage rotation. In some biaxial optic axis figures, only a single isogyre is observed during stage rotation. Typical uniaxial and biaxial interference figures are shown in Part 1 and in color photos 1 to 5. In addition to distinguishing between uniaxial and biaxial materials, the interference figures should be used for *sign determination* and estimation of *optic angle* ($2V$). The presence and character of *dispersion* should be noted.

The *indices of refraction* of the mineral can be approximated either by using oil-immersion techniques or (in thin section) by comparison with the indices of refraction of the mounting medium or of adjacent known minerals. Consideration of *relief* is useful at this stage.

The *color* of the unknown mineral should be noted. If colored, check for *pleochroism* with uncrossed nicols. If pleochroic, the various colors displayed by the mineral should be related to principal optical and crystallographic directions. Table 3 lists minerals with moderate to strong pleochroism.

The presence and quality of *cleavages* should be noted. When working with grain mounts, cleavage determines grain orientation; typical grain shapes and orientations are presented in the mineral descriptions of Part 3, along with the observed indices of refraction. Cleavages are seen in thin sections as parallel cracks within the grains. The angular relationships among various cleavage surfaces and the principal optical and crystallographic directions should be determined; this approach is discussed in more detail in Chapter 10 of Volume 1. The presence of *twinning*—its type, abundance, and orientation—should be noted.

The *birefringence* should be estimated, as described in Part 1. The *crystal habit, geological environment,* and *associated minerals* are often extremely useful in identification.

Optical characteristics of uniaxial and biaxial minerals are given in Tables 1 through 6 and Charts 1 through 6. Note that it is not usually necessary to determine all of the optical and crystallographic parameters in order to identify an unknown material; therefore, obtain only enough information to confirm an identification. Frequently a few features are sufficient. After determining a few characteristics, check the tables and charts in order to limit the possibilities for identification. At this point the distinguishing features are often obvious. If not, then obtain the type of information necessary to distinguish between the various possibilities. In unusual situations it may be necessary to precisely determine one or more of the indices of refraction. In this case, Tables 5 and 6 (listing minerals according to values of ω and β) will prove useful.

Table 1
Colored Minerals Subdivided According to Optical Character and Listed by Average Value of n (Isotropic), ω (Uniaxial), or β (Biaxial)[1]

	Isotropic and Noncrystalline Materials	
Blue	*Green*	*Yellow*
Sodalite (S5), 1.485	Fluorite (F3), 1.434	Opal (S4), ≃1.43
Noselite (S5), 1.490	Hauyne (S5), 1.500	Grossular (G1), 1.734
Hauyne (S5), 1.500	Glass (G2), 1.565	Periclase (P2), 1.740
Brown	Uvarovite (G1), 1.86	Sphalerite (S6), 2.435
Opal (S4), ≃1.43	*Gray*	*Wide Variation in Color*
Glass (G2), 1.565	Glass (G2), 1.565	Fluorite (F3), 1.434
Pyrope (G1), 1.714	*Pink*	Spinel Group (S8), 1.86
Periclase (P2), 1.740	Pyrope (G1), 1.714	
Spessartine (G1), 1.800	Spessartine (G1), 1.800	
Almandine (G1), 1.830	Almandine (G1), 1.830	
Andradite (G1), 1.887	*Violet*	
	Fluorite (F3), 1.434	
	Uniaxial Positive (+) Minerals	
Brown	*Green*	*Yellow*
Vesuvianite (V1), 1.727	Vesuvianite (V1), 1.727	Akermanite (M1), 1.641
Zircon (Z2), 1.941	*Red*	Vesuvianite (V1), 1.727
Cassiterite (C3), 2.000	Cassiterite (C3), 2.000	Cassiterite (C3), 2.000
Rutile (R2), 2.610	Rutile (R2), 2.610	

[1]Many of the minerals listed may also be colorless; the color given may range from strong to weak.

(continued)

Uniaxial Negative (−) Minerals

Black
Anatase (R2), 2.561

Blue
Tourmaline (T3), 1.655
Corundum (C9), 1.769
Anatase (R2), 2.561

Brown
Apatite Group (A3), 1.594
Fluorapatite (A3), 1.641
Tourmaline (T3), 1.655
Ankerite (C2), 1.714
Vesuvianite (V1), 1.727
Siderite (C2), 1.869
Anatase (R2), 2.561

Gray
Apatite Group (A3), 1.594
Fluorapatite (A3), 1.641
Tourmaline (T3), 1.655

Green
Beryl (B1), 1.580
Tourmaline (T3), 1.655
Vesuvianite (V1), 1.727
Corundum (C9), 1.769
Anatase (R2), 2.561

Pink
Tourmaline (T3), 1.655
Corundum (C9), 1.769
Rhodochrosite (C2), 1.802

Yellow
Beryl (B1), 1.580
Tourmaline (T3), 1.655
Gehlenite (M1), 1.660
Vesuvianite (V1), 1.727
Corundum (C9), 1.769
Siderite (C2), 1.869
Anatase (R2), 2.561

Wide Variation in Color
Tourmaline (T3), 1.655

Biaxial Positive (+) Minerals[2]

Blue
Cordierite (C8), 1.549, 75°–90°
Sillimanite (A1), 1.659, 21°–30°
Riebeckite (A2), 1.699, 0°–90°

Brown
Norbergite (H2), 1.572, 44°–50°
Gibbsite (C6), 1.574, 0°–40°
Chondrodite (H2), 1.628, 64°–90°
Humite (H2), 1.636, 64°–84°
Cummingtonite (A2), 1.657, 70°–90°
Sillimanite (A1), 1.659, 21°–30°
Anthophyllite (A2), 1.663, 70°–90°
Hornblende (A2), 1.666, 85°–90°
Gedrite (A2), 1.670, 47°–90°
Clinohumite (H2), 1.672, 52°–90°
Pumpellyite (P6), 1.695, 26°–85°
Riebeckite (A2), 1.699, 0°–90°
Pigeonite (P8), 1.703, 0°–32°
Allanite (E1), 1.757, 57°–90°
Piemontite (E1), 1.768, 50°–86°
Monazite (M3), 1.789, 6°–19°
Sphene (S7), 1.952, 17°–40°
Brookite (R2), 2.584, 0°–30°

Gray
Monazite (M3), 1.789, 6°–19°

Green to Bluish-Green
Chlorite Group (C4), 1.625, 0°–45°
Cummingtonite (A2), 1.657, 70°–90°
Jadeite (P8), 1.660, 68°–72°
Anthophyllite (A2), 1.663, 70°–90°
Hornblende (A2), 1.666, 85°–90°
Gedrite (A2), 1.670, 47°–90°
Diopside (P8), 1.672, 58°
Omphacite (P8), 1.675, 60°–74°
Salite (P8), 1.692, 57°
Pumpellyite (P6), 1.695, 26°–85°
Pigeonite (P8), 1.703, 0°–32°
Augite (P8), 1.706, 25°–60°
Ferrosalite (P8), 1.719, 60°
Aegirine-Augite (P8), 1.726, 70°–90°
Chloritoid (C5), 1.726, 40°–65°
Hedenbergite (P8), 1.739, 63°
Allanite (E1), 1.757, 57°–90°
Orthoferrosilite (P8), 1.771, 35°

Pink
Topaz (T2), 1.622, 48°–68°
Mullite (A1), 1.652, 20°–50°
Zoisite (E1), 1.699, 0°–60°
Rhodonite (R1), 1.727, 61°–76°

Red
Piemontite (E1), 1.768, 50°–86°

Violet
Cordierite (C8), 1.549, 75°–90°
Piemontite (E1), 1.768, 50°–86°

Yellow
Cordierite (C8), 1.549, 75°–90°
Norbergite (H2), 1.572, 44°–50°
Chondrodite (H2), 1.628, 64°–90°
Humite (H2), 1.636, 64°–84°
Gedrite (A2), 1.670, 47°–90°
Clinohumite (H2), 1.672, 52°–90°
Pumpellyite (P6), 1.695, 26°–85°
Zoisite (E1), 1.699, 0°–60°
Staurolite (S10), 1.747, 80°–90°
Allanite (E1), 1.757, 57°–90°

---

[2]The range in values of $2V$ is included for biaxial minerals.

Biaxial Negative (−) Minerals

Blue
- Cordierite (C8), 1.549, 40°–90°
- Glaucophane (A2), 1.631, 32°–50°
- Crossite (A2), 1.666, 42°–90°
- Dumortierite (D1), 1.698, 15°–52°
- Riebeckite (A2), 1.699, 42°–90°
- Sapphirine (S1), 1.717, 51°–69°
- Kyanite (A1), 1.721, 82.5°

Brown
- Montmorillonite (C6), 1.545, 0°–30°
- Vermiculite (C6), 1.56, 0°–8°
- Kaolinite (C6), 1.564, 20°–55°
- Sericite (C6), 1.595, 30°–47°
- Biotite (M2), 1.650, 0°–25°
- Stilpnomelane (S11), 1.660, ≈0°
- Hornblende (A2), 1.666, 23°–90°
- Gedrite (A2), 1.670, 72°–90°
- Axinite (A5), 1.691, 63°–80°
- Grunerite (A2), 1.693, 80°–90°
- Riebeckite (A2), 1.699, 42°–90°
- Oxyhornblende (A2), 1.72, 56°–88°
- Allanite (E1), 1.757, 40°–90°
- Perovskite (P3), 2.34, 0°–90°?

Gray
- Sapphirine (S1), 1.717, 51°–69°

Green
- Serpentine (S3), 1.551, 20°–60°
- Vermiculite (C6), 1.56, 0°–8°
- Lepidolite (M2), 1.568, 0°–58°
- Sericite (C6), 1.595, 30°–47°
- Phlogopite (M2), 1.597, 0°–15°
- Chlorite Group (C4), 1.625, 0°–30°
- Glauconite (M2), 1.63, 0°–20°
- Eckermannite (A2), 1.638, 15°–80°
- Andalusite (A1), 1.638, 71°–86°
- Biotite (M2), 1.650, 0°–25°
- Actinolite (A2), 1.653, 75°–84°
- Stilpnomelane (S11), 1.660, ≈0°
- Hornblende (A2), 1.666, 23°–90°
- Gedrite (A2), 1.670, 72°–90°
- Grunerite (A2), 1.693, 80°–90°
- Arfvedsonite (A2), 1.694, 0°–50°
- Hypersthene (P8), 1.711, 53°
- Chloritoid (C5), 1.726, large
- Ferrohypersthene (P8), 1.737, 53°
- Epidote (E1), 1.740, 64°–90°
- Eulite (P8), 1.756, 77°
- Allanite (E1), 1.757, 40°–90°
- Aegirine (P8), 1.781, 58°–90°
- Fe-rich Olivine (O1), 1.814, 46°–90°

Pink
- Montmorillonite (C6), 1.545, 0°–30°
- Andalusite (A1), 1.638, 71°–86°
- Hypersthene (P8), 1.711, 53°

Red
- Polyhalite (P4), 1.561, 62°–70°
- Sericite (C6), 1.595, 30°–47°
- Biotite (M2), 1.650, 0°–25°
- Perovskite (P3), 2.34, 0°–90°?

Violet
- Cordierite (C8), 1.549, 40°–90°
- Lepidolite (M2), 1.568, 0°–58°
- Glaucophane (A2), 1.631, 32°–50°
- Crossite (A2), 1.666, 42°–90°
- Dumortierite (D1), 1.698, 15°–52°

Yellow
- Montmorillonite (C6), 1.545, 0°–30°
- Cordierite (C8), 1.549, 40°–90°
- Phlogopite (M2), 1.597, 0°–15°
- Glaucophane (A2), 1.631, 32°–50°
- Andalusite (A1), 1.638, 71°–86°
- Stilpnomelane (S11), 1.660, ≈0°
- Crossite (A2), 1.666, 42°–90°
- Gedrite (A2), 1.670, 72°–90°
- Oxyhornblende (A2), 1.72, 56°–88°
- Allanite (E1), 1.757, 40°–90°
- Fe-rich Olivine (O1), 1.814, 46°–90°
- Perovskite (P3), 2.34, 0°–90°?

Table 2
Minerals That Commonly Display Anomalous Interference Colors[1]

Uniaxial Positive (+)
- Brucite (B2), 1.575, dark brown
- Akermanite (M1), 1.642, blue
- Vesuvianite (V1), 1.728, blue or olive-yellow

Uniaxial Negative (−)
- Apophyllite (A4), 1.541, subtle tints
- Gehlenite (M1), 1.660, blue
- Vesuvianite (V1), 1.728, blue or olive-yellow

Biaxial Positive (+)
- Chlorite group (C4), 1.625, brown, violet, blue
- Zoisite (E1), 1.699, blue
- Clinozoisite (E1), 1.701, blue, yellow-brown

Biaxial Negative (−)
- Chlorite group (C4), 1.625, brown, violet, blue
- Epidote (E1), 1.740, subtle blue tints

[1]Minerals within each category are arranged according to average value of ω or β.

Table 3
Minerals with Moderate to Strong Pleochroism[1]

Uniaxial Positive (+)
- Cassiterite (C3), 2.000, (yel, bn, red), $E > O$

Uniaxial Negative (−)
- Apatite Group (A3), 1.635, (bn, gray), $E > O$
- Tourmaline (T3), 1.655, (variable), $O > E$

Biaxial Positive (+)
- Cordierite (C8), 1.549, (yel, bl, viol), $Z > Y > X$
- Norbergite (H2), 1.572, (yel, bn), $X > Y = Z$
- Chlorite (C4), 1.625, (gn), $Z = Y > X$, $X = Y > Z$
- Chondrodite (H2), 1.628, (yel, bn), $X > Y = Z$
- Humite (H2), 1.636, (yel, bn), $X > Y \simeq Z$
- Anthophyllite (A2), 1.663, (bn, gn), $Z > Y \simeq X$, $Z \simeq Y > X$
- Hornblende (A2), 1.666, (gn, bl-gn, yel-bn), $Z > Y > X$, $Y > Z > X$
- Gedrite (A2), 1.670, (gn, bn, yel), $Z > X \simeq Y$
- Clinohumite (H2), 1.672, (yel, bn), $X > Y = Z$
- Pumpellyite (P6), 1.695, (gn, yel, bn), $Y > Z > X$
- Riebeckite (A2), 1.699, (bl, yel-bn), $Z > X \simeq Y$
- Mn-rich Zoisite (E1), 1.699, (pink, yel), $X > Y > Z$
- Pigeonite (P8), 1.703, (gn, bn), $Y > X$ or Z
- Chloritoid (C5), 1.726, (gn, bl-gn), $Y > X > Z$
- Staurolite (S10), 1.747, (golden yel), $Z > Y > X$
- Allanite (E1), 1.757, (yel, bn), $Z > Y > X$
- Piemontite (E1), 1.768, (red, viol, bn), $Z > Y > X$

Biaxial Negative (−)
- Cordierite (C8), 1.549, (yel, bl, viol), $Z > Y > X$
- Chlorite Group (C4), 1.625, (gn), $Z = Y > X$, $X = Y > Z$
- Glaucophane (A2), 1.631, (viol, bl, yel), $Z > Y > X$
- Eckermannite (A2), 1.638, (bl-gn, yel-gn), $X > Y > Z$
- Andalusite (A1), 1.638, (pink, gn, yel), variable
- Chlorite Group (oxidized) (C4), 1.65, (gn), $Z = Y > X$, $X = Y > Z$
- Biotite (M2), 1.650, (bn, red-bn, gn), $Z \simeq Y > X$
- Actinolite (A2), 1.653, (yel-gn, bl-gn), $Z > Y > X$
- Stilpnomelane (S11), 1.660, (gn, bn, yel), $Z = Y >> X$
- Anthophyllite (A2), 1.663, (bn, gn), $Z > Y > X$
- Crossite (A2), 1.666, (yel, bl, viol), $Z > Y > X$
- Hornblende (A2), 1.666, (gn, bl-gn, yel-bn), $Z > Y > X$, $Y > Z > X$
- Gedrite (A2), 1.670, (gn, bn, yel), $Z > Y \simeq X$
- Axinite (A5), 1.691, (clove bn), $Y > X > Z$
- Grunerite (A2), 1.693, (gn, bn), $Z > Y > X$
- Arfvedsonite (A2), 1.694, (bl-gn), $X > Y > Z$
- Dumortierite (D1), 1.698, (bl, viol), $X > Y > Z$
- Riebeckite (A2), 1.699, (bl, yel-bn), $Z > Y \simeq X$
- Kyanite (A1), 1.721, (bl), $Z > Y > X$
- Chloritoid (C5), 1.726, (gn, bl-gn), $Y > X > Z$
- Epidote (E1), 1.740, (yel-gn), $Y > Z > X$
- Allanite (E1), 1.757, (brownish yel, bn, gn), $Z > Y > X$
- Aegirine (P8), 1.781, (gn, yel-gn), $X > Y > Z$

[1]Mineral listings within each category are arranged according to average value of ω or β, and include typical colors and pleochroic formulas.

EXPLANATION OF TABLES 4, 5, AND 6

Minerals are subdivided into the following categories: isotropic, uniaxial positive (+), uniaxial negative (−), biaxial positive (+), biaxial negative (−). If the variation in properties of a mineral exceeds a single category, the mineral is listed in more than one category.

The minerals are arranged on the basis of increasing average values of the index of refraction (ω for uniaxial minerals and β for biaxial minerals). These indices were chosen because they can be determined without the use of interference figures: grains of anisotropic materials that remain in total extinction during stage rotation permit estimation of either ω or β. The symbol * placed next to the average index value of a particular mineral indicates that the range in the value of ω or β is equal to or greater than 0.03.

The colors given are those seen in the microscope.

The abbreviations used are as follows:

acic	acicular	interf	interference	rks	rocks
alb	albite	interm	intermediate	sed	sedimentary, sediments
alt	alteration	lt	light	subh	subhedral
anh	anhedral	max	maximum	T	temperature
anom	anomalous	med	medium	tw	twins
bl	blue	met	metamorphic	uncom	uncommon
bn	brown	min	mineral	var	variety
com	common	mod	moderate	viol	violet
dk	dark	P	pressure	volc	volcanic
elong	elongate	perf	perfect	xls	crystals
euh	euhedral	pleoch	pleochroic	yel	yellow
gdmass	groundmass	poly	polysynthetic	Δ	birefringence
gn	green	prod	product	‖	parallel
hyd	hydrothermal	pshex	pseudohexagonal	⊥	perpendicular
ign	igneous	qtz	quartz		
		reg	regional		

Table 4
Isotropic Materials

Av. *n*	Index of Refraction	Name and Composition	Habit
1.43	1.40–1.46	Opal (S4) SiO_2 + 5–10% H_2O	Amorphous, colloform, tests of microfossils, woody structure
1.434	1.433–1.435	Fluorite (F3) CaF_2	Euh cubes, octahedrons, dodecahedrons, granular less com
1.485	1.483–1.487	Sodalite (S5) $Na_8Al_6Si_6O_{24}Cl_2$	Dodecahedral xls
1.486	1.479–1.493	Analcime (Analcite) (Z1) $NaAlSi_2O_6 \cdot H_2O$	Trapezohedral {211} xls
1.490	1.490	Sylvite (H1) KCl	Cubic, rarely octahedral, granular
1.490	1.485–1.495	Noselite (S5) $Na_8Al_6Si_6O_{24}SO_4$	Dodecahedral xls
1.500	1.496–1.505	Hauyne (S5) $(Na,Ca)_{8-4}Al_6Si_6O_{24}(SO_4)_{1-2}$	Dodecahedral xls, anh
1.544	1.544	Halite (H1) NaCl	Cubes, rarely octahedral, granular
1.565	1.48–1.65	Glass (G2) Rhyolite through basalt	Volc flows to patches in matrix of ign rks
1.714	~1.714	Pyrope (G1) $Mg_3Al_2(SiO_4)_3$	Euh to subh dodecahedral, trapezohedral {112} less com
1.734	~1.734	Grossular (G1) $Ca_3Al_2(SiO_4)_3$	Euh to subh dodecahedral, trapezohedral {112} less com
1.740	1.736–1.745	Periclase (P2) MgO	Euh cubes and octahedrons, anh
1.800	~1.800	Spessartine (G1) $Mn_3Al_2(SiO_4)_3$	Euh to subh dodecahedral, trapezohedral {112} less com
1.830	~1.830	Almandine (G1) $Fe_3Al_2(SiO_4)_3$	Euh to subh dodecahedral {110}, trapezohedral {112} less com
1.86	1.72–2.00	Spinel Group (S8) $(Mg,Fe)(Al,Cr)_2O_4$	Euh to anh, octahedrons and cubes, rounded
1.86	~1.86	Uvarovite (G1) $Ca_3Cr_2(SiO_4)_3$	Euh to subh dodecahedral {110}, trapezohedral {112} less com

Table 4 *(continued)*

Cleavage	Color	Hardness Sp. Grav.	Occurrence	Remarks
	Colorless, rarely pale bn, yel	5½–6½ 1.99–2.25	Secondary from circulating meteoric water in ign rks, sed deposits, and thermal springs	Silica group
{111} perf	Colorless, gn or viol, zoned	4 3.18	Veins and carbonate replacement, hot springs, pegmatites, carbonatites, granites, detrital and carbonate sed	
{110} poor	Colorless to pale gray-bl	5½–6 2.27–2.40	Volc and plutonic Si-poor ign rks, metasomatized carbonate contact rks	Sodalite group, {111} tw rare
{100} poor	Colorless	5½ ~2.25	Vein and cavity fillings in Si-poor ign rks, authigenic, alt prod of feldspars	Zeolite group, Isotropic, but Δ may be <0.002, fine {100}, {110} poly tw
{100} perf	Colorless	2 1.99	Uncom, in evaporite deposits	Halite group
{110} poor	Colorless to pale gray-bl	5½–6 2.27–2.40	Si poor volcanic rks, contact met rks	Sodalite group, {111} tw rare
{110} poor	Bl or bl-gn	5½–6 2.27–2.40	Phonolites or basalts, contact met rks	Sodalite group, {111} tw rare
{100} perf	Colorless	2–2½ 2.165	Evaporite deposits, salt domes	Halite group
None, spherical cracks	Colorless, gray, bn, gn, to opaque	5½–7 <1–2.76	Com in siliceous flows, less com in mafic flows, meteorites	Index is decreased by hydration
	Pale bn to pink	7½ 3.6	Ultramafic rks	Garnet group
	Colorless to pale yel	6½–7 3.6	Contact met rks	Garnet group, Δ = 0.000–0.005
{100} perf	Colorless, pale yel or bn	5½–6 3.56–3.65	Thermally met carbonate rks	{111} tw
	Pale bn to pink	7–7½ 4.2	Rare, Mn-rich skarns and pegmatites	Garnet group, Δ = 0.000–0.004
	Pale bn to pink	7–7½ 4.3	Reg and contact met rks, detrital sed	Garnet group
None, {111} parting rare	Very variable	7½–8 3.55–4.04	High-grade Al-rich met rks, mafic and Si-poor ign rks, com detrital	{111} simple tw com
	Emerald gn	7½ 3.9	Rare, veins in ultramafic rks, with chromite	Garnet group, Δ = 0.000–0.005

Table 4 *(continued)*

Av. *n*	Index of Refraction	Name and Composition	Habit
1.887	~1.887	Andradite (G1) $Ca_3Fe_2(SiO_4)_3$	Euh to subh dodecahedral {110}, trapezohedral {112} less com
2.435	2.37–2.50	Sphalerite (S6) (Zn,Fe)S	Tetrahedral or dodecahedral xls, granular, concretionary

Table 5
Uniaxial Materials

Av. ω	Indices ω, ε	Birefringence $(\varepsilon - \omega)$	Name and Composition	System and Habit
			Uniaxial Positive (+)	
1.483	1.472–1.494 1.470–1.485	0.002–0.010	Chabazite (Z1) $Ca(Al_2Si_4)O_{12} \cdot 6H_2O$	Hexagonal Cubelike rhombohedral $\{10\bar{1}1\}$ xls
1.510	1.508–1.511 1.509–1.511	0.000–0.001 ~isotropic	Leucite (F2) $KAlSi_2O_6$	Tetragonal (pseudoisometric) Euh trapezohedral phenocrysts, rare microlites, also dodecahedrons and cubes
1.534	1.531–1.536 1.533–1.538	0.000–0.003 ~isotropic	Apophyllite (A4) $KCa_4(Si_4O_{10})_2F \cdot 8H_2O$	Tetragonal Anh to euh, {001} plates, elong on *c* with {100} dominant
1.535	1.526–1.544 1.532–1.553	0.005–0.009	Chalcedony (S4) SiO_2	Hexagonal Fine grain size, fibrous, spherulitic, banded, colloform, vuggy pseudomorphic after fossils
1.544	1.544 1.553	0.009	α-Quartz (S4) SiO_2	Hexagonal Euh in volc rks after β-qtz, anh in plutonic and met rks, prisms in low T veins
1.572	1.572 1.592	0.020	Alunite (S12) $KAl_3(SO_4)_2(OH)_6$	Hexagonal Cubelike rhombohedral, tabular {0001}
*1.575	1.559–1.590 1.579–1.600	0.010–0.020	Brucite (B2) $Mg(OH)_2$	Hexagonal {0001} tabular, rarely fibrous

*Range of $\omega \geq 0.03$.

Table 4 *(continued)*

Cleavage	Color	Hardness Sp. Grav.	Occurrence	Remarks
	Pale to deep bn	6½–7 3.9	Contact met, alkaline ign rks, skarns, reg met, serpentinites	Garnet group Δ = 0.000–0.005
{110} perf	Pale yel or bn, colorless	3½–4 3.9–4.1	Hyd veins, contact met rks, Miss. Valley type ore deposits	{111} simple, poly tw com

Table 5 *(continued)*

Twins	Cleavage	Color and Pleochroism	Hardness Sp. Grav.	Occurrence	Remarks
		Uniaxial Positive (+)			
{0001} simple pene-tration com, {1011} simple less com	{10$\bar{1}$1} good	Colorless	4½ 2.1	Cavities and joints in basic and interm volc rks, hyd and hot spring deposits	Zeolite group, usually uni-axial neg (−), may be biaxial
Poly com	{110} poor	Colorless	5½–6 2.47–2.50	Phenocrysts in Si-poor volc rks	Feldspathoid group
{111} rare	{001} perf, {110} poor	Colorless	4½–5 2.33–2.37	Vugs and fissures in mafic volc rks, siliceous ign rks, contact met rks	Rarely uniaxial neg (−)
None		Colorless	6½–7 2.57–2.64	Near surface in cavities and veins, devitrifica-tion of silica gel, sed cement, nodules	Silica group
None		Colorless	7 2.65	Com in ign, met, and sed rks	Silica group
None	{0001} good, {1012} poor	Colorless	3½–4 ~2.71	Hyd alt prod	
None	{0001} perf	Colorless	2½ 2.39	Low-grade marbles from hydration of periclase	Anom dark bn interf color

Table 5 *(continued)*

Av. ω	Indices ω, ε	Birefringence $(\varepsilon - \omega)$	Name and Composition	System and Habit
			Uniaxial Positive (+)	
1.642	1.632–1.651 1.640–1.651	0.000–0.008	Akermanite (M1) $Ca_2MgSi_2O_7$	Tetragonal Sub to euh, tabular {001}
*1.728	1.703–1.752 1.700–1.746	0.001–0.008	Vesuvianite (Idocrase) (V1) $Ca_{10}(Mg,Fe)_2Al_4(SiO_4)_5(Si_2O_7)_2(OH)_4$	Tetragonal Stubby prismatic
*1.941	1.923–1.960 1.968–2.015	0.042–0.065	Zircon (Z2) $ZrSiO_4$	Tetragonal Tiny euh xls with {110} dominant, rounded, embayed
2.000	1.990–2.010 2.093–2.100	0.090–0.103	Cassiterite (C3) SnO_2	Tetragonal Stubby prisms
2.611	2.605–2.616 2.890–2.903	0.285–0.296	Rutile (R2) TiO_2	Tetragonal Euh with {100} and {110} dominant, acic
			Uniaxial Negative (−)	
1.483	1.472–1.494 1.470–1.485	0.002–0.010	Chabazite (Z1) $Ca(Al_2Si_4)O_{12} \cdot 6H_2O$	Hexagonal Cubelike rhombic {10$\bar{1}$1} xls
1.487	1.487 1.484	0.003	α-Cristobalite (S4) SiO_2	Tetragonal Euh octahedral and cubo-octahedral xls, fibrous varieties intergrown with sanidine, submicroscopic
*1.509	1.490–1.528 1.488–1.503	0.002–0.025	Cancrinite (C1) $(Na,Ca,K)_{6-8}(Al_6Si_6O_{24})$ $(CO_3,SO_4,Cl)_{1-2} \cdot 1\text{–}5H_2O$	Hexagonal Anh to subh, may be elong on *c*
1.539	1.529–1.549 1.526–1.544	0.003–0.005	Nepheline (F2) $NaAlSiO_4$	Hexagonal Euh to anh, prisms and basal pinacoid dominant
1.541	1.537–1.545 1.537–1.544	0.000–0.003	Apophyllite (A4) $KCa_4(Si_4O_{10})_2F \cdot 8H_2O$	Tetragonal Anh to euh, {001} plates, elong on *c* with {100} dominant

*Range of $\omega \geq 0.03$.

Table 5 *(continued)*

Twins	Cleavage	Color and Pleochroism	Hardness Sp. Grav.	Occurrence	Remarks
		Uniaxial Positive (+)			
None	{100} poor	Colorless to pale yel Weak $O > E$	5–6 2.94–3.04	Rare, in Si-poor CaAl-rich volc rks	Melilite group, anom bl interf color
Uncom, may be in 4 sectors	{110}, {001} poor	Colorless, may be gn, bn, yel, zoned Weak $O > E$	6–7 3.33–3.45	Contact met rks	Anom bl or olive-yel interf colors, usually uniaxial (−), may be biaxial
{111} uncom	{110} poor	Colorless, may be pale bn Weak $E > O$	7½ 4.6–4.7	Com accessory min in ign and met rks, com detrital min in sed rks	May be en-closed by pleoch haloes
{101} simple or poly com, cyclic	{100}, {110} poor, {111} parting	Yel, bn, red to colorless $E > O$ weak to strong	6–7 6.98–7.02	Siliceous ign rks, hyd veins, pegmatites, gneisses, skarns, low-T colloidal	May be anom biaxial
{011} simple or cyclic	{110}, {100} good, {111} poor, {902}, {101} parting	Deep red-bn, yel-bn, to opaque Weak $E > O$	6–6½ 4.2–5.6	Accessory min in ign and met rks, detrital in sed rks	Rutile group
		Uniaxial Negative (−)			
{0001} simple pene-tration com, {1011} simple less com	{10$\bar{1}$1} good	Colorless	4½ 2.1	Cavities and joints in basic and interm volc rks, hyd and hot spring dep	Zeolite group May be uni-axial (+) or biaxial
{111} poly tw com		Colorless	6½ 2.33	Vugs and fissures in siliceous volc rks, low-T biochem depos-its, and clays	Silica group
Uncommon	{10$\bar{1}$0} perf	Colorless	5–6 2.31–2.51	Nepheline syenite, often with sodalite	Feldspathoid group
	{10$\bar{1}$0} poor, {0001}	Colorless	5½–6 2.56–2.67	Qtz-free alkaline ign rks	Feldspathoid group
{111} rare	{001} perf, {110} poor	Colorless	4½–5 2.33–2.37	Vugs and fissures in mafic volc rks, siliceous ign rks, contact met rks	May have anom interf colors, usually uniaxial (+), ~isotropic

Table 5 *(continued)*

Av. ω	Indices ω, ε	Birefringence (ω − ε)	Name and Composition	System and Habit
			Uniaxial Negative (−)	
*1.570	1.540–1.600 1.535–1.565	0.004–0.037	Scapolite Group (S2) $(Ca,Na)_4[(Al,Si)_3Al_3Si_6O_{24}](Cl,CO_3)$	Tetragonal Granular, prismatic
1.580	1.567–1.594 1.563–1.586	0.004–0.008	Beryl (B1) $Be_3Al_2(SiO_3)_6$	Hexagonal Euh, prismatic with $\{10\bar{1}0\}$ and {0001} dominant, rarely in radial masses, anh
*1.635	1.603–1.667 1.598–1.665	0.003–0.017	Apatite Group (A3) $Ca_5(PO_4)_3(F,OH,Cl)$	Hexagonal Prismatic, oolitic, botryoidal, colloform
1.642	1.633–1.650 1.629–1.646	0.003–0.005	Fluorapatite (A3) $Ca_5(PO_4)_3(F,OH,Cl)$	Hexagonal Prismatic, oolitic, botryoidal, colloform
*1.655	1.635–1.675 1.610–1.650	0.017–0.035	Tourmaline (T3) $Na(Mg,Fe,Li,Al)_3Al_6(Si_6O_{18})(BO_3)_3(OH,F)_4$	Hexagonal Usually euh, stubby 3- to 6-sided prisms
1.658	1.658 1.486	0.172	Calcite (C2) $CaCO_3$	Hexagonal Variable, rhombs and scalehedrons common, usually anh
1.660	1.651–1.670 1.651–1.658	0.000–0.012	Gehlenite (M1) $CaAl_2SiO_7$	Tetragonal Subh to euh, tabular {001}
1.679	1.679 1.500	0.179	Dolomite (C2) $CaMg(CO_3)_2$	Hexagonal Anh, rhombohedral $\{10\bar{1}1\}$
1.704	1.700–1.709 1.509–1.516	0.191–0.193	Magnesite (C2) $MgCO_3$	Hexagonal Anh, $\{10\bar{1}1\}$ rhombs rare

*Range of ω ≥ 0.03.

Table 5 *(continued)*

Twins	Cleavage	Color and Pleochroism	Hardness Sp. Grav.	Occurrence	Remarks
		Uniaxial Negative (−)			
	{100}, {110} good	Colorless	5½–6 2.54–2.81	Met and metasomatic rks with abundant NaCl or CO_2	
	{0001} poor	Colorless, rarely pale yel or gn, $E > O$	7½–8 2.70	Siliceous plutonic rks and pegmatites, less com in met rks and veins in limestone	May be anom biaxial
Rare	{0001} poor	Colorless, rarely pale bn or gray Weak to mod $E > O$	5 2.9–3.5	Ign and met accessory min, detrital in sed, cryptocrystalline aggregates	
Rare	None, or {001} poor	Colorless, rarely pale bn or gray Weak to mod $E > O$	5 2.9–3.5	Ign and met accessory min, detrital in sed, cryptocrystalline in sed	Apatite group
Rare	$\{11\bar{2}0\}$, $\{10\bar{1}1\}$ poor, {0001} parting may be good	Very variable, zoned Strong $O > E$	7–7½ 3.0–3.25	Late stage pneumatolytic accessory min in ign and met rks, com detrital min	Distinguished from biotite by $O > E$
$\{10\bar{1}2\}$ very com, {0001} com	$\{10\bar{1}1\}$ perf	Colorless or turbid	3 2.96	Limestones, marbles, carbonatites, kimberlites, secondary in cavities in ign rks, veins, alt prod of feldspar	Carbonate group, tw ‖ long diagonal, rapid effervescence with HCl
	{100} poor	Colorless to pale yel Weak $O > E$	5–6 2.94–3.04	Rare in Si-poor mafic and CaAl-rich volc rks	Melilite group, anom bl interf color
{0001}, $\{10\bar{1}0\}$ $\{11\bar{2}0\}$ simple, {2021} poly	$\{10\bar{1}1\}$ perf	Colorless	3½–4 2.86–3.10	Dolomite rks, primary in evaporite deposits, dolomitic marbles, hyd veins, secondary in ultramafic rks	Carbonate group, Tw ‖ long and short diagonals, slow effervescence with HCl
{0001} poly rare	$\{10\bar{1}1\}$ perf	Colorless	3½–4½ 3.0	Serpentinites, met schists and contact rks, sed evaporite deposits	Carbonate group effervescence in warm HCl

Table 5 *(continued)*

Av. ω	Indices ω, ε	Birefringence (ω − ε)	Name and Composition	System and Habit
			Uniaxial Negative (−)	
*1.714	1.690–1.738 1.510–1.542	0.180–0.196	Ankerite (C2) $Ca(Mg,Fe)(CO_3)_2$	Hexagonal Anh, rhombohedral $\{10\bar{1}1\}$
*1.728	1.703–1.752 1.700–1.746	0.001–0.008	Vesuvianite (Idocrase) (V1) $Ca_{10}(Mg,Fe)_2Al_4(SiO_4)_5(Si_2O_7)_2(OH)_4$	Tetragonal Stubby prismatic
1.770	1.767–1.772 1.759–1.763	0.007–0.010	Corundum (C9) Al_2O_3	Hexagonal Euh prisms and dipyramids, anh, granular
*1.802	1.769–1.834 1.563–1.608	0.206–0.242	Rhodochrosite (C2) $MnCO_3$	Hexagonal Anh aggregates, botryoidal
1.870	1.864–1.875 1.626–1.633	0.238–0.242	Siderite (C2) $FeCO_3$	Hexagonal Anh aggregates, $\{10\bar{1}1\}$ single xls, radial fibers, botryoidal, earthy, oolitic
2.561	2.561 2.488	0.073	Anatase (R2) TiO_2	Tetragonal Anh, pyramidal, tabular, prismatic

*Range of $\omega \geq 0.03$.

Table 6
Biaxial Materials

Av. β	Indices α, β, γ	Birefringence (γ − α)	Name and Composition	2V and Dispersion	System and Habit
			Biaxial Positive (+)		
1.474	1.469–1.479 1.469–1.480 1.473–1.483	0.002–0.004	α-Tridymite (S4) SiO_2	35°–90°	Orthorhombic Small, euh, pshex basal plates, fan-shaped groups
1.481	1.473–1.483 1.476–1.486 1.485–1.496	0.012–0.013	Natrolite (Z1) $Na_2(Al_2Si_3)O_{10} \cdot 2H_2O$	58°–64° $v > r$ weak	Orthorhombic Prismatic, acic on *c*

Table 5 *(continued)*

Twins	Cleavage	Color and Pleochroism	Hardness Sp. Grav.	Occurrence	Remarks
		Uniaxial Negative (−)			
$\{0001\}$, $\{10\bar{1}0\}$, $\{11\bar{2}0\}$ simple, $\{2021\}$ poly	$\{10\bar{1}1\}$ perf	Colorless, bn with alteration	3½–4 2.86–3.10	Veins and concretions in Fe-rich sed, in hyd ore zones with siderite	Carbonate group, effervescence in warm HCl
Uncom, may be in four sectors	$\{110\}$, $\{001\}$ poor	Colorless, may be gn, bn, yel, zoned Weak $O > E$	6–7 3.33–3.45	Contact met rks	Anom bl or olive-yel interf colors, rarely uniaxial (+) or biaxial
$\{10\bar{1}1\}$ simple or poly, $\{1011\}$, $\{0001\}$ glide com	$\{0001\}$, $\{1011\}$ parting com	Colorless, may be pale yel, bl, pink, gn Weak $O > E$	9 3.98–4.10	Si-poor and Al-rich ign and met rks, a heavy min in sed	May be anom biaxial
Uncom	$\{10\bar{1}1\}$ perf	Colorless, pale pink $O > E$	3½–4 3.2–3.7	High-T metasomatic and hyd veins, alt prod in Mn deposits	Carbonate group
$\{10\bar{1}2\}$ poly uncom	$\{10\bar{1}1\}$ perf	Colorless, pale yel or yel-bn	4–4½ 3.50–3.96	Sed, bog deposits, gangue min in hyd deposits	Carbonate group
$\{112\}$ rare	$\{001\}$, $\{111\}$ perf	Yel, bn, bl, gn, black Weak $E > O$, rarely $O > E$	5½–6 3.82–3.97	Rare, hyd alt prod in ign and met rks, accessory min in detrital sed	Rutile group

Table 6 *(continued)*

Optic Orientation	Cleavage	Color and Pleochroism	Hardness Sp. Grav.	Occurrence	Remarks
		Biaxial Positive (+)			
$a = Y, b = X, c = Z$	None	Colorless	7 2.27	Vugs and fractures in interm volc rks, gdmass and lithophysae in glass	Silica group, pshex $\{10\bar{1}0\}$ twins com
$a = X, b = Y, c = Z$	$\{110\}$ perf, $\{010\}$ parting	Colorless	5 2.20–2.26	Cavities and veins in mafic volc rks and serpentinites, alt prod of feldspathoids	Zeolite group

Table 6 *(continued)*

Av. β	Indices α, β, γ	Birefringence $(\gamma - \alpha)$	Name and Composition	$2V$ and Dispersion	System and Habit
			Biaxial Positive (+)		
1.494	1.487–1.499 1.487–1.501 1.488–1.505	0.001–0.007	Heulandite (Z1) $(Ca,Na_2)(Al_2Si_7)O_{18} \cdot 6H_2O$	≃30° $r > v$ mod	Monoclinic Tabular {010}, equant, anh
1.523	1.497–1.530 1.513–1.533 1.518–1.544	0.006–0.021	Thomsonite (Z1) $NaCa_2[(Al,Si)_5O_{10}]_2 \cdot 6H_2O$	42°–75° $r > v$ strong	Orthorhombic Bladed, acic on *c*
1.524	1.519–1.521 1.522–1.526 1.529–1.531	0.010	Gypsum (S12) $CaSO_4 \cdot 2H_2O$	58°	Monoclinic Euh, elong on *c*, stubby prismatic or acic, massive
1.531	1.527 1.531 1.538	0.010	Low Albite (An_0) (F1) $NaAlSi_3O_8$	77°	Triclinic Euh to subh, {010} plates, or elong on *a*
*1.549	1.522–1.560 1.524–1.574 1.527–1.578	0.005–0.018	Cordierite (C8) $(Mg,Fe)_2Al_3(Si_5Al)O_{18}$	75°–90°(+), usually 40°–90°(−) $v > r$ weak	Orthorhombic (pshex) Euh to anh, poikiloblastic
1.553	1.549 1.553 1.557	0.008	Andesine (An_{40}) (F1) $(NaSi)_{0.6}(CaAl)_{0.4}AlSi_2O_8$	83° $r > v$ weak	Triclinic Euh to subh, {010} plates or elong on *a*
1.564	1.560 1.564 1.568	0.008	Labradorite (An_{60}) (F1) $(NaSi)_{0.4}(CaAl)_{0.6}AlSi_2O_8$	80° $r > v$ weak	Triclinic Euh to subh, {010} plates, elong on *a*
1.572	1.561–1.567 1.566–1.579 1.587–1.593	0.026–0.027	Norbergite (H2) $Mg_2SiO_4 \cdot Mg(OH,F)_2$	44°–50°	Orthorhombic Anh, massive, pinacoidal {100}, {010}, or {001} plates
1.574	1.568–1.580 1.568–1.580 1.587–1.600	0.019	Gibbsite (C6) $Al(OH)_3$	0°–40° $r > v$ strong, occ $v > r$	Monoclinic {001} plates
1.576	1.570 1.576 1.614	0.044	Anhydrite (S12) $CaSO_4$	≃43° $v > r$ strong	Orthorhombic {100}, {010}, {001} dominant, rarely fibrous
1.592	1.586 1.592 1.614	0.028	Colemanite (C7) $Ca_2B_6O_{11} \cdot 5H_2O$	55° $v > r$ weak	Monoclinic Prismatic, elong on *c* with {110} dominant

*Range of $\beta \geq 0.03$.

Table 6 *(continued)*

Optic Orientation	Cleavage	Color and Pleochroism	Hardness Sp. Grav.	Occurrence	Remarks
		Biaxial Positive (+)			
$b = Z, c \wedge Y = -8°$ to $-32°$	{010} perf	Colorless	3½–4 2.2	Cavities and veins in mafic volc rks	Zeolite group
$a = X, b = Z, c = Y$	{010} perf, {100} good	Colorless	5–5½ 2.1–2.39	Cavities and veins in mafic volc rks	Zeolite group
$b = Y, c \wedge Z \simeq 52°$	{010} perf, {100}, {111} good	Colorless	2 2.31	Evaporites, salt dome caps, some limestones, oxidation of sulfides	{100} simple, poly tw com
On {001} $a \wedge \alpha' = 3°$ on {010} $a \wedge \alpha' = 20°$	{001} perf, {010} good, {110} poor	Colorless	6–6½ 2.63	Felsic and interm ign rks, pegmatites, hyd veins, low-grade and contact met rks, authigenic, detrital sed rks	Feldspar group, poly tw com
$a = Y, b = Z, c = X$	{010} good, {100}, {001} poor, {001} parting	Colorless, yel, bl, viol Usually none, may be strong $Z > Y > X$	7–7½ 2.53–2.78	Contact met rks, hornfelses, less com in reg met and ign rks, rare in sed rks	Pleoch haloes, resembles qtz and feldspar, {110} cyclic and poly com
On {001} a $\wedge \alpha' = 2.5°$ on {010} a $\wedge \alpha' = 10°$	{001} perf, {010} good, {110} poor	Colorless	6–6½ 2.68	Felsic and interm ign rks, med to high grade met rks, detrital sed	Feldspar group, poly tw com
On {001} a $\wedge \alpha' = 10°$ on {010} a $\wedge \alpha' = 23°$	{001} perf, {010} good, {110} poor	Colorless	6–6½ 2.71	Mafic ign rks, detrital sed	Feldspar group, poly tw com
$a = X, b = Z, c = Y$	{001} poor	Colorless, yel, or bn $X > Y = Z$ mod	6–6½ 3.1–3.4	Rare, contact met rks	Humite group
$b = X, c \wedge Z = -21°$	{001} perf	Colorless to pale bn	2½–3½ 2.38–2.42	Lateritic clays and soils, bauxite, low-T hyd veins	Clay group
$a = Y, b = X, c = Z$	{010} perf, {001}, {100} good	Colorless	3–3½ 2.96	Evaporites, oxidized zones with ores, fumaroles	Sulfate group, {011} simple and poly tw com
$b = X, c \wedge Z = 84°$	{010} perf, {001} poor	Colorless	4½ 2.42	Rare, boron-rich playas	

Table 6 *(continued)*

Av. β	Indices α, β, γ	Birefringence $(\gamma - \alpha)$	Name and Composition	$2V$ and Dispersion	System and Habit
			Biaxial Positive (+)		
1.610	1.595–1.610 1.605–1.615 1.632–1.645	0.030–0.038	Pectolite (P1) $Ca_2NaH(SiO_3)_3$	50°–63° $r > v$ weak	Triclinic Acic on *c*, massive
1.622	1.609–1.630 1.612–1.632 1.618–1.640	0.009–0.010	Topaz (T2) $Al_2(SiO_4)(F,OH)_2$	48°–68° $r > v$ mod	Orthorhombic Euh, elong on *c*
1.6235	1.621–1.622 1.623–1.624 1.630–1.632	0.009–0.010	Celestite (S12) $SrSO_4$	50° $v > r$ mod	Orthorhombic Tabular with {001} and {110} dominant, elong prismatic on *a*, fibrous, granular
1.625	1.57–1.67 1.57–1.68 1.59–1.68	0.000–0.010	Chlorite Group (C4) $(Mg,Al,Fe)_{12}$ $[(Si,Al)_8O_{20}](OH)_{16}$	0°–45°(+), 0°–30°(−) $v > r$ strong	Monoclinic {001} dominant, pshex plates, scaly, massive
*1.628	1.592–1.643 1.602–1.655 1.619–1.675	0.025–0.037	Chondrodite (H2) $2Mg_2SiO_4 \cdot Mg(OH,F)_2$	64°–90° $r > v$ weak to strong	Monoclinic Massive, granular, pinacoidal plates on {100}, {010}, or {001}
1.629	1.611–1.630 1.617–1.641 1.632–1.669	0.021–0.039	Prehnite (P5) $Ca_2Al(AlSi_3)O_{10}(OH)_2$	65°–69° $r > v$ weak, may be $v > r$ wk to str	Orthorhombic Tabular {001}, spherulitic
*1.636	1.607–1.643 1.619–1.653 1.639–1.675	0.028–0.036	Humite (H2) $3Mg_2SiO_4 \cdot Mg(OH,F)_2$	65°–84° $r > v$ weak	Orthorhombic Massive, granular, pinacoidal plates
1.638	1.636–1.637 1.637–1.639 1.647–1.649	0.011–0.012	Barite (S12) $BaSO_4$	36°–37° $v > r$ weak	Orthorhombic Tabular plates on {001}
1.652	1.651 1.652 1.660	0.009	Enstatite (En_{100}) (P8) $MgSiO_3$	35° $r > v$ weak	Orthorhombic Stubby prismatic on *c*
*1.652	1.634–1.666 1.635–1.670 1.644–1.690	0.010–0.024	Mullite (Al) $(Al_2O_3)_{3-2}(SiO_2)_{2-1}$	20°–50° $r > v$ weak	Orthorhombic Prismatic fibrous ‖ to *c*, cross section ~ square

*Range of $\beta \geq 0.03$.

Table 6 *(continued)*

Optic Orientation	Cleavage	Color and Pleochroism	Hardness Sp. Grav.	Occurrence	Remarks
		Biaxial Positive (+)			
$a \wedge Y = -10°$ to $-16°$ $b \wedge Z \simeq 2°$ $c \wedge X \simeq -5°$ to $-11°$	{100}, {001} perf	Colorless	4½–5 2.75–2.90	Cavities and veins in mafic volc rks	{100} simple tw com
$a = X, b = Y, c = Z$	{001} perf	Colorless, detrital grains are yel-pink $X, Y =$ yel, $Z =$ pink	8 3.49–3.57	Vugs and fissures in siliceous ign rks, pegmatites, greisen, detrital sed	
$a = Z, b = Y, c = X$	{001} perf, {210} good, {010} poor	Colorless	3–3½ ~3.97	Evaporites, carbonate sed	Sulfate group
$b = Y, c \wedge Z = 0°–9°$, also $c \wedge X = 0°–7°$	{001} perf	Colorless to lt or dk gn Weak to mod, $Z = Y > X$ or $X = Y > Z$	1–2½ 2.64–3.21	Fine-grained sed, low-grade met rks, deuteric alt prod	Anom bn, viol, bl interf colors, {001} poly tw com
$b = Z, c \wedge Y = 22°–31°$	{001} poor	Colorless to yel or bn Mod $X > Y = Z$	6–6½ 3.1–3.4	Contact met rks, carbonatites	Humite group, {001} simple or poly tw com
$a = X, b = Y, c = Z$	{001} good, {110} poor	Colorless	6–6½ 2.90–2.95	Veins and cavities in ultramafic, mafic, and interm ign rks	Poly tw
$a = X, b = Z, c = Y$	{001} poor	Colorless to yel or bn $X > Y \simeq Z$ mod	6–6½ 3.1–3.4	Limestone or dolomite contact met rks with CaMg silicates	Humite group
$a = Z, b = Y, c = X$	{001}, {210} perf, {010} good to poor	Colorless	2½–3½ 4.5	Gangue min in hyd veins in limestones and dolomites	Sulfate group, {110} poly tw com
$a = X, b = Y, c = Z$	{210} good, {100}, {010} parting	Colorless	5 3.21	Ultramafic ign rks, volc ign rks, high-grade met rks, pyroxenites, dunites, peridotites, meteorites	Pyroxene group, {100} simple or poly tw com
$a = X, b = Y, c = Z$	{010} good	Colorless, or pinkish $X > Y = Z$ weak	6–7 ~3.19	Rare, high-T, low-P met rks, xenoliths	Aluminosilicate group

Table 6 *(continued)*

Av. β	Indices α, β, γ	Birefringence $(\gamma - \alpha)$	Name and Composition	$2V$ and Dispersion	System and Habit
			Biaxial Positive (+)		
1.655	1.64–1.65 1.65–1.66 1.65–1.67	~0.015	Boehmite (C6) γ–AlO(OH)	~80°	Orthorhombic {010} plates
*1.657	1.632–1.663 1.638–1.677 1.655–1.697	0.022–0.035	Cummingtonite (A2) $(Mg_{0.69-0.3}Fe_{0.31-0.7})_7$ $Si_8O_{22}(OH)_2$	70°–90° $v > r$ weak	Monoclinic Prismatic, acic asbestiform, elong on *c*
1.659	1.657–1.660 1.658–1.661 1.677–1.682	0.020–0.022	Sillimanite (A1) Al_2SiO_5	21°–30° $r > v$ strong	Orthorhombic Prismatic, acic, elong on *c*
1.660	1.654–1.658 1.657–1.663 1.665–1.674	0.011–0.016	Jadeite (P8) $NaAl(SiO_3)_2$	68°–72° $v > r$ mod	Monoclinic Fibrous aggregates, anh, granular
1.662	1.648–1.663 1.655–1.669 1.662–1.679	0.014–0.027	Spodumene (P8) $LiAl(SiO_3)_2$	55°–70° $v > r$ weak	Monoclinic Euh prismatic on *c*
*1.663	1.604–1.694 1.617–1.710 1.628–1.722	0.024–0.028	Anthophyllite (A2) $(Mg_{0.9-0.1}Fe_{0.1-0.9})_7Si_8O_{22}$ $(OH)_2$	70°–90°(+), 82°–90°(−) $r > v$, $v > r$ weak to mod	Orthorhombic Prismatic, fibrous, elong on *c*
*1.666	1.615–1.705 1.618–1.714 1.632–1.730	0.014–0.026	Hornblende (A2) $Ca_2(Mg,Fe)_4(Al,Fe)$ $(Si_7Al)O_{22}(OH)_2$	85°–90°(+), 23°–90°(−) $r > v$, $v > r$ mod	Monoclinic Prismatic, elong on *c*
*1.670	1.627–1.690 1.635–1.705 1.644–1.717	0.017–0.027	Gedrite (A2) $(Mg_{0.9-0.1}Fe_{0.1-0.9})_5Al_2$ $Si_6Al_2O_{22}(OH)_2$	47°–90°(+), 72°–90°(−) $v > r$ weak	Orthorhombic Prismatic, fibrous, elong on *c*
1.672	1.664 1.672 1.694	0.030	Diopside (Di_{100}) (P8) $CaMg(SiO_3)_2$	58° $r > v$ weak	Monoclinic Stubby prismatic, elong on *c*, granular

*Range of $\beta \geq 0.03$.

Table 6 *(continued)*

Optic Orientation	Cleavage	Color and Pleochroism	Hardness Sp. Grav.	Occurrence	Remarks
		Biaxial Positive (+)			
$a = Z, b = Y, c = X$	{010} good to perf	Colorless	3½–4 3.01–3.06	Bauxite, lateritic clay and soil	Clay group
$b = Y, c \wedge Z =$ 16°–21°	{110} good	Colorless, pale gn or bn Weak, $Z > Y \geq X$	5–6 3.3–3.6	Contact and reg met rks, interm and mafic ign rks	Amphibole group, {100} simple or poly tw com
$a = X, b = Y, c = Z$	{010} perf	Colorless, rarely bn or bluish	6–7 3.24	High-T met rks, detrital, rare in pegmatites and qtz veins	Aluminosilicate group
$b = Y, c \wedge Z =$ 33°–36°	{110} good {100} parting	Colorless, greenish Weak, X = pale gn, Y = colorless, Z = pale yel	6 3.25–3.40	High-P, low-T met rks	Pyroxene group, {100} fine poly tw
$b = Y, c \wedge Z =$ 22°–26°	{110} good, {010} parting	Colorless	6½–7 3.0–3.2	Granitic pegmatites	Pyroxene group
$a = X, b = Y, c = Z$	{210} perf, {010} good, {100} poor {001} parting	Colorless, pale bn, pale gn weak to mod $Z > Y \simeq X$, $Z \simeq Y > X$	5½ ~3.1	High-grade met rks	Amphibole group
$b = Y, c \wedge Z =$ 13°–24°	{110} perf, {100}, {001} parting rare	Gn, bl-gn, yel-bn Mod $Z > Y > X$, less com $Y > Z > X$	5–6 3.04–3.45	Common in ign and met rks	Amphibole group, {100} simple tw com
$a = X, b = Y, c = Z$	{210} perf, {010} good, {100} poor, {001} parting	Pale gn, bn, yel Weak to mod $Z > Y \simeq X$	5½ ~3.3	Medium-grade met rks	Amphibole group
$b = Y, c \wedge Z =$ 38°	{110} good, {100} parting	Pale gn, colorless Weak to none	5½ 3.2	Ca- and Mg-rich met rks, granulites, gneisses, alkali-rich mafic ign rks, peridotites, anorthosites, pegmatites	Pyroxene group, {100}, {001} simple or poly tw com

Table 6 *(continued)*

Av. β	Indices α, β, γ	Birefringence $(\gamma - \alpha)$	Name and Composition	2V and Dispersion	System and Habit
			Biaxial Positive (+)		
*1.672	1.623–1.702 1.636–1.709 1.651–1.728	0.028–0.045	Clinohumite (H2) $4Mg_2SiO_4 \cdot Mg(OH,F)_2$	52°–90° $r > v$ weak to strong	Monoclinic Granular, pinacoidal plates on {100}, {010}, or {001}
1.674	1.665 1.672–1.676 1.684–1.686	0.019–0.021	Lawsonite (L1) $CaAl_2Si_2O_7(OH)_2 \cdot H_2O$	76°–87° $r > v$ strong	Orthorhombic Euh, tabular on {001}
1.675	1.667–1.669 1.674–1.676 1.689–1.693	0.022–0.024	Omphacite (P8) $(Ca,Na)(Mg,Fe^{+2},Fe^{+3},Al)(SiO_3)_2$	60°–74° $r > v$ mod	Monoclinic Coarse granular
1.692	1.685 1.692 1.714	0.029	Salite (Di_{70}) (P8) $Ca(Mg_{0.7}Fe_{0.3})SiO_3$	57° $r > v$ mod	Monoclinic Stubby, prismatic on c, granular
*1.695	1.674–1.702 1.675–1.715 1.688–1.722	0.012–0.022	Pumpellyite (P6) $Ca_4(Mg,Fe,Mn)(Al,Fe,Ti)_5 O(OH)_3 [Si_2O_7]_2[SiO_4]_2 \cdot 2H_2O$	26°–85°(+), rarely (−) $v > r$ strong	Monoclinic Fibrous, bladed, acic, elong on b
*1.699	1.677–1.701 1.684–1.714 1.689–1.720	0.012–0.018	Riebeckite (A2) $Na_2(Fe,Mg)_3(Fe,Al)_2(Si_8O_{22})(OH)_2$	0°–90°(+) $v > r$ strong 42°–90°(−) $r > v$	Monoclinic Prismatic, fibrous, elong on c
1.699	1.685–1.707 1.688–1.711 1.697–1.725	0.005–0.020	Zoisite (E1) $Ca_2Al_3[Si_2O_7][SiO_4](OH)$	0°–60° $v > r$ strong $r > v$ strong	Orthorhombic Elong on b, granular
*1.703	1.682–1.722 1.684–1.722 1.705–1.751	0.023–0.029	Pigeonite (P8) $Ca(Mg,Fe,Al)((Si,Al)O_3)_2$	0°–32° $r > v$ or $v > r$ weak to mod	Monoclinic Euh to anh, stubby prismatic, elong on c
1.705	1.635–1.731 1.651–1.760 1.670–1.775	0.035–0.043	Mg-rich Olivine (Fo_{100-50}) (O1) $(Mg_{2-1},Fe_{0-1})SiO_4$	82°–90°(+) 73°–90°(−) $v > r$ weak $r > v$ weak	Orthorhombic Anh to euh, elong on c, skeletal
*1.706	1.671–1.735 1.672–1.741 1.703–1.761	0.018–0.030	Augite (P8) $Ca(Mg,Fe,Al)((Si,Al)O_3)_2$	25°–60° $r > v$ weak to mod	Monoclinic Stubby prismatic, elong on c, microlites, anh

*Range of $\beta \geq 0.03$.

Table 6 *(continued)*

Optic Orientation	Cleavage	Color and Pleochroism	Hardness Sp. Grav.	Occurrence	Remarks
		Biaxial Positive (+)			
$b = Z, c \wedge Y = 7°–15°$	{001} poor	Colorless, yel to bn Mod $X > Y = Z$	6–6½ 3.1–3.4	Contact met rks, uncom in altered peridotites	Humite group, {100} simple or poly tw com
$a = Y, b = Z, c = X$	{100}, {010} perf, {101} poor	Colorless	7–8 ~3.1	Rare, low-T, high-P met rks	{110} simple or poly tw com
$b = Y, c \wedge Z =$ 39°–43°	{110} good	Colorless to pale gn Weak, $X = Y =$ pale gn, $Z =$ colorless	5–6 3.29–3.37	Uncommon, in high-grade met rks, eclogite	Pyroxene group, {100} poly tw
$b = Y, c \wedge Z = 42°$	{110} good, {100} parting	Brownish gn Mod	5½–6½ 3.3	Ca- and Mg-rich met rks, granulites, gneisses, alkali-rich mafic ign rks	Pyroxene group, {100}, {001} simple or poly tw com
$b = Y, c \wedge Z = 4°$ to $-24°$, rarely $b = Z$, $c \wedge Y \simeq -50°$	{001}, {100} good	Pale gn, yel, or bn Strong $Y > Z > X$	6 3.18–3.23	Greenschist and blueschist met rks	{100}, {001} tw relatively com
$b = Z, c \wedge Y = 8°–9°$ $b = Y, c \wedge X = 7°–8°$	{110} perf, {010}, {001} parting	Bl to yel-bn Mod to strong $Z > Y \simeq X$	6 3.0–3.4	Na-rich plutonic ign rks, occasionally in met rks	Amphibole group
$a = Y, b = X, c = Z$ (α-Zois) $a = X, b = Y, c = Z$ (β-Zois)	{100} perf, {001} poor	Mainly colorless, Mn var pink to yel $X > Y > Z$ mod	6 3.15–3.36	Med-grade met rks, from deuteric alt of plagioclase	Epidote group, anom bl interf color
$b = X, c \wedge Z =$ 37°–44°, $b = Y, c \wedge Z =$ 40°–44°	{110} good, {001} parting	Colorless, pale gn or bn Mod $Y > X$ or Z	6 3.3–3.46	Matrix phase in andesites, basalts, and diabases	Pyroxene group, {100} simple or poly tw com
$a = Z, b = X, c = Y$	{010}, {100} poor	Colorless	7 3.22–3.75	Ultramafic and mafic ign rks, contact met rks	Olivine group, {100}, {011}, {012} simple tw mod com
$b = Y, c \wedge Z =$ 35°–50°	{110} good, {100}, {001} parting	Colorless to pale gn Weak Z or $X > Y$	5–6 3.2–3.6	Com in mafic and ultramafic ign rks, met gneisses and granulites	Pyroxene group, {100} simple or poly tw com

Table 6 *(continued)*

Av. β	Indices α, β, γ	Birefringence $(\gamma - \alpha)$	Name and Composition	$2V$ and Dispersion	System and Habit
			Biaxial Positive (+)		
1.710	1.697–1.714 1.699–1.722 1.702–1.729	0.005–0.015	Clinozoisite (E1) $Ca_2(Al,Fe)Al_2O[Si_2O_7]$ $[SiO_4](OH)$	14°–90° $v > r$ strong	Monoclinic Elong on *b*
1.715	1.682–1.706 1.705–1.725 1.730–1.752	0.046–0.052	Diaspore (C6) α-AlO(OH)	84°–86° $v > r$ weak	Orthorhombic {010} tabular
1.719	1.713 1.719 1.739	0.026	Ferrosalite (Di_{30}) (P8) $Ca(Fe_{0.7},Mg_{0.3})(SiO_3)_2$	60° $r > v$ mod	Monoclinic Stubby prismatic, elong on *c*, granular
*1.726	1.700–1.722 1.710–1.742 1.730–1.758	0.030–0.036	Aegirine-augite (P8) (Na,Ca) (Fe^{+3},Fe^{+2},Mg,Al) $(SiO_3)_2$	70°–90° $r > v$ mod to strong	Monoclinic Stubby prismatic, elong on *c*
1.726	1.713–1.728 1.719–1.734 1.723–1.740	0.010–0.012	Chloritoid (C5) $(Fe,Mg,Mn)_2$ $(Al,Fe)Al_3O_2$ $(SiO_4)_2(OH_4)$	40°–65°(+) rarely large (−) $r > v$ strong	Monoclinic, triclinic less com Tabular {001}
1.727	1.711–1.738 1.714–1.741 1.724–1.751	0.011–0.014	Rhodonite (R1) $(Mn,Ca,Fe)SiO_3$	61°–76° $v > r$ weak	Triclinic Euh to anh, tabular on {001}
1.739	1.732 1.739 1.757	0.025	Hedenbergite (P8) $CaFe(SiO_3)_2$	63° $r > v$ strong	Monoclinic Stubby prismatic, elong on *c*, granular
1.747	1.736–1.747 1.740–1.754 1.745–1.762	0.009–0.015	Staurolite (S10) $(Fe^{+2})_2Al_9Si_4O_{22}$ $(O,OH)_2$	80°–90° $r > v$ weak to mod	Monoclinic Large euh xls, elong on *c*

*Range of $\beta \geq 0.03$.

Table 6 *(continued)*

Optic Orientation	Cleavage	Color and Pleochroism	Hardness Sp. Grav.	Occurrence	Remarks
		Biaxial Positive (+)			
$b = Y, c \wedge X = 0°–90°$	{001} perf, {100} good	Colorless	6–6½ 3.21–3.49	Low- to med-grade met rks, rarely primary in ign rks, alt prod of plagioclase	Epidote group, {100} poly tw com, anom interf colors
$a = Z, b = Y, c = X$	{010} good to perfect	Colorless	6½–7 3.3–3.5	Bauxite, lateritic clays and soils, Al-rich met rks, hyd alt prod of Al-rich minerals	Clay group
$b = Y, c \wedge Z = 45°$	{110} good, {001}, {100} parting	Brownish gn Moderate	5½–6½ 3.4	Fe-rich contact met rks, alkali granites	Pyroxene group, {100}, {001} simple and poly tw com
$b = Y, c \wedge X = -12°$ to $-20°$	{110} good, {100}, {001} parting	Pale gn, yel-gn Weak $X > Y > Z$	6 3.4–3.6	Alkali ign rks, Na-rich met rks	Pyroxene group, {100} simple tw com
$b = X, c \wedge Z =$ 14°–42°	{001} perf, {110} poor, {001} parting poor	Colorless to gn or bl-gn Mod $Y > X > Z$	6½ 3.26–3.80	Low-grade met rks, uncommon in hyd veins	{001} poly or simple tw com
$a \wedge X \simeq 5°, b \wedge Y \simeq 20°, c \wedge Z \simeq 25°$	{110}, $\{1\bar{1}0\}$ perf, {001} good	Colorless, rarely pale pink Weak, $X =$ orange, $Y =$ rose-pink, $Z =$ pale yel-orange	5½–6½ 3.55–3.76	Hyd deposits, contact skarns, and marbles	{010} poly tw com
$b = Y, c \wedge Z = 48°$	{110} good, {100} parting	Brownish gn Strong	6½ 3.5	Fe-rich contact met rks, alkali granites	Pyroxene group, {100}, {001} simple and poly tw com
$a = Y, b = X, c = Z$	{010} poor	Golden yel Mod $Z > Y > X$	7–7½ 3.74–3.83	Common as porphyroblasts in med-grade met rks	{023}, {232} penetration and cruciform tw com

Table 6 *(continued)*

Av. β	Indices α, β, γ	Birefringence $(\gamma - \alpha)$	Name and Composition	$2V$ and Dispersion	System and Habit
			Biaxial Positive (+)		
*1.757	1.690–1.791 1.700–1.815 1.706–1.828	0.013–0.036	Allanite (Orthite) (E1) $(Ca,Ce)_2(Fe^{+2},Fe^{+3})Al_2O[Si_2O_7][SiO_4](OH)$	40°–90° $r > v$ strong, rarely $v > r$	Monoclinic Elong on b, {100} tablets, anh
*1.768	1.725–1.794 1.730–1.807 1.750–1.832	0.025–0.082	Piemontite (E1) $Ca_2(Mn,Fe,Al)_2AlO[Si_2O_7][SiO_4](OH)$	50°–86° $r > v$ strong, $v > r$ rare	Monoclinic Elong on b
1.771	1.768 1.771 1.788	0.022	Orthoferrosilite (P8) $FeSiO_3$	35° $r > v$ strong	Orthorhombic Stubby prismatic, elong on c
1.789	1.770–1.800 1.777–1.801 1.828–1.851	0.045–0.075	Monazite (M3) $(Ca,La,Th)PO_4$	6°–19° $v > r$ or $r > v$ weak	Monoclinic Small euh xls, elong on b, tabular {100}
*1.952	1.843–1.950 1.870–2.034 1.943–2.110	0.100–0.192	Sphene (Titanite) (S7) $CaTiSiO_5$	17°–40° $r > v$ strong	Monoclinic Euh with rhombic cross-section, anh
2.584	2.583 2.584 2.700	0.117	Brookite (R2) TiO_2	0°–30° $v > r$ strong crossed	Orthorhombic Euh, with {010} dominant, or elong on c
			Biaxial Negative (−)		
1.472	1.454 1.472 1.488	0.034	Kernite (K1) $Na_2B_4O_7 \cdot 4H_2O$	80° $r > v$ mod	Monoclinic Anh to euh, or elong ‖ to c
1.499	1.482–1.500 1.491–1.507 1.493–1.513	0.010	Stilbite (Z1) $(Ca,Na_2,K_2)(Al_2Si_7)O_{18} \cdot 7H_2O$	30°–49° $v > r$	Monoclinic Elong on c, tabular {010}
1.517	1.502–1.514 1.512–1.522 1.514–1.525	0.008–0.016	Laumontite (Z1) $Ca(Al_2Si_4)O_{12} \cdot 4H_2O$	25°–47° $v > r$ strong	Monoclinic Columnar to prismatic on c

*Range of $\beta \geq 0.03$.

Table 6 *(continued)*

Optic Orientation	Cleavage	Color and Pleochroism	Hardness Sp. Grav.	Occurrence	Remarks
		Biaxial Positive (+)			
$b = Y$, $c \wedge X = -1°$ to $-42°$	{001} good to poor, {100}, {110} poor	Brownish yel, bn, or gn Mod $Z > Y > X$	5–6½ 3.4–4.2	Accessory mineral in granites, granodiorites, monzonites, nepheline syenites, limestone skarns, and pegmatites	Epidote group, may be metamict isotropic, or surrounded with pleoch haloes, {100} tw com
$b = Y$, $c \wedge X = -2°$ to $-9°$	{001} perf	Deep reds, viol, bn Strong $Z > Y > X$	6–6½ 3.40–3.52	Low-grade meta rks, hyd and metasomatic, vesicles and fissures in volc rks	Epidote group, interf color often masked by min color, {100} poly tw com
$a = X$, $b = Y$, $c = Z$	{210} good, {100}, {010} parting	Pale gn Weak or none	6 3.96	Metamorphosed Fe-rich sed	Pyroxene group, {100} simple or poly tw com
$b = X$, $c \wedge Z = 2°–7°$	{100} good, {001} parting	Yel-bn, gray, colorless Weak $Y > X = Z$	5 5.0–5.3	Rare, accessory mineral in siliceous ign and met rks, large xls in pegmatites, stable in detrital rks	
$b = Y$, $c \wedge Z \simeq 51°$	{110} good, {221} parting	Colorless, pale bn or yel-bn Weak $Z > Y > X$	5–5½ 3.45–3.56	Common accessory in ign, met, and sed rks	
$a = X$, $b = Z$, $c = Y$ (yel light) $a = Y$, $b = Z$, $c = X$ (gn to bl light)	{120} poor, {001} very poor	Yel-gn, red-bn, deep bn	5½–6 4.12	Accessory mineral in pegmatites, schists, gneisses, and hyd veins	Rutile group
		Biaxial Negative (−)			
$b = Z$, $c \wedge Y = -20°$	{100}, {001} perf, {$\bar{2}$01} good	Colorless	~2½ ~1.93	Playa lake evaporites	
$b = Y$, $c \wedge X = 0°–7°$	{010} perf, {100} poor	Colorless	3½–4 2.1–2.2	Cavities and veins in volc ign rks, thermal springs, hyd veins	Zeolite group, {001} cruciform penetration tw com
$b = Y$, $c \wedge Z = -8°$ to $-40°$	{010}, {110} perf	Colorless	3–4 2.23–2.41	Cavities and veins in ign rks, low-grade met rks	Zeolite group, {100} simple tw com

Table 6 *(continued)*

Av. β	Indices α, β, γ	Birefringence $(\gamma - \alpha)$	Name and Composition	$2V$ and Dispersion	System and Habit
			Biaxial Negative (−)		
1.518	1.507–1.513 1.516–1.520 1.517–1.521	0.008–0.010	Scolecite (Z1) $Ca(Al_2Si_3)O_{10} \cdot 3H_2O$	36°–56° $v > r$ strong	Monoclinic Acic on *c*
1.518	1.514–1.516 1.518–1.519 1.521–1.522	0.007	Microcline (F1) $(K_{1-0.92},Na_{0-0.08})AlSi_3O_8$	66°–68° $r > v$ mod	Triclinic Anh, blocky prismatic, elong on *c*
1.523	1.518–1.520 1.522–1.524 1.522–1.525	0.005	Orthoclase (F1) $(K_{1-0.85},Na_{0-0.15})AlSi_3O_8$	35°–50° $r > v$ mod	Monoclinic Anh, blocky prismatic, elong on *a*, microlites, spherulites
1.524	1.518–1.521 1.523–1.525 1.524–1.526	0.006	High Sanidine (F1) $(K_{1-0.70},Na_{0-0.30})AlSi_3O_8$	15°–63° $v > r$ weak	Monoclinic Euh phenocrysts, tabular {010}
1.525	1.518–1.524 1.522–1.529 1.522–1.530	0.005–0.006	Sanidine (F1) $(K_{1-0.77},Na_{0-0.33})AlSi_3O_8$	18°–24° $r > v$ weak	Monoclinic {010} tablets, elong on *a*, subh microlites
1.535	1.524–1.526 1.529–1.532 1.530–1.534	0.006–0.007	Anorthoclase (F1) $(K_{0.37-0.10},Na_{0.63-0.90})$ $AlSi_3O_8$	42°–52° $r > v$ weak	Triclinic Rhomb-shaped, with pinacoids dominant
1.542	1.538 1.542 1.546	0.008	Oligoclase (An_{20}) (F1) $(NaSi)_{0.8}(CaAl)_{0.2}$ $AlSi_2O_8$	87° $r > v$ weak	Triclinic Euh to subh, {010} plates or elong on *a*
*1.545	1.48–1.57 1.50–1.59 1.50–1.60	0.02–0.03	Montmorillonite (C6) $(0.5Ca,Na)_{0.67}$ $(Al_{3.33}Mg_{0.67})$ $Si_8O_{20}(OH)_4 \cdot nH_2O$	0°–30°	Monoclinic {001} flakes
*1.549	1.522–1.560 1.524–1.574 1.527–1.578	0.005–0.018	Cordierite (C8) $(Mg,Fe)_2Al_3(Si_5Al)O_{18}$	40°–90°(−), may be 75°–90°(+)	Orthorhombic (pshex) Euh to anh, poikiloblastic

*Range of $\beta \geq 0.03$.

Table 6 *(continued)*

Optic Orientation	Cleavage	Color and Pleochroism	Hardness Sp. Grav.	Occurrence	Remarks
		Biaxial Negative (−)			
$b = Z$, $c \wedge X \simeq 18°$	{110} perf	Colorless	5 2.25–2.29	Cavities and veins in mafic volc rks	Zeolite group, {100} poly tw com
Optic axial plane near {001}, $c \wedge Y \simeq 18°$, $a \wedge X \simeq 18°$, $b \simeq Z$	{001} perf, {010} ≃perf, {110} poor	Colorless	6–6½ 2.56–2.63	Met rks, siliceous plutonic ign rks, pegmatites, detrital and authigenic in sed	Feldspar group, alb and pericline poly tw com
$b = Z$, $c \wedge Y =$ 13°–21°	{001} perf, {010} ≃perf, {110} poor	Colorless	6–6½ 2.55–2.63	Felsic plutonic and volc rks, high-grade met rks, detrital and authigenic in sed	Feldspar group, Carlsbad tw com
$b = Y$, $c \wedge Z = 20°$	{001} perf, {010} ≃perf, {110} poor	Colorless	~6 2.56–2.62	Uncom, phenocrysts or matrix xls in felsic volc rks	Feldspar group, Carlsbad tw com
$b = Z$, $c \wedge Y \simeq 21°$	{001} perf, {010} ≃perf	Colorless	6 2.56–2.62	Phenocryst or matrix xls in felsic volc rks, spherulites in rhyolite or obsidian	Feldspar group, Carlsbad tw com
$b \wedge Z \simeq 5°$, $c \wedge Y \simeq$ 20° optic axial plane $\simeq\perp${010}	{001} perf {010} ≃perf, {110} poor	Colorless	6 2.56–2.62	Uncom, phenocryst or matrix xls in Na-rich volc rks	Feldspar group, fine poly tw com, Carlsbad tw rare
On {001}, $a \wedge \alpha' = 1°$, on {010}, $a \wedge \alpha' = 5°$	{001} perf, {010} good, {110} poor	Colorless	6–6½ 2.66	Felsic and interm ign rks, detrital in sed	Feldspar group, alb tw very com, Carlsbad and pericline less so, $\alpha' \wedge$ alb tw = 1° max
$b = Y$, $c \wedge X =$ small, $a \wedge Z =$ small	{001} perf	Colorless, pale yel, bn, or pink	1–2 2.0–2.7	From weathering of ign rks, bentonite clay, in fine-grained sed	Clay min group,
$a = Y$, $b = Z$, $c = X$	{010} good, {100}, {001} poor, {001} parting	Usually colorless, may be yel, bl, viol Usually none, may be strong $Z > Y > X$	7–7½ 2.53–2.78	Contact met rks, hornfelses, less com in reg met and ign rks, rare in sed	Pleoch haloes, resembles qtz and feldspar, {110} cyclic and poly tw com

Table 6 *(continued)*

Av. β	Indices α, β, γ	Birefringence $(\gamma - \alpha)$	Name and Composition	$2V$ and Dispersion	System and Habit
			Biaxial Negative (−)		
*1.551	1.529–1.567 1.530–1.573 1.537–1.574	0.004–0.009	Serpentine Group (S3) $Mg_3Si_2O_5(OH)_4$	20°–60° $r > v$ weak	Monoclinic Crysotile is asbestiform, elong on *a;* antigorite and lizardite have undulatory and mottled extinction and are netlike and leaflike
*1.56	1.52–1.56 1.54–1.58 1.54–1.58	0.02–0.03	Vermiculite (C6) (Mg,Ca) $[(Mg,Fe)_5(Fe,Al)]$ $(Si_5Al_3)O_{20}(OH)_4 \cdot 8H_2O$	0°–8° $v > r$ weak	Monoclinic {001} flakes
1.561	1.547–1.548 1.560–1.562 1.567	0.019–0.020	Polyhalite (P4) $K_2Ca_2Mg(SO_4)_4 \cdot 2H_2O$	62°–70° $v > r$	Triclinic Massive, fibrous, elong on *b*, foliate on {010}, xls rare
1.564	1.553–1.563 1.559–1.569 1.560–1.570	0.006–0.007	Kaolinite (C6) $Al_2Si_2O_5(OH)_4$	20°–55° $r > v$ weak	Monoclinic {001} flakes
*1.568	1.525–1.548 1.551–1.585 1.554–1.587	0.018–0.038	Lepidolite (M2) $K_2(Li,Al)_{5-6}$ $[Si_{6-7}Al_{2-1}O_{20}](OH,F)_4$	0°–58°, (usually 30°–50°)	Monoclinic Pshex {001} plates
1.574	1.569 1.574 1.579	0.010	Bytownite (An_{80}) (F1) $(NaSi)_{0.2}(CaAl)_{0.8}$ $AlSi_2O_8$	85° $v > r$ weak	Triclinic Euh to sub, {010} plates, elong on *a*
1.585	1.577 1.585 1.590	0.013	Anorthite (An_{100}) (F1) $CaAl_2Si_2O_8$	78° $v > r$ weak	Triclinic Euh to sub, {010} plates, elong on *a*
1.587	1.534–1.556 1.586–1.589 1.596–1.601	0.045–0.062	Pyrophyllite (P7) $Al_2Si_4O_{10}(OH)_2$	53°–62° $r > v$ weak	Monoclinic Tabular {001}, fine-grained aggregates

*Range of $\beta \geq 0.03$.

Table 6 *(continued)*

Optic Orientation	Cleavage	Color and Pleochroism	Hardness Sp. Grav.	Occurrence	Remarks
		Biaxial Negative (−)			
$b = Y$, $c \wedge X = 0°–7°$	{001} perf	Colorless to pale gn Weak $Z > Y = X$	2½–3½ 2.5–2.6	Alt prod of olivine and pyroxene in ultramafic rks, met contact rks	Pseudomorphs after olivine and pyroxene, mod com
$b = Y$, $c \wedge X = 3°–6°$	{001} perf	Colorless, pale bn or gn Weak $Z = Y > X$	~1½ ~2.4	Alt prod of Fe- and Mg-rich micas, in soils and fine sed	Clay min group
Uncertain	{100} good, {010} parting good	Colorless to pale red	3½ 2.78	Evaporite deposits	{010} poly tw very com, {001} poly tw com
$b = Z$, $X \wedge \{001\} = 3°$	{001} perf	Colorless, pale bn to opaque None, weak $Z = Y > X$	2–3 2.5–2.7	Alt prod of feldspar, detrital in sed rks, hyd alt prod	Clay min group, {001} tw
$b = Y$, $c \wedge X = 3°–10°$	{001} perf	Colorless, rarely viol or gn $Y = Z > X$	2½–4 2.80–2.90	Granite pegmatites	Mica group, {001} complex tw
On {001} $a \wedge \alpha' = 23°$, on {010} $a \wedge \alpha' = 34°$	{001} perf, {010} good, {110} poor	Colorless	6–6½ 2.74	Mafic ign rks, meteorites, rare in detrital sed	Feldspar group, alb tw very com, Carlsbad and pericline tw less so, $\alpha' \wedge$ alb tw = 45° max
On {001} $a \wedge \alpha' = 43°$, on {010} $a \wedge \alpha' = 39°$	{001} perf, {010} good, {110} poor	Colorless	6–6½ 2.76	Uncom, olivine norites, basalts, calc-silicate met rks	Feldspar group, alb tw very com, Carlsbad and pericline tw less so, $\alpha' \wedge$ alb tw = 63° max
$b = Y$, $c \wedge X = 10°$	{001} perf	Colorless	1–1½ 2.84	Uncom, low-grade met rks, hyd alt prod	

Table 6 *(continued)*

Av. β	Indices α, β, γ	Birefringence $(\gamma - \alpha)$	Name and Composition	$2V$ and Dispersion	System and Habit
			Biaxial Negative (−)		
*1.59	1.54–1.57 1.57–1.61 1.57–1.61	0.03	Illite (C6) $K_{1-1.5}Al_4(Si_{7-6.5}Al_{1-1.5})O_{20}(OH)_4$	0°–10°	Monoclinic {001} flakes
1.591	1.539–1.550 1.589–1.594 1.589–1.600	0.05	Talc (T1) $Mg_3Si_4O_{10}(OH)_2$	0°–30° $r > v$ mod	Monoclinic Aggregates of {001} plates
*1.595	1.55–1.57 1.58–1.61 1.59–1.62	0.036–0.049	Sericite (C6) Usually $K_2Al_4[Si_6Al_2O_{20}](OH,F)_4$	30°–47° $r > v$ mod	Monoclinic Fine-grained {001} flakes
1.596	1.552–1.574 1.582–1.610 1.587–1.616	0.036–0.049	Muscovite (M2) $K_2Al_4[Si_6Al_2O_{20}](OH,F)_4$	30°–47° $r > v$ mod	Monoclinic Pshex {001} plates
*1.597	1.530–1.590 1.557–1.637 1.558–1.637	0.028–0.049	Phlogopite (M2) $K_2(Mg,Fe)_6[Si_6Al_2O_{20}](OH,F)_4$	0°–15° $v > r$ weak to mod	Monoclinic Pshex {001} tablets
1.601	1.564–1.580 1.594–1.609 1.600–1.609	0.028–0.038	Paragonite (M2) $Na_2Al_4[Si_6Al_2O_{20}](OH)_4$	0°–40° $r > v$ mod	Monoclinic Scaly aggregates with {001} dominant
*1.625	1.57–1.67 1.57–1.68 1.59–1.68	0.000–0.010	Chlorite Group (C4) $(Mg,Al,Fe)_{12}[(Si,Al)_8O_{20}](OH)_{16}$	0°–30°(−), 0°–45°(+) $v > r$ strong	Monoclinic {001} dominant, pshex plates, scaly, massive
1.625	1.606–1.621 1.618–1.632 1.630–1.644	0.023–0.024	Tremolite (A2) $Ca_2(Mg_{1-0.8},Fe_{0-0.2})_5Si_8O_{22}(OH)_2$	84°–88° $v > r$ weak	Monoclinic Acic to fibrous, elong on *c*
*1.63	1.56–1.61 ~1.61–1.65 ~1.61–1.65	0.014–0.032	Glauconite (M2) (C6) $(K,Na)(Al,Fe^{+3},Mg)_2(Al,Si)_4O_{10}(OH)_2$	0°–20° $r > v$	Monoclinic Fine-grained granules, pellets, or pseudomorphs after foraminifera

*Range of $\beta \geq 0.03$.

Table 6 *(continued)*

Optic Orientation	Cleavage	Color and Pleochroism	Hardness Sp. Grav.	Occurrence	Remarks
		Biaxial Negative (−)			
$b = Z, c \wedge X =$ small, $a \wedge Y =$ small	{001} perf	Colorless	1–2 2.6–2.9	Alt prod of feldspars and muscovite, shales and clays, hyd alt prod	Clay min group
$b = Y, c \wedge X \simeq 10°$, $a \simeq Z$	{001} perf	Colorless	1 2.6–2.8	Met dolomites, alt prod in ultramafic rks	Tw rare
$b = Z, c \wedge X = 0°–5°$, $a \simeq Y$	{001} perf	Colorless, pale bn, red, or gn None	2½–3 2.77–2.88	Retrograde metamorphism of feldspars, hyd alt prod, argillaceous sed rks	Clay min group, {001}, {110} tw com
$b = Z, c \wedge X = 0°–5°$, $a \simeq Y$	{001} perf	Colorless	2½–3 2.77–2.88	Very com in met rks, intrusive rks, pegmatites, deuteric or hyd alt prod, fine-grained sed	Mica group, {001} poly tw, {110} simple tw
$b = Y, c \wedge X = 5°–10°$	{001} perf	Colorless, pale yel or gn Weak $Z = Y > X$	2–2½ 2.76–2.90	Mod com in ultramafic rks, impure marbles	Mica group {001} tw
$b = Z, c \wedge X \simeq 5°$	{001} perf	Colorless	2½ 2.85	Low- to med-grade met rks, fine-grained sed	Mica group
$b = Y, c \wedge Z = 0°–9°$, also $c \wedge X = 0°–7°$	{001} perf	Colorless to lt or dk gn Weak to mod $Z = Y > X$ or $X = Y > Z$	1–2½ 2.64–3.21	Fine-grained sed rks, low-grade met rks, deuteric alt prod	{001} poly tw com, anom bn, viol, bl interf colors
$b = Y, c \wedge Z =$ 17°–21°	{110} perf	Colorless	5–6 3.0–3.2	Contact and reg low-grade met rks, hyd alt prod in mafic and ultramafic ign rks	Amphibole group, {100} simple and poly tw com, {001} poly tw rare
$b = Y, c \wedge X \simeq 10°$	{001} perf	Yel-gn to olive-gn, or colorless	2 2.4–3.0	Marine detrital sands	Mica group, also Clay mineral group

Table 6 *(continued)*

Av. β	Indices α, β, γ	Birefringence $(\gamma - \alpha)$	Name and Composition	$2V$ and Dispersion	System and Habit
			Biaxial Negative (−)		
*1.631	1.594–1.631 1.614–1.648 1.619–1.649	0.019–0.024	Glaucophane (A2) $Na_2(Mg,Fe)_3(Al,Fe)_2$ $(Si_8O_{22})(OH)_2$	32°–50° $v > r$ weak	Monoclinic Prismatic, elong on c
1.638	1.612–1.638 1.625–1.652 1.630–1.654	0.009–0.020	Eckermannite (A2) $Na(Na_{2.0-1.5},Ca_{0-0.5})$ $(Mg,Fe)_4(Fe,Al)(Si_8O_{22})$ $(OH)_2$	15°–80° $r > v$ strong	Monoclinic Prismatic, elong on c
1.638	1.629–1.640 1.633–1.644 1.638–1.650	0.009–0.011	Andalusite (A1) Al_2SiO_5	71°–86° $v > r$ weak, rarely $r > v$	Orthorhombic Euh, prismatic, elong on c, inclusions com and often symmetrically arranged
1.639	1.616–1.640 1.628–1.650 1.631–1.653	0.013–0.014	Wollastonite (W1) $CaSiO_3$	38°–60° $r > v$ weak	Triclinic Columnar, bladed, elong on b
1.645	1.630–1.638 1.642–1.648 1.644–1.650	0.012–0.014	Margarite (M2) $Ca_2Al_4[Si_4Al_4O_{22}]$ $(OH)_4$	40°–67° $v > r$ mod	Monoclinic pshex {001} plates
*1.65	1.60–1.67 1.61–1.69 1.61–1.69	0.00–0.02	Chlorite Group (oxidized) (C4) $(Mg,Al,Fe)_{12}$ $[(Si,Al)_8O_{20}]$ $(OH)_{16}$	0°–30°(−), 0°–45°(+) $v > r$ strong	Monoclinic {001} dominant, pshex plates, scaly, massive
*1.650	1.565–1.625 1.605–1.696 1.605–1.696	0.04–0.07	Biotite (M2) $K_2(Mg,Fe)_{6-4}$ $(Fe,Al,Ti)_{0-2}$ $[Si_{6-5}Al_{2-3}O_{20}](OH,F)_4$	0°–25°, usually <10° $r > v$ or $v > r$ weak	Monoclinic {001} tabular, pshex cross section
*1.653	1.621–1.670 1.632–1.675 1.644–1.688	0.017–0.022	Actinolite (A2) $Ca_2(Mg_{0.8-0.2}, Fe_{0.2-0.8})_5$ $Si_8O_{22}(OH)_2$	75°–84° $v > r$ weak	Monoclinic Acic to fibrous, elong on c
*1.660	1.543–1.634 1.576–1.745 1.576–1.745	0.030–0.110	Stilpnomelane (S11) $(K,Na,Ca)_{0-1.4}$ $(Fe^{+3},Fe^{+2},Mg,Al,Mn)_{5.9-8.2}$ $[Si_8O_{20}](OH)_4$ $(O,OH,H_2O)_{3.6-8.5}$	≃0°, may be up to 40°	Monoclinic Micaceous {001} plates

*Range of $\beta \geq 0.03$.

Table 6 *(continued)*

Optic Orientation	Cleavage	Color and Pleochroism	Hardness Sp. Grav.	Occurrence	Remarks
		Biaxial Negative (−)			
$b = Y, c \wedge Z = 6°–8°$	{110} perf, {010}, {001} parting	Viol, bl, yel Mod to strong $Z > Y > X$	6 3.0–3.4	High-P met rks	Amphibole group, tw uncom
$b = Y, c \wedge X = -30°$ to $-57°$	{110} perf, {010} parting	Bl-gn to yel-gn Strong $X > Y > Z$	5–6 3.0–3.5	Na-rich plutonic ign rks, rare in alkali-rich volc rks	Amphibole group, {100} simple or poly tw com
$a = Z, b = Y, c = X$	{110} good	Colorless, pink, yel, gn Weak to mod	6½–7½ 3.13–3.16	Low- to med-grade met rks, detrital in sed rks	Aluminosilicate group, {101} tw rare
$b \wedge Y = 0°–5°, c \wedge X = -33°$ to $-44°$, optic axial plane $\simeq$ {010}	{100} perf, {001}, {$\bar{1}$02} good	Colorless	4½–5 2.9–3.1	Contact met rks	{100} poly tw com
$b = Z, c \wedge X = 11°–13°$	{001} ~perf	Colorless	3½–4½ 3.0–3.1	Uncom, met rks with corundum and diaspore	Mica group, {001} poly tw
$b = Y, c \wedge Z = 0°–9°$, also $c \wedge X = 0°–7°$	{001} perf	Colorless to lt or dk gn Weak to mod $Z = Y > X$ or $X = Y > Z$	1–2½ 2.64–3.21	Fine-grained sed, low-grade met rks, deuteric alt prod	{001} poly tw com
$b = Y, c \wedge X = 0°–9°$	{001} perf	Bn, red-bn, gn Strong $Z \simeq Y > X$	2½–3 2.7–3.3	Com in ign and met rks, detrital sed	Mica group, {001} complex tw
$b = Y, c \wedge Z = 13°–17°$	{110} perf	Pale yel-gn or bl-gn Weak $Z > Y > X$	5–6 3.0–3.2	Contact and reg low-grade and high-P met rks, hyd alt prod of pyroxene	Amphibole group, {100} simple and poly tw com, {001} poly tw rare
$b = Y, c \wedge X = 7°, a \simeq Z$	{001} perf, {010} poor	Gn, bn, yel Strong $Z = Y >> X$	3–4 2.59–2.96	Low-grade Fe- and Mn-rich schists, high-P met rks	Easily confused with biotite

Table 6 *(continued)*

Av. β	Indices α, β, γ	Birefringence $(\gamma - \alpha)$	Name and Composition	$2V$ and Dispersion	System and Habit
			Biaxial Negative (−)		
*1.663	1.604–1.694 1.617–1.710 1.628–1.722	0.024–0.028	Anthophyllite (A2) $(Mg_{0.9-0.1}Fe_{0.1-0.9})_7$ $Si_8O_{22}(OH)_2$	82°–90°(−), 0°–90°(+) $r > v$, $v > r$ weak to mod	Orthorhombic Prismatic, fibrous, elong on *c*
*1.666	1.631–1.677 1.648–1.684 1.649–1.689	0.012–0.019	Crossite (A2) $Na_2(Mg,Fe)_3(Al,Fe)_2$ $(Si_8O_{22})(OH)_2$	42°–90° $r > v$ very strong	Monoclinic Prismatic, fibrous, elong on *c*
*1.666	1.615–1.705 1.618–1.714 1.632–1.730	0.014–0.026	Hornblende (A2) $Ca_2(Mg,Fe)_4(Al,Fe)$ $(Si_7Al)O_{22}(OH)_2$	23°–90°(−), 85°–90°(+) $r > v$, $v > r$ mod	Monoclinic Prismatic, elong on *c*
*1.670	1.627–1.690 1.635–1.705 1.644–1.717	0.017–0.027	Gedrite (A2) $(Mg_{0.9-0.1}Fe_{0.1-0.9})_5$ $Al_2Si_6Al_2O_{22}(OH)_2$	72°–90°(−), 47°–90°(+) $v > r$ weak	Orthorhombic Prismatic, fibrous, elong on *c*
1.674	1.637–1.641 1.672–1.676 1.676–1.681	0.039–0.040	Spurrite (S9) $Ca_5(SiO_4)_2CO_3$	35°–41° $r > v$ weak	Monoclinic Usually anh
1.682	1.527–1.542 1.670–1.695 1.676–1.699	0.149–0.157	Aragonite (C2) $CaCO_3$	18°–23° $v > r$ weak	Orthorhombic ‖ or radial aggregates, elong on *c*
1.685	1.677 1.685 1.690	0.011	Bronzite (En_{80}) (P8) $(Mg_{0.8},Fe_{0.2})SiO_3$	77° $r > v$ mod	Orthorhombic Stubby prismatic on *c*
1.691	1.674–1.693 1.681–1.701 1.684–1.704	0.009–0.011	Axinite (A5) $Ca_2(Fe,Mn,Mg)Al_2BO_3$ $(Si_4O_{12})OH$	63°–80° $v >$ r	Triclinic Wedge-shaped with {010} and {011} dominant
*1.693	1.663–1.688 1.677–1.709 1.697–1.729	0.035–0.045	Grunerite (A2) $(Fe_{1-0.7},Mg_{0-0.3})_7$ $Si_8O_{20}(OH)_2$	80°–90° $r > v$ weak, inclined	Monoclinic Prismatic, acic, asbestiform, elong on *c*
*1.694	1.674–1.700 1.679–1.709 1.686–1.710	0.005–0.012	Arfvedsonite (A2) $Na(Na_{2.0-1.5},Ca_{0-0.5})$ $(Fe,Mg)_4(Fe,Al)(Si_8O_{22})$ $(OH)_2$	0°–50° $r > v$ very strong	Monoclinic Prismatic, elong on *c*

*Range of $\beta \geq 0.03$.

Table 6 *(continued)*

Optic Orientation	Cleavage	Color and Pleochroism	Hardness Sp. Grav.	Occurrence	Remarks
		Biaxial Negative (−)			
$a = X, b = Y, c = Z$	{210} perf, {010} good, {100} poor, {001} parting	Colorless, pale bn or gn Weak to mod $Z > Y > X$	5½ ~3.1	High-grade met rks	Amphibole group
$b = Y, c \wedge Z = 8°$, $b = Z, c \wedge Y = 9°$	{110} perf, {010}, {001} parting	Yel, bl, viol Mod to strong $Z > Y > X$	6 3.0–3.4	High-P met rks	Amphibole group, tw uncom
$b = Y, c \wedge Z =$ 13°–24°	{110} perf, {100}, {001} parting rare	Gn, bl-gn, yel-bn Mod with $Z \geq Y > X$, less com $Y > Z > X$	5–6 3.04–3.45	Very com in ign and met rks	Amphibole group, {100} simple tw com
$a = X, b = Y, c = Z$	{210} perf, {010} good, {100} poor, {001} parting	Pale gn, bn, yel Weak to mod $Z > Y = X$	5½ ~3.3	Med-grade met rks	Amphibole group
$b = X, c \wedge Y \simeq 33°$, $a \simeq Z$	{001} good, {100} poor	Colorless	5 3.01	Uncom, high-grade contact met rks	$\{20\bar{5}\}$ simple tw, {001} poly tw
$a = Y, b = Z, c = X$	{010} poor, {110} rare	Colorless	3½–4 2.95	In sed as oolites, needles or hard parts of marine organisms, hot springs, amyg-dules in mafic volc rks, high-P met rks	Carbonate group, {110} poly and cyclic tw com, pshex outline on *c*
$a = X, b = Y, c = Z$	{210} good, {100}, {010} parting	Colorless	5–6 3.36	Mafic and ultramafic ign rks, high-grade met rks, charnockites	Pyroxene group, {100} simple and poly tw com
$X \simeq \perp\{111\}$	{100} good, {001}, {110}, {011} poor	Clove-bn to colorless $Y > X > Z$ mod	6½–7 3.26–3.36	Calc-silicate met rks, veins in ign rks	Axinite group, {110} poly or simple tw rare
$b = Y, c \wedge Z =$ 12°–16°	{110} good	Pale gn or bn Weak to mod $Z > Y \geq X$	5–6 3.3–3.6	Fe-rich siliceous met rks	Amphibole group, {110} simple, poly tw com
$b = Z, c \wedge X = -2°$ to $-25°$	{110} perf, {010} parting	Bl-gn Strong $X > Y > Z$	5–6 3.0–3.5	Na-rich plutonic rks, rare in alkali-rich volc rks	Amphibole group, {100} simple or poly tw com

Table 6 *(continued)*

Av. β	Indices α, β, γ	Birefringence $(\gamma - \alpha)$	Name and Composition	2V and Dispersion	System and Habit
			Biaxial Negative (−)		
*1.695	1.674–1.702 1.675–1.715 1.688–1.722	0.012–0.022	Pumpellyite (P6) $Ca_4(Mg,Fe,Mn)$ $(Al,Fe,Ti)_5O(OH)_3$ $[Si_2O_7]_2[SiO_4]_2 \cdot 2H_2O$	rarely (−), 26°–85°(+) $v > r$	Monoclinic Fibrous, bladed, acic, elong on *b*
*1.698	1.655–1.686 1.675–1.722 1.684–1.723	0.011–0.027	Dumortierite (D1) $(Al,Fe)_7O_3(BO_3)(SiO_4)_3$	15°–52° $v > r$ strong, $r > v$ rare	Orthorhombic Fibrous, acic, elong on *c*, euh xls rare
*1.699	1.677–1.701 1.684–1.714 1.689–1.720	0.012–0.018	Riebeckite (A2) $Na_2(Fe,Mg)_3\ (Fe,Al)_2$ $(Si_8O_{22})(OH)_2$	42°–90°(−), 0°–90°(+) $r > v$ (−), $v > r$ (+) strong	Monoclinic Prismatic, fibrous, elong on *c*
1.705	1.635–1.731 1.651–1.760 1.670–1.775	0.035–0.043	Mg-rich Olivine (Fo_{100-50}) (O1) $(Mg_{2-1},Fe_{0-1})SiO_4$	73°–90°(−), 82°–90°(+) $v > r$ weak, $r > v$ weak	Orthorhombic Anh to euh, elong on *c*, skeletal
1.711	1.699 1.711 1.714	0.014	Hypersthene (En_{60}) (P8) $(Mg_{0.6},Fe_{0.4})SiO_3$	53° $r > v$ weak	Orthorhombic Stubby prismatic on *c*
1.717	1.701–1.729 1.703–1.732 1.705–1.734	0.004–0.006	Sapphirine (S1) $(Mg,Fe)_2Al_4O_6(SiO_4)$	51°–69° $v > r$ strong, inclined	Monoclinic Euh to anh, tabular {010}, may be elong on *c*
*1.720	1.650–1.700 1.670–1.770 1.680–1.800	0.018–0.083	Oxyhornblende (A2) Ca_2Na $(Mg,Fe^{+2},Fe^{+3},Al,Ti)_5$ $[(Si_3Al)O_{11}]_2(O,OH)_2$	56°–88° $v > r$ weak, $r > v$ strong	Monoclinic Short prismatic, elong on *c*
1.721	1.710–1.713 1.720–1.722 1.727–1.729	0.016–0.017	Kyanite (A1) Al_2SiO_5	82.5° $r > v$ weak	Triclinic Elong on *c*, bladed, fibrous

*Range of $\beta \geq 0.03$.

Table 6 *(continued)*

Optic Orientation	Cleavage	Color and Pleochroism	Hardness Sp. Grav.	Occurrence	Remarks
		Biaxial Negative (−)			
$b = Y$, $c \wedge Z = 4°$ to $-24°$, rarely $b = Z$, $c \wedge Y \simeq -50°$	{001}, {100} good	Pale gn, yel, or bn Strong $Y > Z > X$	6 3.18–3.23	Greenschist and blue-schist met rks	{100}, {001} tw relatively com
$a = Z$, $b = Y$, $c = X$	{100} good, {110} poor	Colorless to bl and viol Strong $X > Y > Z$	7–8½ 3.26–3.41	Hyd or metasomatic rks, granite peg-matites, aplites, gneisses, altered ign rks, qtz veins	{110} cyclic tw, trillings, and pshex forms
$b = Z$, $c \wedge Y = 8°–9°$, $b = Y$, $c \wedge X = 7°–8°$	{110} perf, {010}, {001} parting	Bl to yel-bn Mod to strong $Z > Y \simeq X$	6 3.0–3.4	Na-rich plutonic ign rks, occasionally in met rks	Amphibole group, tw uncom
$a = Z$, $b = X$, $c = Y$	{010}, {100} poor	Colorless	7 3.22–3.75	Ultramafic and mafic ign rks	Olivine group, {100}, {011}, {012} simple tw mod com
$a = X$, $b = Y$, $c = Z$	{210} good, {100}, {010} parting	Lt pink to gn Strong	5–6 3.51	Norites and gabbros	Pyroxene group, {100} simple or poly tw com
$b = Y$, $c \wedge Z = -6°$ to $-15°$	{100}, {010}, {001} poor	Pale bl or gray Mod $Z > Y > X$	7½ 3.4–3.58	Regional or contact met rks with low SiO_2 and high MgO and Al_2O_3, granulites, emery deposits	{010}, {100} poly tw uncom
$b = Y$, $c \wedge Z = 0°–19°$	{110} good, {100}, {001} parting poor	Yel to very dk bn	5–6 3.2–3.3	Mafic volc rks	Amphibole group, {100} simple or poly tw mod com
On {100}, $c \wedge Z' =$ 27°–30°, on {010}, $c \wedge Z' =$ 5°–8°, on {001}, $c \wedge X' =$ 0°–2.5°, $X \simeq \perp\{100\}$	{100} perf, {010} good, {001} weak	Colorless, rarely bluish Weak $Z > Y > X$	4–4½ ∥ c 6–7 ∥ b 3.59	Med- to high-P met rks	Aluminosili-cate group, {100} simple or poly tw com, {001} poly tw less so

Table 6 *(continued)*

Av. β	Indices α, β, γ	Birefringence $(\gamma - \alpha)$	Name and Composition	$2V$ and Dispersion	System and Habit
			Biaxial Negative (−)		
1.726	1.713–1.728 1.719–1.734 1.723–1.740	0.010–0.012	Chloritoid (C5) $(Fe,Mg,Mn)_2$ $(Al,Fe)Al_3O_2$ $(SiO_4)_2(OH)_4$	Large (−) uncom, 40°–65°(+) $r > v$ strong	Monoclinic, triclinic less com Tabular {001}
1.737	1.722 1.737 1.739	0.017	Ferrohypersthene (En_{40}) (P8) $(Mg_{0.4},Fe_{0.6})SiO_3$	53° $v > r$ weak	Orthorhombic Stubby prismatic on *c*
*1.740	1.714–1.728 1.722–1.758 1.729–1.776	0.015–0.048	Epidote (E1) $Ca_2(Al,Fe)Al_2O[Si_2O_7]$ $[SiO_4](OH)$	64°–90° $r > v$ strong	Monoclinic Elong on *b*
1.756	1.745 1.756 1.764	0.019	Eulite (En_{20}) (P8) $(Mg_{0.2}Fe_{0.8})SiO_3$	77° $v > r$ mod	Orthorhombic Stubby prismatic on *c*
1.757	1.690–1.791 1.700–1.815 1.706–1.828	0.013–0.036	Allanite (Orthite) (E1) $(Ca,Ce)_2(Fe^{+2},Fe^{+3})Al_2O$ $[Si_2O_7][SiO_4](OH)$	40°–90°(−), 57°–90°(+) $r > v$ strong, rarely $v > r$	Monoclinic Elong on *b*, {100} tablets, anh
*1.781	1.722–1.776 1.742–1.820 1.758–1.836	0.036–0.060	Aegirine (P8) $NaFe^{+3}(SiO_3)_2$	58°–90° $r > v$ mod to strong	Monoclinic Elong or acic xls on *c*
1.814	1.731–1.827 1.760–1.869 1.775–1.879	0.043–0.052	Fe-rich Olivine (Fo_{50-0}) (O1) $(Mg_{1-0},Fe_{1-2})SiO_4$	46°–90° $r > v$ weak	Orthorhombic Anh to euh, elong on *c*, skeletal
2.34	2.30–2.38 2.30–2.38 2.30–2.38	0.000–0.002	Perovskite (P3) $CaTiO_3$	Indeterminate	Orthorhombic, nearly isotropic, tiny euh cubes or octahedrons

*Range of $\beta \geq 0.03$.

Table 6 *(continued)*

Optic Orientation	Cleavage	Color and Pleochroism	Hardness Sp. Grav.	Occurrence	Remarks
		Biaxial Negative (−)			
$b = X$, $c \wedge Z =$ 14°–42°	{001} perf, {110} poor, {001} parting poor	Colorless to gn or bl-gn Mod $Y > X > Z$	6½ 3.26–3.80	Low-grade met rks, uncom in hyd veins	{001} poly or simple tw com
$a = X$, $b = Y$, $c = Z$	{210} good, {100}, {010} parting	Pale gn Weak or none	5–6 3.66	In diorites, monzonites, and granites	Pyroxene group, {100} simple and poly tw com
$b = Y$, $c \wedge X = 0°–5°$	{001} perf, {100} good	Yel-gn Mod $Y > Z > X$	6–6½ 3.21–3.49	Low- to med-grade reg and contact met rks, rarely primary in ign rks	Epidote group, {100} poly tw com, subtle bl anom interf color tints
$a = X$, $b = Y$, $c = Z$	{210} good, {100}, {010} parting	Pale gn Weak or none	5–6 3.80	In diorites, monzonites, and granites	Pyroxene group, {100} simple and poly tw com
$b = Y$, $c \wedge X = 0°$ to $-42°$	{001} mod to poor, {100}, {110} poor	Brownish yel, bn or gn Mod $Z > Y > X$	5–6½ 3.4–4.2	Accessory mineral in granites, granodiorites, monzonites, nepheline syenites, limestones, skarns, and pegmatites	Epidote group, may be metamict isotropic, may be surrounded by pleoch haloes, {100} tw com
$b = Y$, $c \wedge X =$ 10°–12°	{110} good, {100}, {001} parting	Bright gn to yel gn Strong $X > Y > Z$	6 3.4–3.6	Alkaline ign rks, occasionally in Na-rich met rks	Pyroxene group, {100} simple tw com
$a = Z$, $b = X$, $c = Y$	{010}, {100} poor	Yel-gn, pale yel Weak $Y > X = Z$	6½ 3.75–4.39	Alkaline ign rks, cavities in volc rks, Fe-rich met rks	Olivine group, {100}, {011}, {012} simple tw mod com
$a = X$, $b = Z$, $c = Y$	{100} poor cubic	Yel, bn, lt bn-red Weak $Z > X$	5½ 3.98–4.26	Accessory in Si-poor mafic, ultramafic and alkaline ign rks, marbles	{111} poly tw very com

Chart 1: Uniaxial, (+) and (−)

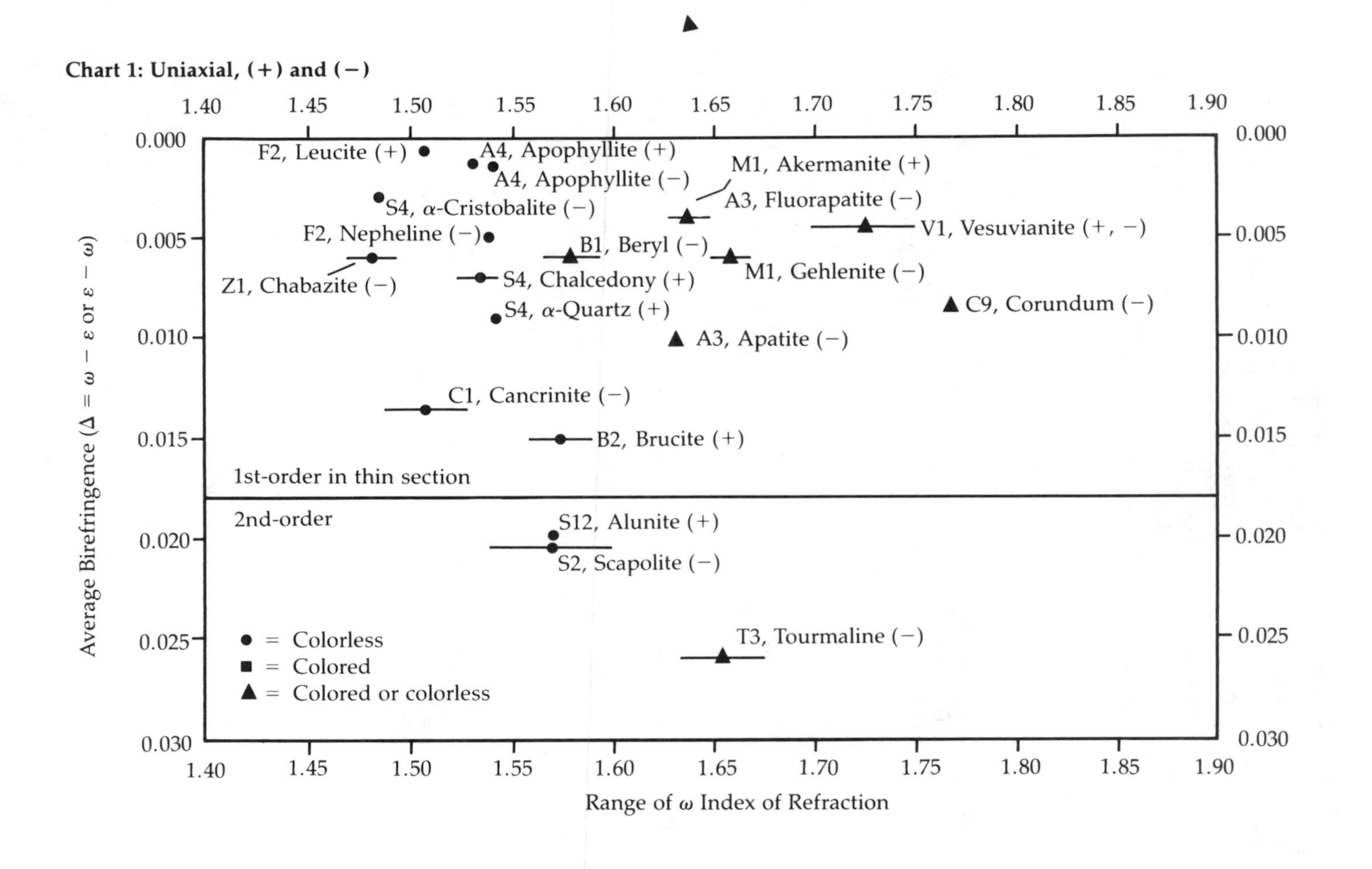

Chart 2: Uniaxial, (+) and (−)

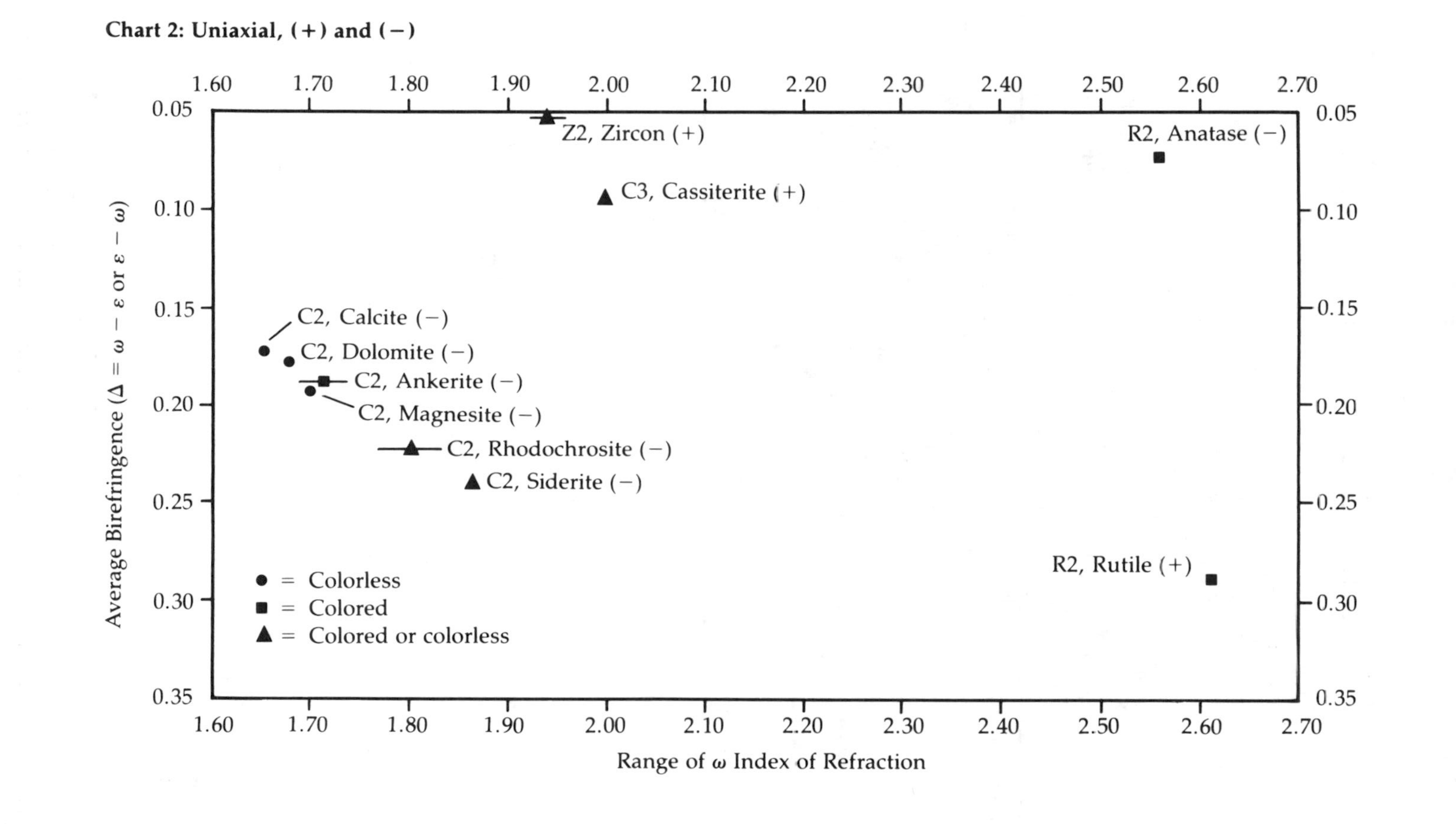

Chart 3: Biaxial (+) Minerals

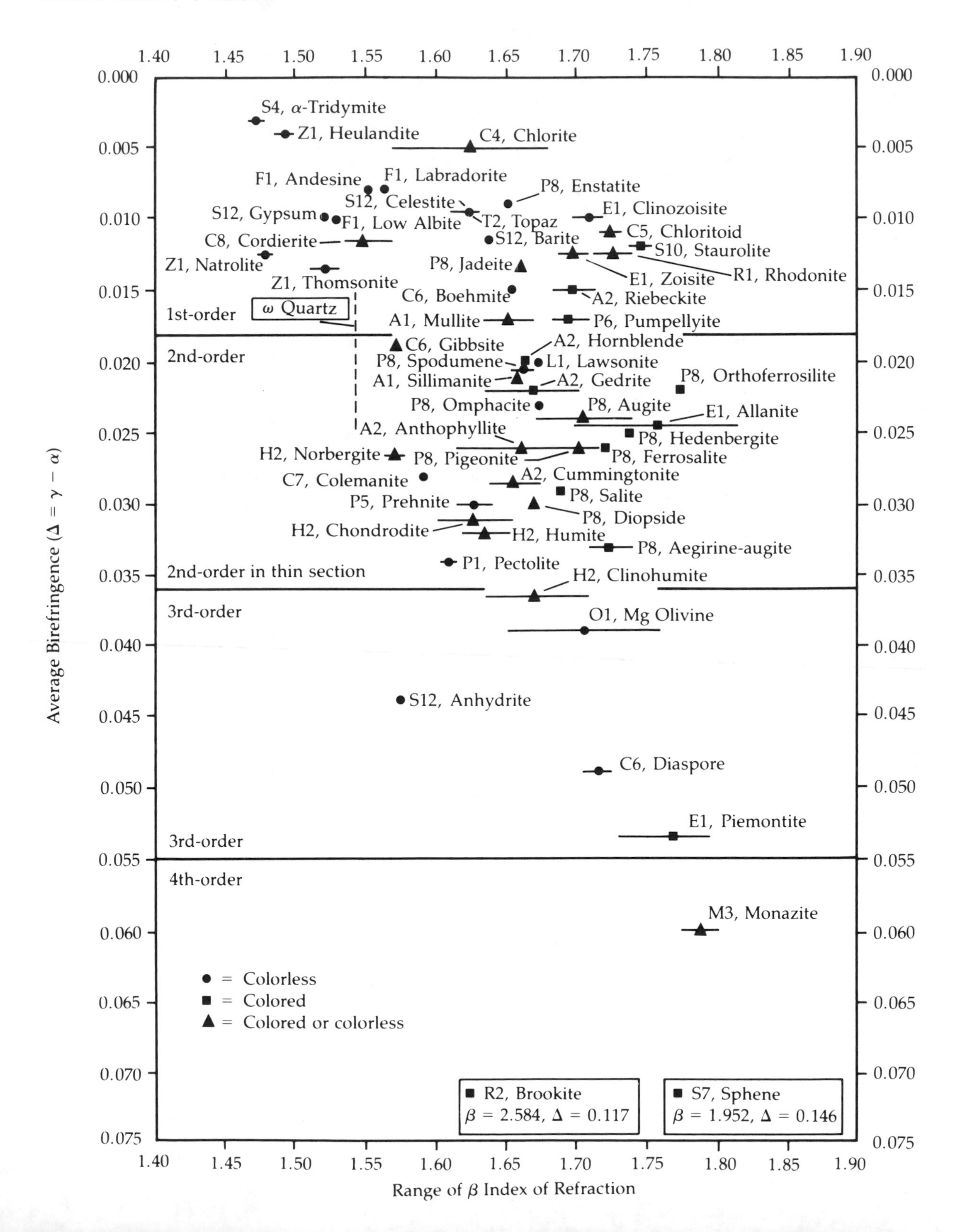

Chart 4: Biaxial (−) Minerals

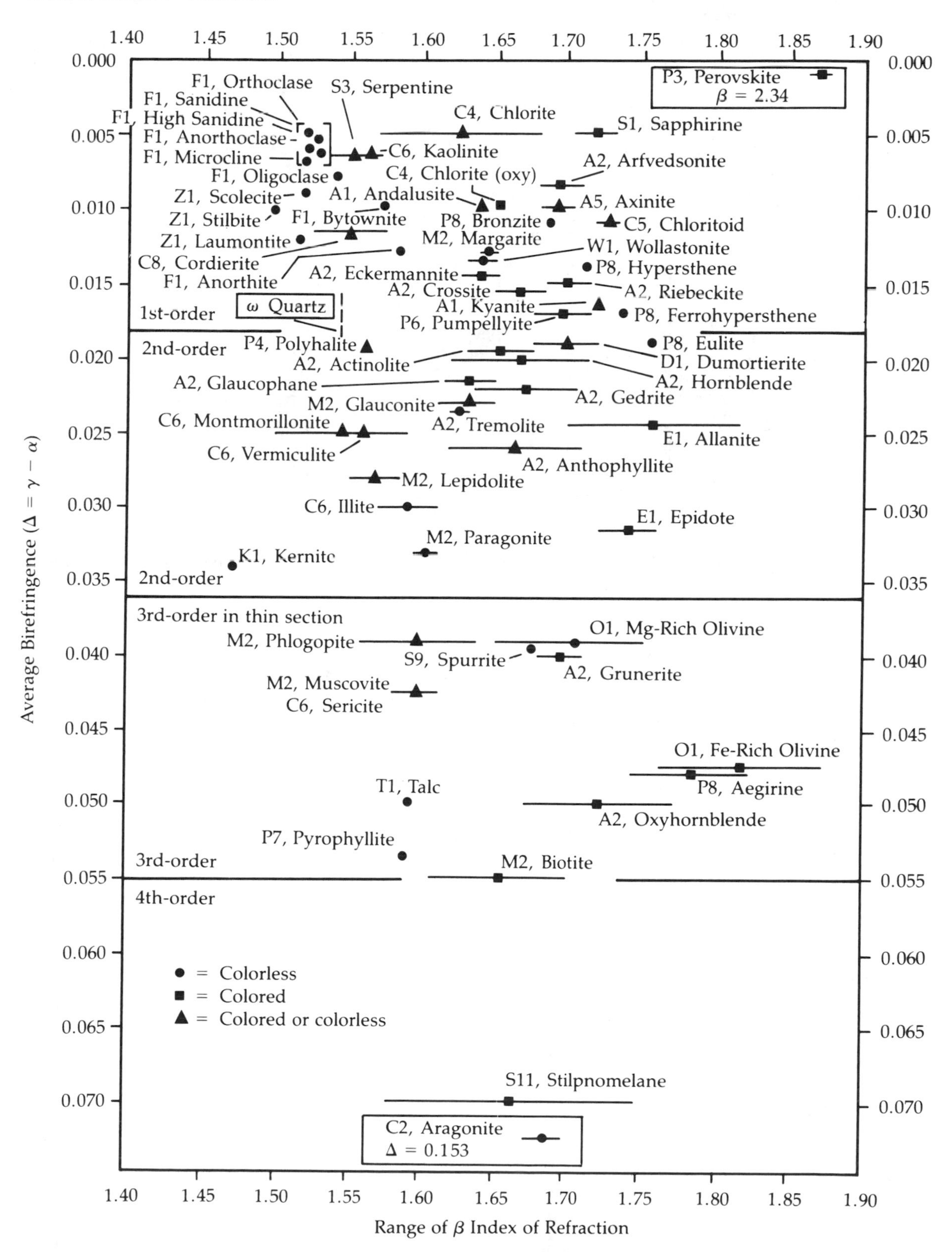

Chart 5: Biaxial Positive (+)

Listed According to Increasing Minimum Value of $2V$

Range of $2V$

0° 10° 20° 30° 40° 50° 60° 70° 80° 90°

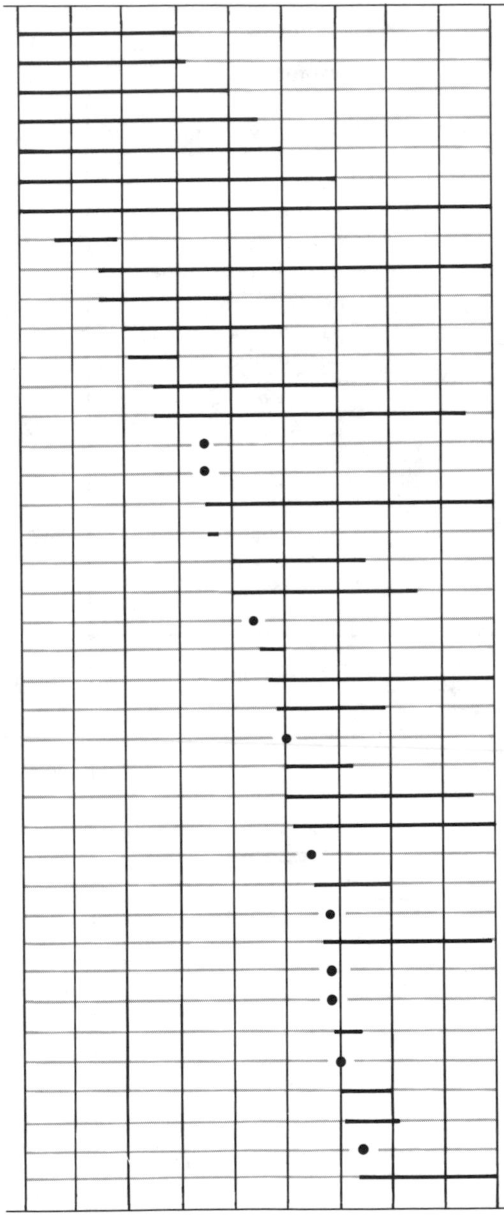

Mineral Name	Average Value of β	Range in Birefringence	Twins[1]	Number of Good Cleavages
Brookite (R2)	2.584	0.117	—	0
Pigeonite (P8)	1.703	0.023–0.029	S, P	2
Gibbsite (C6)	1.574	~0.019	S	1
Chlorite Group (C4)	1.625	0.000–0.010	P	1
Heulandite (Z1)	1.494	0.001–0.007	—	1
Zoisite (E1)	1.699	0.005–0.020	—	1
Riebeckite (A2)	1.699	0.012–0.018	—	2
Monazite (M3)	1.789	0.045–0.075	S	1
Clinozoisite (E1)	1.710	0.005–0.015	P	2
Sphene (S7)	1.952	0.100–0.192	S	2
Mullite (A1)	1.652	0.010–0.024	—	1
Sillimanite (A1)	1.659	0.020–0.022	—	1
Augite (P8)	1.706	0.018–0.030	S, P	2
Pumpellyite (P6)	1.695	0.012–0.022	S	2
Orthoferrosilite (P8)	1.771	0.022	S, P	2
Enstatite (P8)	1.652	0.009	S, P	2
α-Tridymite (S4)	1.474	0.002–0.004	S, P, Cy	0
Barite (S12)	1.638	0.011–0.012	P	4
Chloritoid (C5)	1.726	0.010–0.012	P, S	1
Thomsonite (Z1)	1.523	0.006–0.021	—	2
Anhydrite (S12)	1.576	0.044	S, P	3
Norbergite (H2)	1.572	0.026–0.027	—	0
Gedrite (A2)	1.670	0.017–0.027	—	3
Topaz (T2)	1.622	0.009–0.010	—	1
Celestite (S12)	1.623	0.009–0.010	—	3
Pectolite (P1)	1.610	0.030–0.038	S	2
Piemontite (E1)	1.768	0.025–0.082	P	1
Clinohumite (H2)	1.672	0.028–0.045	S, P	0
Colemanite (C7)	1.592	0.028	—	1
Spodumene (P8)	1.662	0.014–0.027	—	2
Salite (P8)	1.692	0.029	S, P	2
Allanite (E1)	1.757	0.013–0.036	S	1
Gypsum (S12)	1.524	0.010	S, P	4
Diopside (P8)	1.672	0.030	S, P	2
Natrolite (Z1)	1.481	0.012–0.013	—	2
Ferrosalite (P8)	1.719	0.026	S, P	2
Omphacite (P8)	1.675	0.022–0.024	P	2
Rhodonite (R1)	1.727	0.011–0.014	P	3
Hedenbergite (P8)	1.739	0.025	S, P	2
Chondrodite (H2)	1.628	0.025–0.037	S, P	0

[1] *Twin Types*

S = Simple

P = Polysynthetic

C = Cruciform

Cy = Cyclic

Chart 5: Biaxial Positive (+)
Listed According to Increasing Minimum Value of $2V$

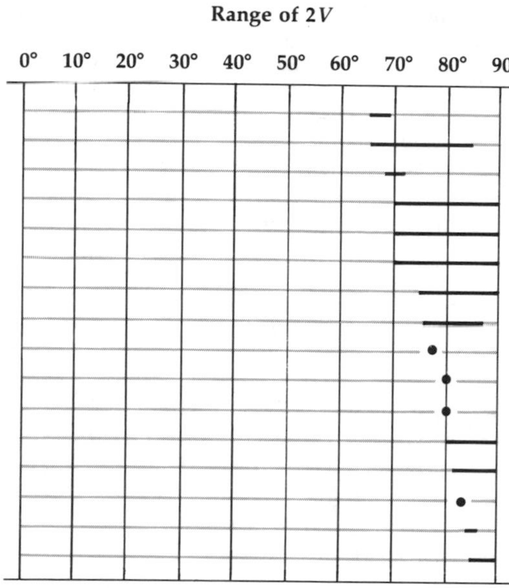

Range of $2V$ (0° 10° 20° 30° 40° 50° 60° 70° 80° 90°)	Mineral Name	Average Value of β	Range in Birefringence	Twins[1]	Number of Good Cleavages
	Prehnite (P5)	1.629	0.021–0.039	P	1
	Humite (H2)	1.636	0.028–0.036	—	0
	Jadeite (P8)	1.660	0.011–0.016	P	2
	Anthophyllite (A2)	1.663	0.024–0.028	—	3
	Aegirine-augite (P8)	1.726	0.030–0.036	S	2
	Cummingtonite (A2)	1.657	0.022–0.035	S, P	2
	Cordierite (C8)	1.549	0.005–0.018	P, Cy	1
	Lawsonite (L1)	1.674	0.019–0.021	S, P	2
	Low Albite (F1)	1.531	0.010	P, S	2
	Labradorite (F1)	1.564	0.008	P, S	2
	Boehmite (C6)	1.655	~0.015	—	1
	Staurolite (S10)	1.747	0.009–0.015	C	0
	Mg-Rich Olivine (O1)	1.705	0.035–0.043	S	0
	Andesine (F1)	1.553	0.008	P, S	2
	Diaspore (C6)	1.715	0.046–0.052	—	1
	Hornblende (A2)	1.666	0.014–0.026	S	2

[1] *Twin Types*
S = Simple
P = Polysynthetic
C = Cruciform
Cy = Cyclic

Chart 6: Biaxial Negative (−)
Listed According to Increasing Minimum Value of $2V$

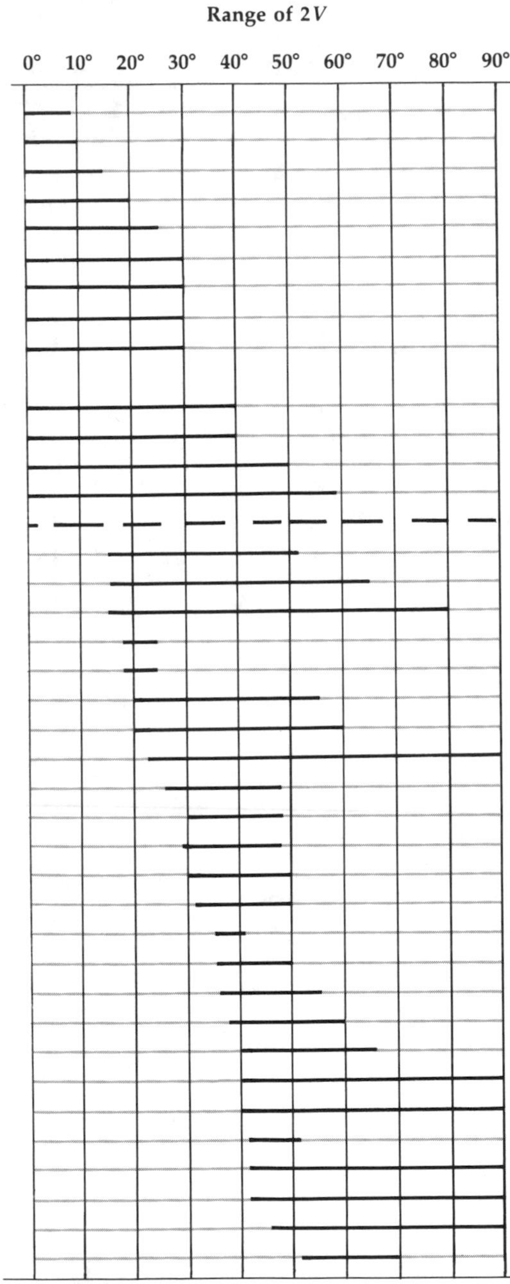

Range of $2V$

0° 10° 20° 30° 40° 50° 60° 70° 80° 90°

Mineral Name	Average Value of β	Range in Birefringence	Twins[1]	Number of Good Cleavages
Vermiculite (C6)	1.56	0.02–0.03	—	1
Illite (C6)	1.59	0.03	—	1
Phlogopite (M2)	1.597	0.028–0.049	S	1
Glauconite (M2)	1.63	0.014–0.032	—	1
Biotite (M2)	1.650	0.04–0.07	S	1
Montmorillonite (C6)	1.545	0.02–0.03	—	1
Talc (T1)	1.591	~0.05	—	1
Chlorite Group (C4)	1.625	0.000–0.010	P	1
Chlorite Group (C4) (oxidized)	1.65	0.00–0.02	P	1
Paragonite (M2)	1.601	0.028–0.038	—	1
Stilpnomelane (S11)	1.660	0.030–0.110	—	1
Arfvedsonite (A2)	1.694	0.005–0.012	S, P	2
Lepidolite (M2)	1.568	0.018–0.038	S	1
Perovskite (P3)	2.34	0.000–0.002	P	0
Dumortierite (D1)	1.698	0.011–0.027	Cy	1
High Sanidine (F1)	1.524	0.006	S	2
Eckermannite (A2)	1.638	0.009–0.020	S, P	2
Aragonite (C2)	1.682	0.149–0.157	P, Cy	0
Sanidine (F1)	1.525	0.005–0.006	S	2
Kaolinite (C6)	1.564	0.006–0.007	S	1
Serpentine Group (S3)	1.551	0.004–0.009	—	1
Hornblende (A2)	1.666	0.014–0.026	S	2
Laumontite (Z1)	1.517	0.008–0.016	S	3
Sericite (C6)	1.595	0.036–0.049	—	1
Muscovite (M2)	1.596	0.036–0.049	—	1
Stilbite (Z1)	1.499	~0.010	C	1
Glaucophane (P8)	1.613	0.019–0.024	—	2
Spurrite (S9)	1.674	0.039–0.040	S, P	1
Orthoclase (F1)	1.523	0.005	S	2
Scolecite (Z1)	1.518	0.008–0.010	P	2
Wollastonite (W1)	1.639	0.013–0.014	P	3
Margarite (M2)	1.645	0.012–0.014	P	1
Cordierite (C8)	1.549	0.005–0.018	P, Cy, S	1
Allanite (E1)	1.757	0.013–0.036	S	1
Anorthoclase (F1)	1.535	0.006–0.007	P, S	2
Crossite (A2)	1.666	0.012–0.019	—	2
Riebeckite (A2)	1.699	0.012–0.018	—	2
Fe-Rich Olivine (O1)	1.814	0.043–0.052	S	0
Sapphirine (S1)	1.717	0.004–0.006	—	0

[1] *Twin Types*
S = Simple
P = Polysynthetic
C = Cruciform
Cy = Cyclic

Chart 6: Biaxial Negative (−)
Listed According to Increasing Minimum Value of $2V$

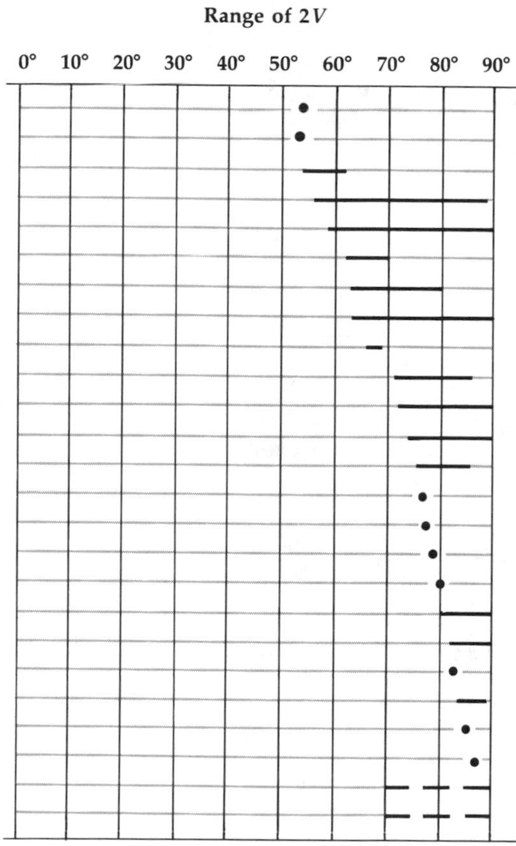

Range of $2V$: 0° 10° 20° 30° 40° 50° 60° 70° 80° 90°

Mineral Name	Average Value of β	Range in Birefringence	Twins[1]	Number of Good Cleavages
Ferrohypersthene (P8)	1.737	0.017	S, P	2
Hypersthene (P8)	1.711	0.014	S, P	2
Pyrophyllite (P7)	1.587	0.045–0.062	—	1
Oxyhornblende (A2)	1.720	0.018–0.083	S, P	2
Aegirine (P8)	1.781	0.036–0.060	S	2
Polyhalite (P4)	1.561	0.019–0.020	P	1
Axinite (A5)	1.691	0.009–0.011	—	1
Epidote (E1)	1.740	0.015–0.048	P	2
Microcline (F1)	1.518	0.007	P, S	2
Andalusite (A1)	1.638	0.009–0.011	—	2
Gedrite (A2)	1.670	0.017–0.027	—	3
Mg-Rich Olivine (O1)	1.705	0.035–0.043	S	0
Actinolite (A2)	1.653	0.017–0.022	S, P	2
Bronzite (P8)	1.685	0.011	S, P	2
Eulite (P8)	1.756	0.019	S, P	2
Anorthite (F1)	1.585	0.013	P, S	2
Kernite (K1)	1.472	0.034	S	3
Grunerite (A2)	1.693	0.035–0.045	S, P	2
Anthophyllite (A2)	1.663	0.024–0.028	—	3
Kyanite (A1)	1.721	0.016–0.017	S, P	2
Tremolite (A2)	1.625	0.023–0.024	S, P	2
Bytownite (F1)	1.574	0.010	P, S	2
Oligoclase (F1)	1.542	0.008	P, S	2
Pumpellyite (P6)	1.695	0.012–0.022	S	2
Chloritoid (C5)	1.726	0.010–0.012	P, S	1

[1] *Twin Types*
S = Simple
P = Polysynthetic
C = Cruciform
Cy = Cyclic

PART 3

Mineral Descriptions

When you steal from one author, it's plagiarism;
if you steal from many, it's research.
WILSON MIZNER

The mineral descriptions depart from tradition, inasmuch as they are arranged alphabetically rather than on the basis of increasing structural complexity. It has been my experience that the average student "could care less" about a mineral's structure when the task at hand is identification. In addition, the mineral species or group name is located in upper page corners to facilitate the locating of a particular mineral.

For the most part, the mineral species are arranged in structural or compositional groups (for example, amphiboles, carbonates, silica minerals) within the alphabetical framework. Exceptions are the opaque minerals, which include a variety of compositions. In addition, each mineral species is alphabetically cross-referenced to mineral group. It would be wise to examine the alphabetic mineral listing that follows this section in order to become familiar with the general arrangement.

Some of the described minerals are pictured in *color photomicrographs* as well as in black and white. These are reproduced in the color insert.

The *color* (or colors) of each mineral is given in the mineral description. Note that this is the color as seen through the microscope, not the color in hand specimen. The *interference color,* listed after the birefringence, is the maximum interference color that can be observed in a thin section of standard 0.03 mm thickness.

The *optic orientation* uses the following convention: a positive (unlabeled) angle indicates that a principal optical vibration direction is in front of or below the positive end of a crystallographic reference axis. Thus, in the monoclinic system, "$c \wedge Z = 32°$" means that the optical direction Z is 32° in front of the positive end of the c axis, that is, between $+c$ and $+a$;

"$c \wedge Z = -32^\circ$" means that Z lies 32° behind the positive side of the *c* axis, between $+c$ and $-a$.

The section of the mineral description labeled *Extinction in section* refers to the type of extinction observed for randomly oriented grains in thin section. If an extinction type for a specific grain orientation is needed, this can be deduced from sketches of cleavage fragments or from consideration of the mineral's block diagram and optical orientation.

The meanings of textural terms are generally self-evident. An exception is *bird's-eye maple structure,* which refers to a granular

Alphabetical Arrangement of Mineral Descriptions

- Aluminosilicate Group (A1)
  - Andalusite
  - Kyanite
  - Sillimanite
  - Mullite
- Amphibole Group (A2)
  - Anthophyllite
  - Gedrite
  - Cummingtonite-grunerite series
  - Tremolite-actinolite series
  - Hornblende
  - Oxyhornblende
  - Glaucophane-crossite-riebeckite series
  - Eckermannite-arfvedsonite series
- Apatite Group (A3)
- Apophyllite (A4)
- Axinite (A5)
- Beryl (B1)
- Brucite (B2)
- Cancrinite (C1)
- Carbonate Group (C2)
  - Calcite
  - Dolomite-ankerite series
  - Magnesite
  - Rhodochrosite
  - Siderite
  - Aragonite
- Cassiterite (C3)
- Chlorite Group (C4)
- Chloritoid (C5)
- Clays and clay minerals (C6)
  - Kaolinite
  - Montmorillonite
  - Illite
  - Vermiculite
  - Sericite
  - Glauconite
  - Boehmite
  - Diaspore
  - Gibbsite
- Colemanite (C7)
- Cordierite (C8)
- Corundum (C9)
- Dumortierite (D1)
- Epidote Group (E1)
  - Zoisite
  - Clinozoisite-epidote series
  - Piemontite
  - Allanite
- Feldspar Group (F1)
  - Sanidine
  - High sanidine
  - Anorthoclase
  - Orthoclase
  - Microcline
  - Plagioclase series
- Feldspathoid Group (F2)
  - Leucite
  - Nepheline
- Fluorite (F3)
- Garnet Group (G1)
  - Pyralspite garnets
  - Ugrandite garnets
- Glass (G2)
- Halite Group (H1)
  - Halite
  - Sylvite
- Humite Group (H2)
  - Norbergite
  - Chondrodite
  - Humite
  - Clinohumite
- Kernite (K1)
- Lawsonite (L1)
- Melilite Group (M1)
  - Gehlenite-akermanite series
- Mica Group (M2)
  - Biotite
  - Glauconite
  - Lepidolite
  - Margarite
  - Muscovite
  - Paragonite
  - Phlogopite
- Monazite (M3)
- Olivine Group (O1)
  - Forsterite-fayalite series
  - Monticellite

or pebbled appearance, observed between crossed nicols near the extinction position.

Included in the description of each mineral is a schematic sketch showing the general shape of cleavage fragments as seen in the microscope. The quality of the cleavages shown is usually somewhat better than might be expected. This has been done deliberately in order to facilitate the descriptions of extinction angles, grain shapes, and types of interference figures. Indices of refraction calculated for cleavage surfaces generally give a ranged value as a result of compositional variability.

Alphabetical Arrangement of Mineral Descriptions *(continued)*

Acmite. See Pyroxene Group (P8).
Actinolite. See Amphibole Group (A2).
Adularia. See Feldspar Group (F1).
Aegirine. See Pyroxene Group (P8).
Aegirine-augite. See Pyroxene Group (P8).
Akermanite. See Melilite Group (M1).
Albite. See Feldspar Group (F1).
Allanite. See Epidote Group (E1).
Almandine. See Garnet Group (G1).

ALUMINOSILICATE GROUP (A1)

Andalusite

Al_2SiO_5
Orthorhombic
$\alpha = 1.629-1.640$, $\beta = 1.633-1.644$, $\gamma = 1.638-1.650$
$\gamma - \alpha = 0.009-0.011$ (1st-order yellow)
$2V = 71°-86° (-)$
$a = Z$, $b = Y$, $c = X$

Pleochroism and color: commonly colorless, but may be pleochroic with X = pink to yellow (color photo 6), Y = colorless, pale yellow, or green, Z = colorless, pale yellow, or olive-green.
Dispersion: $v > r$ weak, rarely $r > v$.
Habit: euhedral prismatic {110} crystals common.
Cleavage: good {110} at nearly 90° (Fig. 1).
Extinction in section: parallel to dominant faces and cleavage traces in prismatic section, and length low (fast); symmetric to crystal outline and cleavage traces when viewed along c.
Twinning: rare on {101}, may be cruciform.
Hardness: 6½–7½.
Specific gravity: 3.13–3.16.
Occurrence: low- to moderate-grade metamorphic rocks; detrital grains in sediments; rare in siliceous igneous rocks and veins; commonly associated minerals are cordierite, quartz, biotite, garnet, staurolite, and sillimanite.
Alteration: commonly to sericite; with increasing grade of metamorphism may be completely or partially inverted to sillimanite (often as pseudomorphs).

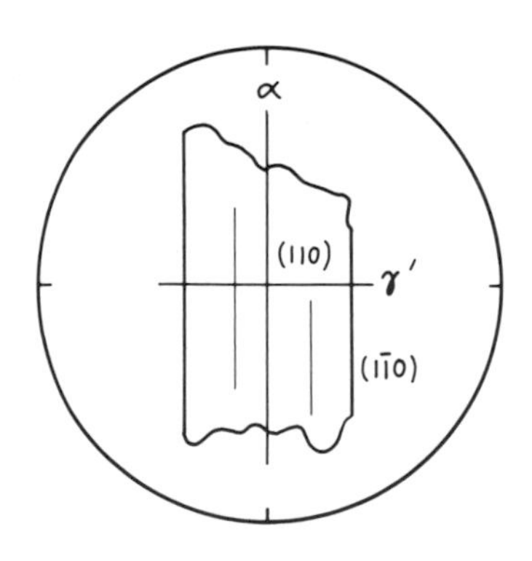

Figure 1. Andalusite grains on {110} have parallel extinction, and permit estimation of α and γ' (1.635–1.647). The figure is monosymmetric O.N.

Figure 2. Porphyroblasts of the chiastolite variety of andalusite in micaceous schist. Plane-polarized light. Photo length is 3.3 mm. (From J. Hinsch, E. Leitz, Inc.)

Characteristic features: euhedral rectangular to square outline common when viewed along *c;* cleavage is parallel to length when viewed in prismatic section; symmetrically arranged carbonaceous inclusions may be present, as in Figure 2 and color photo 7 (variety chiastolite); moderate relief; weak birefringence; length low (fast).

Look-alikes: sillimanite is length high (slow); topaz is (+) with a smaller 2*V;* hypersthene is length high (slow).

Kyanite

Al_2SiO_5
Triclinic
$\measuredangle\alpha \simeq 90°$, $\measuredangle\beta \simeq 101°$, $\measuredangle\gamma \simeq 106°$
$\alpha = 1.710–1.713$, $\beta = 1.720–1.722$, $\gamma = 1.727–1.729$
$\gamma - \alpha = 0.016–0.017$ (1st-order red)
$2V = 82.5° (-)$

Optic orientation: in {100}, $Z' \wedge c = 27°–30°$; in {010}, $Z' \wedge c = 5°–8°$; in {001}, $X' \wedge a = 0°–2½°$; $X \sim \perp$ {100}.

Pleochroism and color: commonly colorless, may be weakly pleochroic with X = colorless, Y = pale violet-blue, Z = pale cobalt-blue; $Z > Y > X$.

Dispersion: $r > v$, very weak.

Habit: bladed to columnar, elongated on *c*, narrow elongate sections are parallel to {010}, broad elongate sections are parallel to {100}.

Cleavage: {100} perfect (Fig. 3), {010} good, {001} weak (parting).

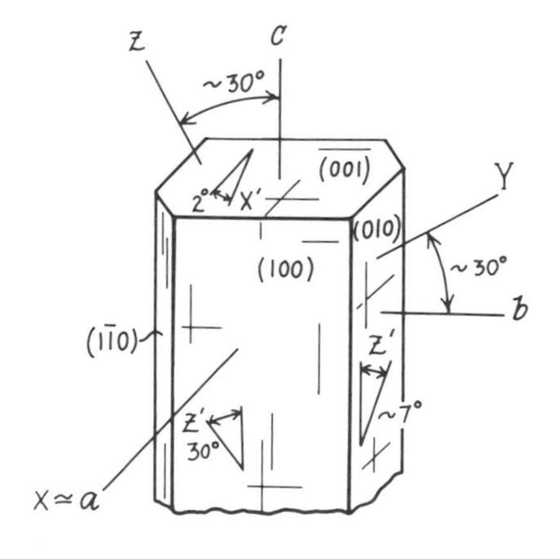

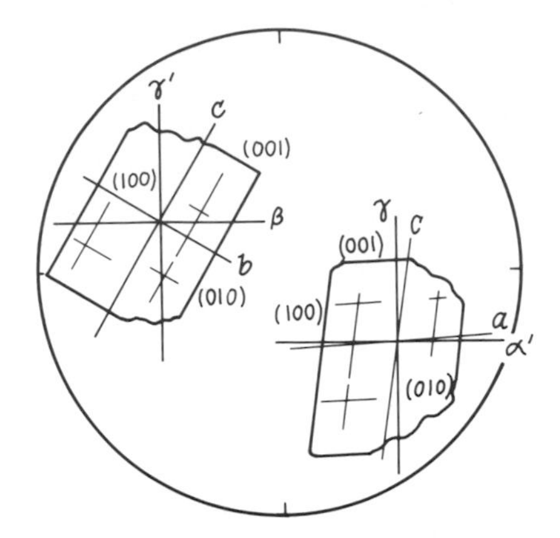

Figure 3. Kyanite fragments lying on {100} permit estimation of β' ($\simeq\beta$) and γ' ($\simeq\gamma$). Extinction angles are $\gamma' \wedge \{010\} \simeq 30°$ and $\gamma' \wedge \{001\} \simeq 60°$. The figure is almost bisymmetric Bxa. Fragments lying on {010} permit estimation of α' ($\simeq 1.713$) and γ' ($\simeq 1.726$). Extinction angles are $\gamma' \wedge c = 5°-8°$ and $\alpha' \wedge a \simeq$ parallel. The figure is nonsymmetric O.N. Less common fragments (not shown) lying on {001} have an extinction angle of $\alpha' \wedge \{010\} = 0°-2.5°$.

Extinction in section: inclined or parallel, depending on orientation.
Twinning: simple or polysynthetic {100} common, polysynthetic {001} less so.
Hardness: 4–4½ ‖ *c*, 6–7 ‖ *b*.
Specific gravity: 3.59.
Occurrence: metamorphic rocks with moderate to high pressure; with quartz, muscovite, almandine, staurolite, biotite, and sillimanite (Fig. 4); rare in pegmatites and quartz veins; common as a detrital mineral.
Alteration: commonly to sericite; less commonly to pyrophyllite and chlorite; may invert to other aluminosilicates.
Characteristic features: high relief; weak birefringence; cleavage; inclined extinction; large $2V$ (−); generally blocky grains. See color photo 8.
Look-alikes: sillimanite has small $2V$ (+); andalusite has parallel extinction; rhodonite is (+).

Figure 4. Irregular grain of kyanite (center). Gray grains are mainly biotite. Small grains at right are mainly sillimanite. Crossed nicols. Photo length is 2.22 mm.

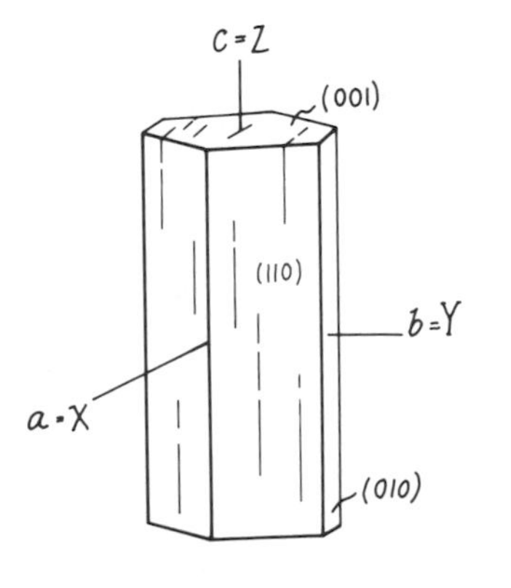

Sillimanite

Al_2SiO_5
Orthorhombic
$\alpha = 1.657-1.660$, $\beta = 1.658-1.661$, $\gamma = 1.677-1.682$
$\gamma - \alpha = 0.020-0.022$ (2nd-order blue)
$2V = 21°-30°$ (+)
$a = X$, $b = Y$, $c = Z$
Pleochroism and color: usually colorless, rarely brownish or bluish, with $Z > Y > X$.
Dispersion: $r > v$ strong.

Habit: prismatic or needlelike parallel to *c* (color photos 9–11); sometimes felted or fibrous; length high (slow).
Cleavage: perfect {010} (Fig. 5).
Extinction in section: generally parallel to direction of elongation and length high (slow); when viewed along *c*, extinction is symmetrical to {110} faces and parallel to {010} cleavage trace.
Twinning: none.
Hardness: 6–7.
Specific gravity: 3.24.
Occurrence: high-temperature metamorphic rocks (Figs. 6 and 7); rare in pegmatites and quartz veins, and as a detrital mineral.
Alteration: to sericite, pyrophyllite, kaolinite, or montmorillonite; may invert to other aluminosilicate minerals.
Characteristic features: occurrence; habit; high relief; small $2V$ (+).

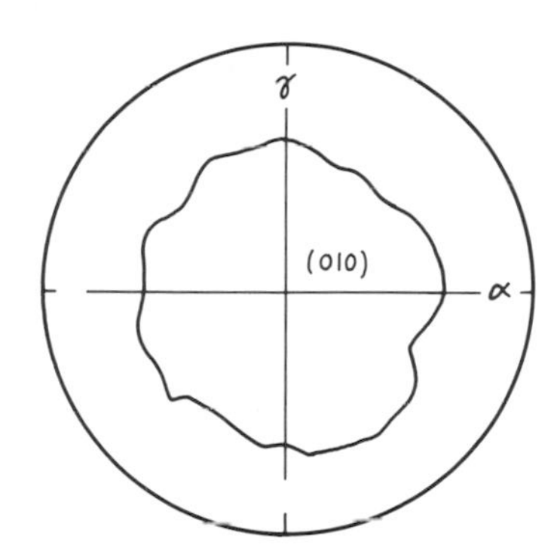

Figure 5. Sillimanite fragments on the {010} cleavage permit estimation of α and γ. Grain outline is irregular and figure is bisymmetric O.N. If habit is retained, grains may be acicular.

A

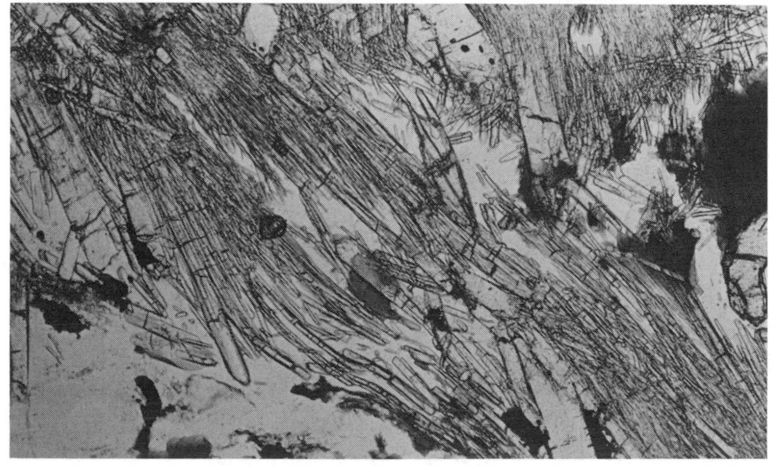

B

Figure 6. (A) Sillimanite aggregate in quartz-mica-garnet schist. Crossed nicols. (B) The same, viewed in plane-polarized light. Note the high relief of sillimanite and garnet (right and bottom) against quartz. Photo length is 2.06 mm. (From J. Hinsch, E. Leitz, Inc.)

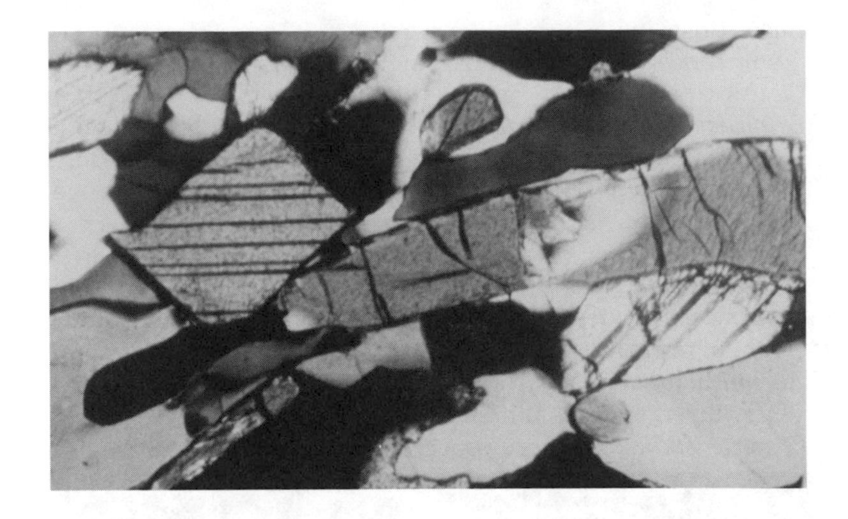

Figure 7. Sillimanite, quartz, and garnet. The blocky sillimanite grain (left) is viewed along the *c* axis, whereas the elongate grain of sillimanite (center right) is viewed almost normal to the *c* axis. Crossed nicols. Photo length is 0.91 mm. (From C. Shultz, Slippery Rock University)

Look-alikes: andalusite has large $2V$ (−) and is length low (fast); kyanite has large $2V$ (−) and usually inclined extinction; zoisite is length low (fast) and (−); mullite is indistinguishable optically but is quite rare.

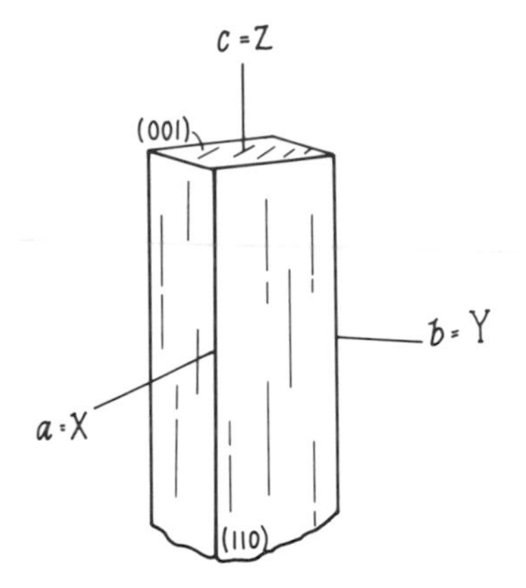

Mullite

$(Al_2O_3)_{3-2}(SiO_2)_{2-1}$
Orthorhombic
$\alpha = 1.634–1.666$, $\beta = 1.635–1.670$, $\gamma = 1.644–1.690$
$\gamma - \alpha = 0.010–0.024$ (2nd-order blue)
$2V = 20°–50°$ (+)
$a = X$, $b = Y$, $c = Z$
Pleochroism and color: colorless, or X = pale pink, Y = colorless, Z = colorless.
Dispersion: $r > v$ weak.
Habit: prismatic or acicular parallel to *c;* almost square cross-section.
Cleavage: {010} good (Fig. 8).
Extinction in section: generally parallel to cleavage or larger grain edges, and length high (slow); symmetric to {110} faces when viewed along *c* axis.
Twinning: none.
Hardness: 6–7.
Specific gravity: 3.19.
Occurrence: high-temperature low-pressure metamorphic rocks; xenoliths or contact zones of mafic magmas.
Alteration: to other aluminosilicates or clay minerals.

Characteristic features: occurrence; resemblance to sillimanite.
Look-alikes: essentially indistinguishable from sillimanite by optical methods; andalusite has {110} cleavage, lower birefringence, and a high 2*V*.

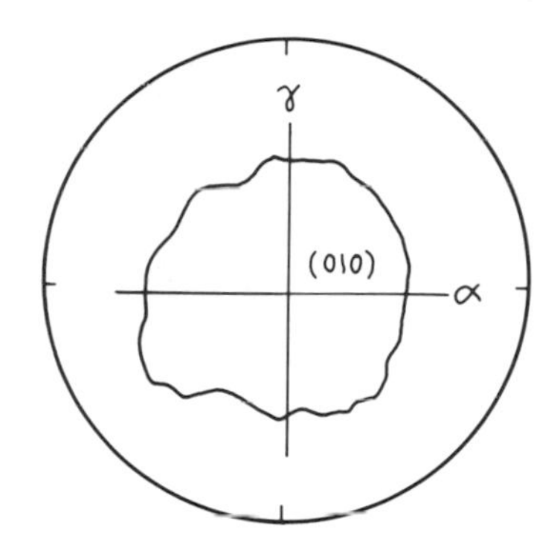

Figure 8. Most mullite fragments lie on {010}, permit estimation of α and γ, and yield a bisymmetric O.N. figure. If habit is retained, grains may be acicular.

Alunite. See Sulfate Group (S12).

Amesite. See Chlorite Group (C4).

THE AMPHIBOLE GROUP (A2)

Compositional variation within the amphibole group is based on a standard formula:

$$A_{0-1}B_2C_5{}^{VI}T_8{}^{IV}O_{22}(OH,F,Cl)_2$$

where A = (Na, K),
B = (Fe^{+2}, Mn, Mg, Ca, Na),
C = (Al, Cr, Ti, Fe^{+3}, Mg, Mn), and
T = (Si, Al, Cr^{+3}, Fe^{+3}, Ti^{+4}).

The superscript Roman numerals refer to octahedral and tetrahedral coordinations.

Amphiboles have been classified into four principal groups, based upon the number of atoms of Ca + Na and Na in the B sites of the structure; these are shown in Figure 9. Mineral names, assigned according to subdivision within each group, are based on the number of Si atoms and the ratio $Mg/(Mg + Fe^{+2})$. Prefixes may be added to the mineral names if one or more elements that are not essential constituents of the end members are present in substantial amounts. Amphibole nomenclature used will be that described by Leake (1978).

Due to the wide range in composition, as well as large differences in abundance, only the most common amphiboles will be described below (see Table A2).

The calcic amphiboles are the most abundant. If a calcic amphibole is identified solely or largely on its physical properties, and cannot be confidently assigned to an end member composition, it should be referred to as *hornblende.* If the optical (but not chemical) properties of any amphibole are known, the assigned name should be made into an adjective (for example, tremolitic amphibole, or edenitic hornblende); this is prudent in that optical determinations of amphiboles are notoriously inaccurate because of wide compositional substitutions.

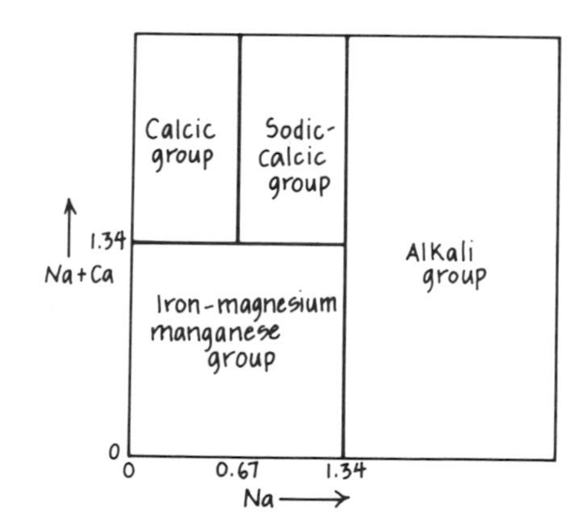

Figure 9. Amphibole classification based upon the number of atoms of Na + Ca, and Na in the B sites of the structure (after Leake, 1978).

Table A2
The Common Amphiboles

Mineral and System	Formula
Iron-magnesium-manganese amphiboles	
Orthorhombic system	
Anthophyllite	$(Mg_{0.9-0.1}Fe^{+2}_{0.1-0.9})_7Si_8O_{22}(OH)_2$
Gedrite	$(Mg_{0.9-0.1}Fe^{+2}_{0.1-0.9})_5Al_2Si_6Al_2O_{22}(OH)_2$
Monoclinic system	
Cummingtonite to grunerite	$(Mg_{0.69-0}Fe^{+2}_{0.31-1})_7Si_8O_{22}(OH)_2$
Calcic amphiboles (all monoclinic)	
Tremolite to ferro-actinolite	$Ca_2(Mg,Fe^{+2})_5Si_8O_{22}(OH)_2$
Edenite to ferro-edenite	$NaCa_2(Mg,Fe^{+2})_5Si_7AlO_{22}(OH)_2$
Pargasite to hastingsite	$NaCa_2(Mg,Fe^{+2})_4AlSi_6Al_2O_{22}(OH)_2$
Tschermakite to ferro-alumino-tschermakite	$Ca_2(Mg,Fe^{+2})_3Al_2Si_6Al_2O_{22}(OH)_2$
Hornblende is composed of mixtures of various calcic amphibole end-member components. Oxyhornblende is similar, but with larger amounts of Fe^{+3} and less Fe^{+2} and Mg^{+2}, such that $(OH + F + Cl) < 1.0$.	
Sodic-calcic amphiboles (all monoclinic)	
Katophorite (rare)	$NaCaNa(Mg,Fe^{+2})_4(Al,Fe^{+3})Si_7AlO_{22}(OH)_2$
Alkali amphiboles (all monoclinic)	
Glaucophane to riebeckite	$Na_2(Mg,Fe^{+2})_3(Al,Fe^{+3})_2Si_8O_{22}(OH)_2$
Eckermannite to arfvedsonite	$Na(Na_{2.0-1.5},Ca_{0-0.5})(Mg,Fe^{+2})_4(Al,Fe^{+3})Si_8O_{22}(OH)_2$

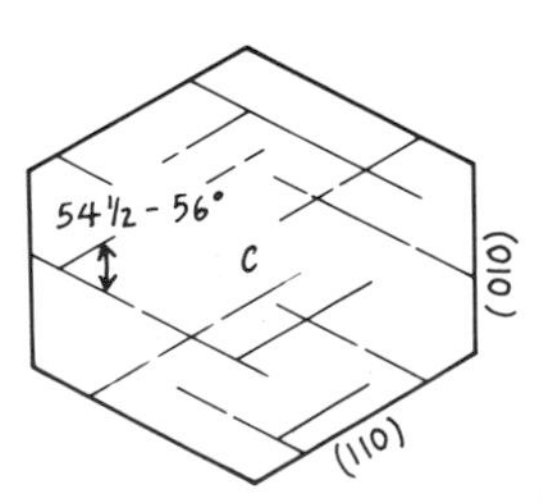

Figure 10. The common six-sided cross-section shown by members of the amphibole group when viewed along the *c* axis.

The amphiboles are monoclinic, with the exception of anthophyllite and gedrite, which are orthorhombic. The habit of amphiboles is generally elongate prismatic (parallel to *c*), but may be asbestiform or stubby prismatic. Prismatic crystals have dominant {110} and {010} forms and are generally four- or six-sided (Fig. 10). Monoclinic amphiboles possess perfect {110} cleavage (color photo 12). Corresponding to this, orthorhombic amphiboles have a perfect {210} cleavage. The angles between the two corresponding cleavages are 56° and 124°, approximately. Simple or polysynthetic {100} twinning is common in monoclinic varieties. A flow chart for tentative amphibole identification is given in Figure 11.

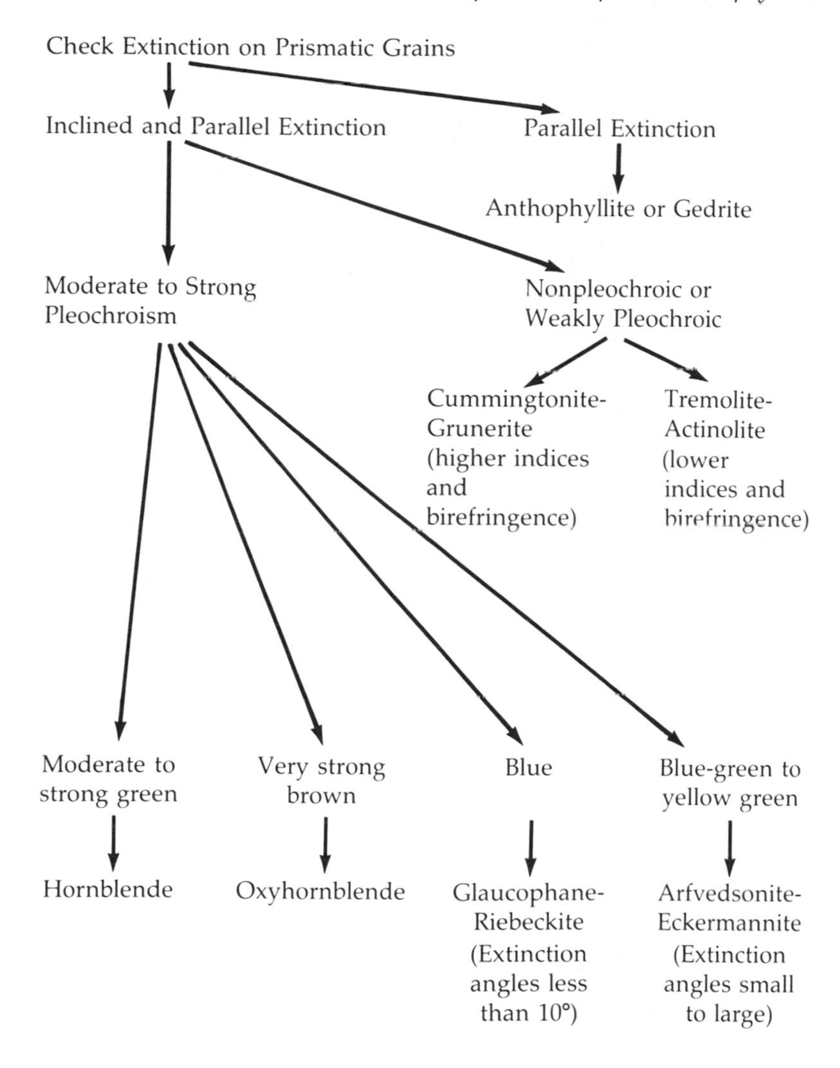

Figure 11. Flowchart for tentative amphibole identification. Detailed optical and chemical data are necessary for firm identification.

Anthophyllite

$(Mg_{0.9-0.1}Fe^{+2}_{0.1-0.9})_7Si_8O_{22}(OH)_2$
Orthorhombic
$\alpha = 1.604-1.694$, $\beta = 1.617-1.710$, $\gamma = 1.628-1.722$ (Fig. 12)
$\gamma - \alpha = 0.024-0.028$ (2nd-order green to green-yellow)
$2V = 82°-90°(-)$, $70°-90°(+)$
$a = X, b = Y, c = Z$
Pleochroism and color: colorless to pale brown or green, weak to moderate pleochroism, with $Z > Y \simeq X$ or $Z \simeq Y > X$.
Dispersion: $r < v$ or $r > v$, weak to moderate.
Habit: prismatic; fibrous; asbestiform parallel to c, with {210}, {100}, and {010} dominant.

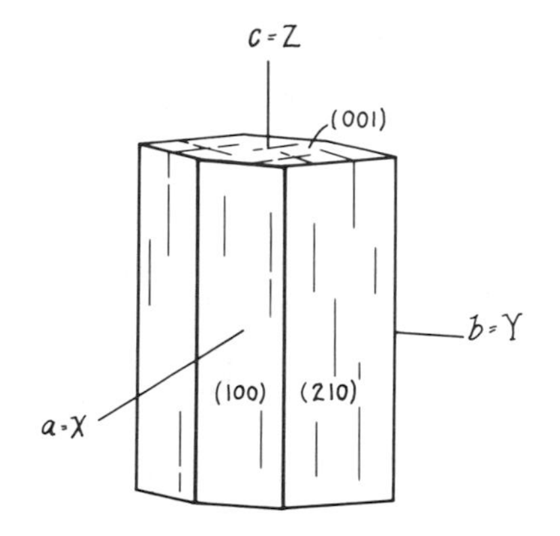

Figure 12. Variation in indices of refraction and $2V$ in the magnesio-anthophyllite to ferro-anthophyllite series based on data in Deer, Howie, and Zussman (1963), Troeger (1979), and Fleischer, Wilcox, and Matzko (1984). Ferro-anthophyllite is theoretical; in nature the series only exists to approximately 40 percent substitution of Fe for Mg.

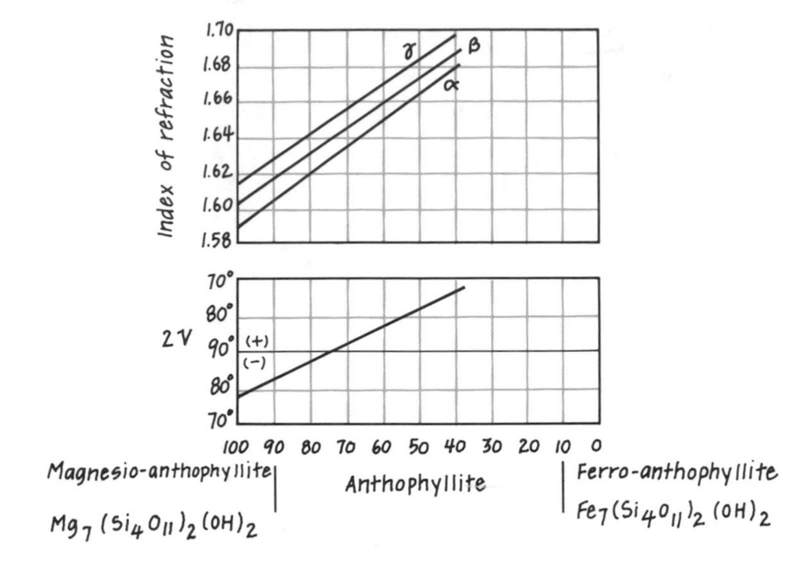

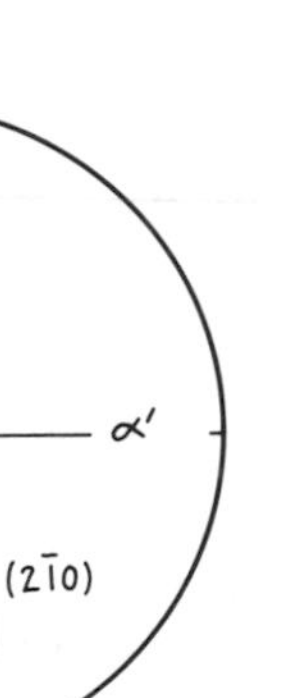

Figure 13. Anthophyllite fragments on {210} have parallel extinction, are length high (slow), and permit estimation of α' (1.614–1.707) and γ. The figure is monosymmetric Bxa or Bxo.

Cleavage: {210} perfect at 54°30′ and 125°30′ (Fig. 13), {010} good, {100} poor, {001} parting.

Extinction in section: usually parallel to elongate {210} faces and cleavages in prismatic section; length high (slow); symmetrical to {210} faces and cleavage traces when viewed along *c*.

Twinning: none.

Hardness: 5½.

Specific gravity: ≃ 3.1.

Occurrence: high-grade metamorphism of ultramafic or impure argillaceous rocks; commonly with cordierite, hornblende, plagioclase, sillimanite, talc, or serpentine.

Alteration: commonly to talc or serpentine.

Characteristic features: usually parallel extinction; perfect {210} cleavage; large $2V$ with (+) sign most common; occurrence.

Look-alikes: zoisite has higher indices and lower birefringence; the angle between {110} faces is about 90° in sillimanite; monoclinic amphiboles usually show inclined extinction; optically indistinguishable from gedrite, with which it forms an isomorphous series by replacement of Al^{+3} for Si^{+4} (see Figures 12 and 14).

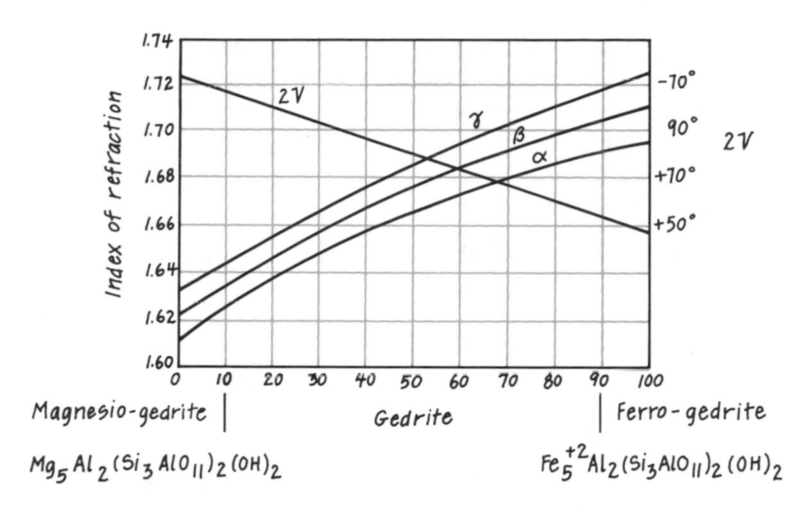

Figure 14. Variation in indices of refraction and $2V$ in the magnesio-gedrite to ferro-gedrite series, based on data in Deer, Howie, and Zussman (1963), Troeger (1979), and Fleischer, Wilcox, and Matzko (1984).

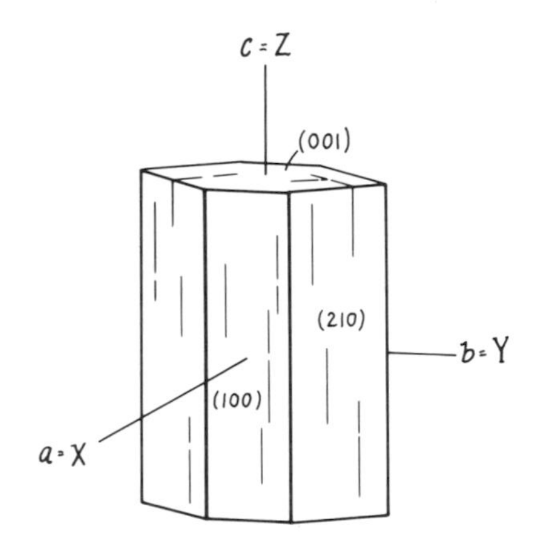

Gedrite

$(Mg_{0.9-0.1}Fe^{+2}_{0.1-0.9})_5Al_2Si_6Al_2O_{22}(OH)_2$
Orthorhombic
$\alpha = 1.627-1.690$, $\beta = 1.635-1.705$, $\gamma = 1.644-1.717$ (Fig. 14)
$\gamma - \alpha = 0.017-0.027$ (1st-order red to 2nd-order green-yellow)
$2V = 72°-90°(-)$, $47°-90°(+)$
$a = X, b = Y, c = Z$
Pleochroism and color: weakly pleochroic in various shades of pale green-brown and yellow, with $Z > Y \simeq X$; X = yellow-brown to greenish yellow, Y = brownish yellow to greenish yellow, Z = yellowish brown to grayish green.
Dispersion: $r < v$ weak.
Habit: prismatic; fibrous; asbestiform parallel to c, with {210}, {100}, and {010} most common.
Cleavage: {210} perfect at 54°30′ and 125°30′ (Fig. 15), {010} good, {100} poor, {001} parting.
Extinction in section: usually parallel to elongate {210} faces and cleavages, and length high (slow); symmetrical to {210} faces and cleavages when viewed along c.
Twinning: none.
Hardness: 5½.
Specific gravity: ~ 3.3.
Occurrence: medium-grade metamorphic rocks (amphibolite, eclogite, gneiss, and hornfels) with garnet and cordierite.
Alteration: commonly to talc or serpentine.
Characteristic features: usually parallel extinction; perfect {210} cleavage; large $2V$, commonly with (+) sign; occurrence.

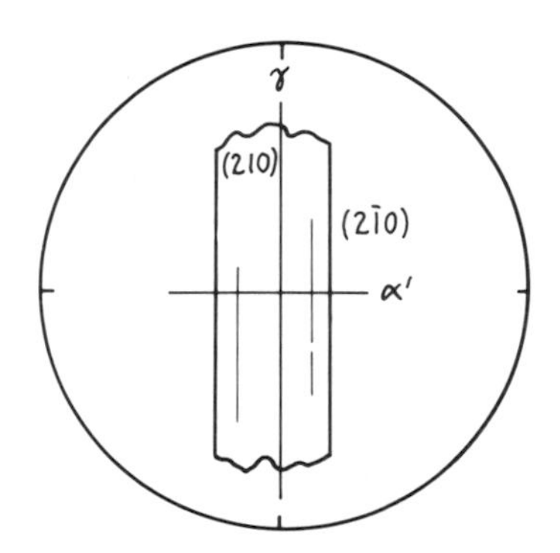

Figure 15. Gedrite grain on {210} has parallel extinction, and permits estimation of α' (1.633–1.702) and γ. The figure is monosymmetric Bxa or Bxo.

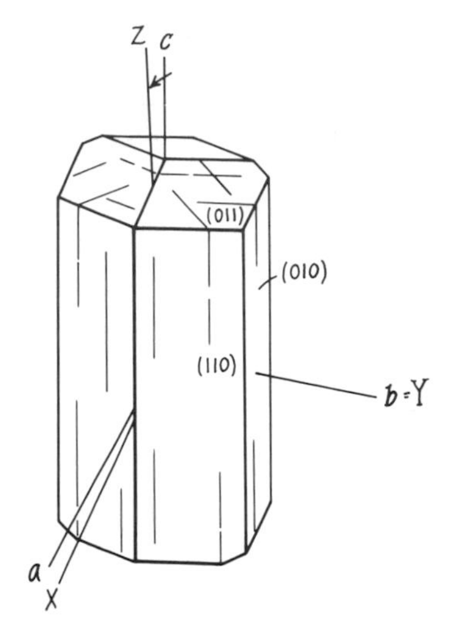

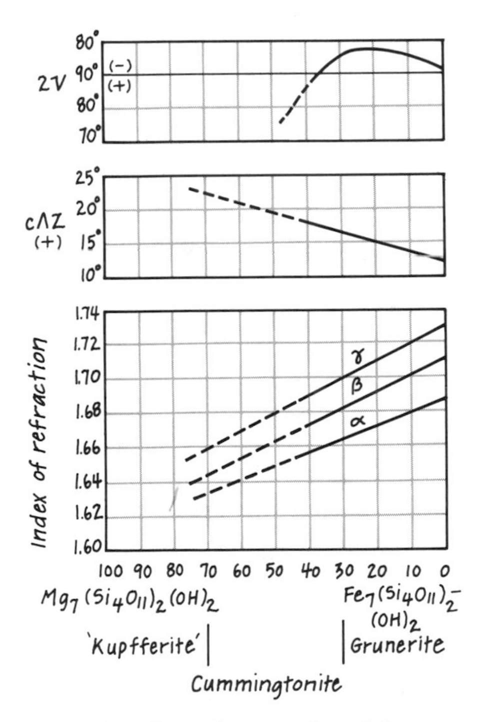

Figure 16. Optical properties of the cummingtonite-grunerite series (after Klein, 1964). The Mg-rich end-member, called kupfferite, does not occur in nature.

Look-alikes: zoisite has higher indices and lower birefringence; monoclinic amphiboles usually show inclined extinction; optically indistinguishable from anthophyllite, with which it forms an isomorphous series.

Cummingtonite-Grunerite Series

$(Mg_{0.69-0}Fe^{+2}_{0.31-1})Si_8O_{22}(OH)_2$
Monoclinic
$\measuredangle\beta \simeq 102°$
$\alpha = 1.632-1.688$, $\beta = 1.638-1.709$, $\gamma = 1.655-1.729$ (Fig. 16)
$\gamma - \alpha = 0.022-0.045$ (2nd-order blue to 3rd-order green)
$2V = 80°-90°(-)$, $70°-90°(+)$
$b = Y, c \wedge Z = 12°-21°$
Pleochroism and color: colorless, pale green, yellow, or brown; weak pleochroism, with $Z > Y \geq X$.
Dispersion: $r > v$ or $v > r$, weak, inclined.
Habit: prismatic; elongate parallel to *c* with {110} and {010} dominant; bladed, acicular, asbestiform, or fibrous.
Cleavage: {110} good at ~55° and ~125° (Fig. 17).
Extinction in section: inclined or parallel to elongate principal faces and cleavage traces, and length high (slow); symmetrical to {110} faces and cleavages when viewed along *c*.
Twinning: {100} simple and polysynthetic common.
Hardness: 5–6.
Specific gravity: 3.3–3.6.
Occurrence: cummingtonite is most common in contact or regionally metamorphosed rocks and less common in intermediate or mafic igneous rocks; typically associated with hornblende

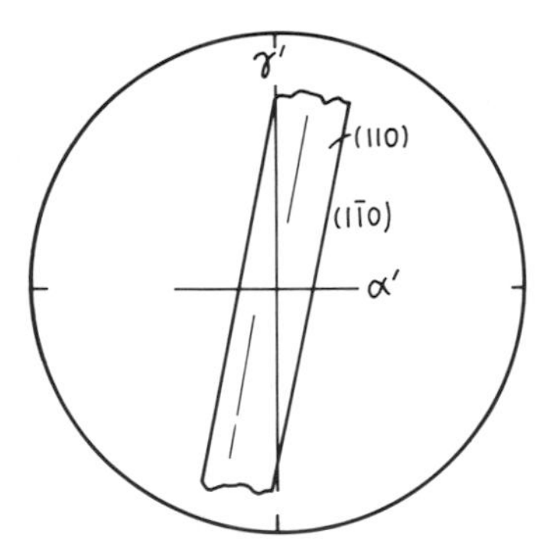

Figure 17. Fragments on {110} have extinction of $\gamma' \wedge \{110\} = 10°$ for 100 percent Fe grunerite; the indices are α' (1.708) and γ' (1.728). Values range to Mg-rich cummingtonite with $\gamma' \wedge \{110\} = 16°$; the indices are α' (1.637) and γ' (1.653), and the figure is nonsymmetric.

or anthophyllite; grunerite occurs in metamorphosed iron-rich sediments.

Alteration: to hornblende by reaction, or to talc or serpentine by weathering.

Characteristic features: large 2*V*(+ or −); inclined extinction common; twinning; pale or colorless; {110} cleavage and habit.

Look-alikes: anthophyllite and gedrite generally show parallel extinction and are untwinned; hornblende has strong pleochroism; tremolite and actinolite have lower indices.

Tremolite-Actinolite Series

$Ca_2Mg_5Si_8O_{22}(OH)_2$ to $Ca(Mg_{0.5}Fe^{+2}_{0.5})_5Si_8O_{22}(OH)_2$
Monoclinic
$\measuredangle\beta \simeq 105°$
$\alpha = 1.608–1.647$, $\beta = 1.618–1.659$, $\gamma = 1.630–1.667$ (Fig. 18)
$\gamma - \alpha = 0.020–0.022$ (2nd-order blue)
$2V = 79°–85°(-)$
$b = Y, c \wedge Z = 12°–15°$

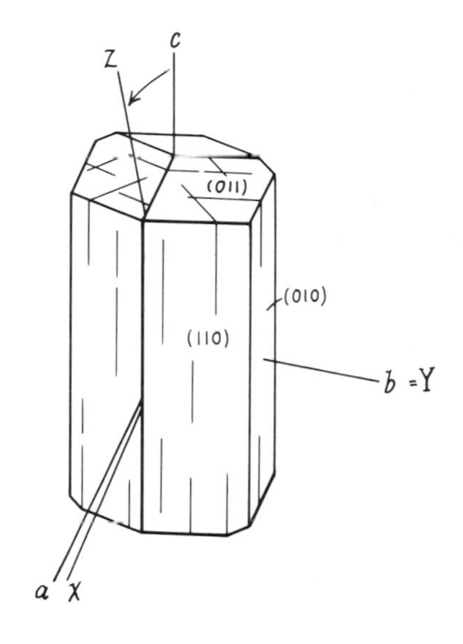

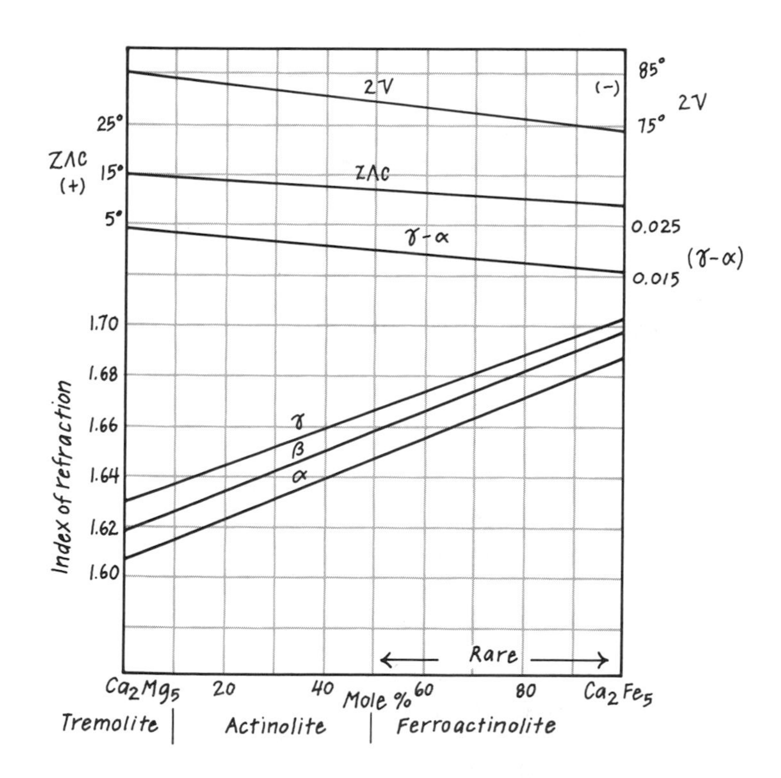

Figure 18. Optical properties of the tremolite-actinolite series (after Troeger, 1971, p. 93).

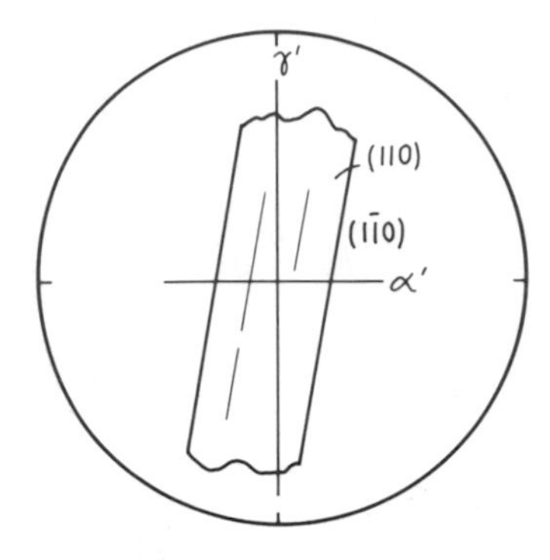

Figure 19. Fragment on {110}. Tremolite has $\gamma' \wedge \{110\} = 13°$, and permits estimation of α' (1.616) and γ' (1.629). Values range to $\gamma' \wedge \{110\} = 11°$, α' (1.656) and γ' (1.666) for Fe-rich actinolite. The figure is nonsymmetric.

Pleochroism and color: tremolite is colorless; actinolite is pale yellow-green or blue-green, with X = pale yellow-green, Y = light green, Z = light bluish green; $Z > Y > X$; pleochroism weak.

Dispersion: $r < v$ weak.

Habit: acicular to fibrous parallel to *c*, with {110}, {010}, and {011} common.

Cleavage: {110} perfect at ≃56° and ≃124° (Fig. 19).

Extinction in section: parallel or inclined to elongate {110} faces and cleavages; symmetrical to {110} faces and cleavages when viewed along *c*.

Twinning: {100} simple and polysynthetic common; {001} polysynthetic rare.

Hardness: 5–6.

Specific gravity: 3.0–3.2.

Occurrence: contact and low-grade metamorphism of ultramafic or carbonate rocks (color photos 13 and 14); hydrothermal alteration of mafic and ultramafic rocks; pseudomorphs after igneous pyroxenes by deuteric alteration.

Alteration: tremolite commonly alters to talc; actinolite alters to chlorite and/or calcite-dolomite.

Characteristic features: large $2V(-)$; inclined extinction common; twinning; pale or colorless; {110} cleavage and habit; may be weakly pleochroic.

Look-alikes: cummingtonite and grunerite have higher indices and more commonly exhibit polysynthetic twins; hornblende has strong pleochroism; wollastonite has lower birefringence and smaller $2V$.

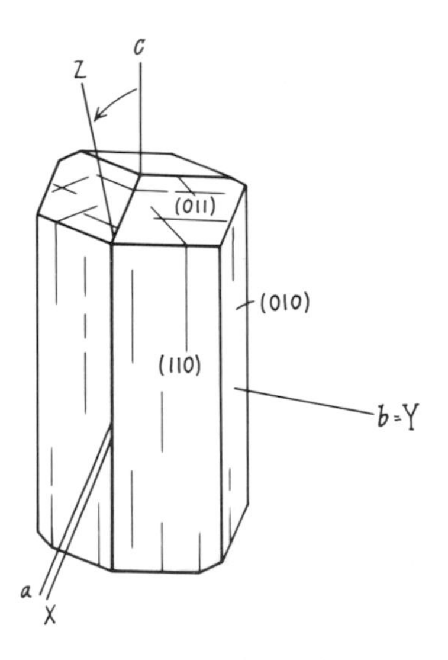

Hornblende

Composition consists of edenite, pargasite, and tschermakite Mg-rich components and their Fe-rich equivalents; approximately

$Ca_2(Mg,Fe^{+2})_4(Al,Fe^{+3})(Si_7Al)O_{22}(OH)_2$ (see Table A2)

Monoclinic

$\measuredangle\beta = 105°30'$

$\alpha = 1.615–1.705$, $\beta = 1.618–1.714$, $\gamma = 1.632–1.730$ (Fig. 20)

$\gamma - \alpha = 0.014–0.026$ (1st-order yellow-orange to 2nd-order green)

$2V = 23°–90°(-)$, $85°–90°(+)$; large negative values common

$b = Y$, $c \wedge Z = 13°–34°$

Pleochroism and color: color is strong and variable in green, blue-green, and yellow-brown; moderate pleochroism, with $Z > Y > X$ or less commonly $Y > Z > X$; examples of pleochroism are from Deer, Howie, and Zussman (1963):

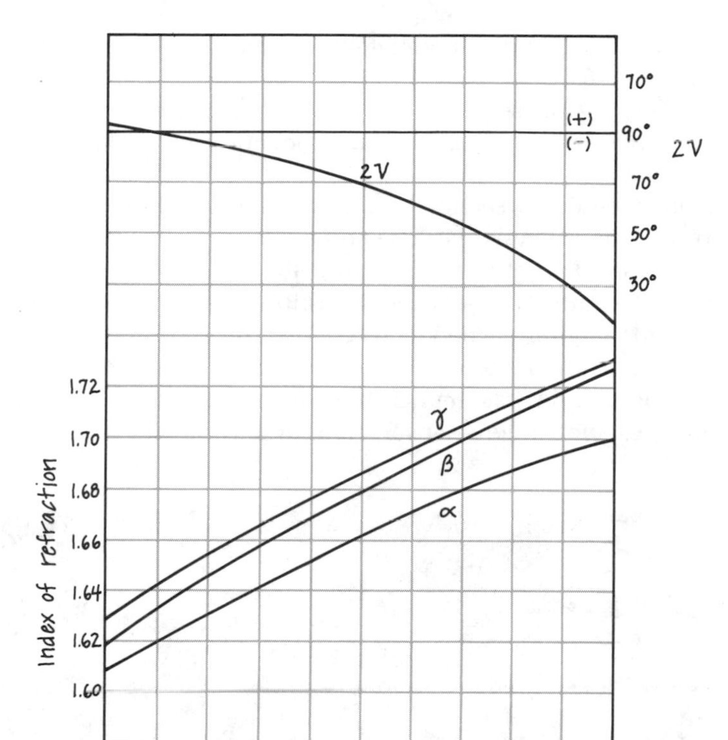

Figure 20. The relationship between chemical composition and optical properties of the common hornblendes (modified from Deer, Howie, and Zussman, 1963, Fig. 76).

X	Y	Z
colorless	pale yellow-green	pale green
colorless	pale brown-green	pale apple-green
yellowish green	yellowish green	bluish green
pale yellow	light green	light blue-green
yellow-green	pale brown	brown
pale yellow-brown	pale yellow	yellow-brown
yellow-green	green	bluish green
greenish brown	reddish brown	red-brown

Dispersion: $r > v$ or $v > r$ moderate.

Habit: prismatic with {110} and {010} common; pseudohexagonal cross-section when c is vertical.

Cleavage: {110} perfect at 56° and 124°; parting uncommon on {100} or {001}. See color photos 15 and 16 and Figure 21.

Extinction in section: parallel or inclined to elongate {110} faces or cleavages; length high (slow); symmetrical to {110} faces and cleavages when viewed along c.

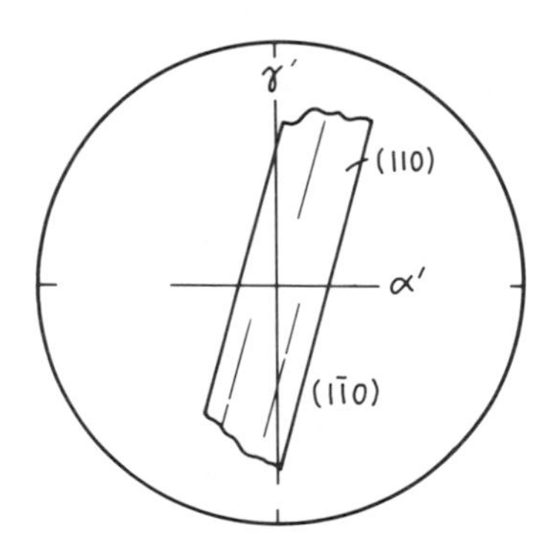

Figure 21. Hornblende fragment on {110}. Extinction of $\gamma' \wedge \{110\}$ = 4°–23°. The indices of refraction are highly variable, and are probably in the range of α' (1.62–1.70) and γ' (1.63–1.72). The figure is nonsymmetric.

Twinning: {100} simple, moderately common.
Hardness: 5–6.
Specific gravity: 3.04–3.45.
Occurrence: common in igneous and metamorphic rocks (Figs. 22 and 23).
Alteration: deuteric reaction to form biotite; alteration commonly to chlorite, often with calcite and epidote.
Characteristic features: strong color; strong pleochroism; large $2V$ (−) common; characteristic {110} cleavage and pseudo-hexagonal outline when viewed along the c axis.
Look-alikes: cummingtonite-grunerite usually twinned on {100}; oxyhornblende has higher birefringence; many other amphiboles show intense absorption, or may be blue.

Figure 22. Aggregate of hornblende grains in amphibolite. Grains show either two, one, or no cleavages as a function of orientation. Crossed nicols. Photo length is 2.22 mm.

Figure 23. Hornblende with plagioclase (twinned) and quartz (clear). Crossed nicols. Photo length is 2.22 mm.

Oxyhornblende

$Ca_2Na(Mg,Fe^{+2},Fe^{+3},Al,Ti)_5[(Si_3Al)O_{11}]_2(O,OH)_2$, with (OH + F + Cl) < 1.0

Monoclinic

$\measuredangle\beta \simeq 106°$

$\alpha = 1.650\text{–}1.700$, $\beta = 1.670\text{–}1.770$, $\gamma = 1.680\text{–}1.800$

$\gamma - \alpha = 0.018\text{–}0.083$ (from 2nd-order through white of the higher orders)

$2V = 56°\text{–}88° (-)$

$b = Y, c \wedge Z = 0°\text{–}19°$

Pleochroism and color: very strong pleochroism, with $Z > Y > X$; X = pale yellow to yellow-brown, Y = brown to red-brown, Z = dark red-brown to dark green-brown.

Dispersion: $v > r$ weak, or $r > v$ strong.

Habit: euhedral short prismatic phenocrysts with {110} and {010} common.

Cleavage: {110} good at ≃56° and ≃124° (Fig. 24); some parting on {100} or {001}.

Extinction in section: inclined or parallel to elongate {110} faces and cleavages; length high (slow); symmetrical to {110} faces or cleavage when viewed along the c axis.

Twinning: simple or polysynthetic on {100} fairly common.

Hardness: 5–6.

Specific gravity: 3.2–3.3.

Occurrence: phenocrysts in volcanic or hypabyssal rocks that vary in composition from basalt to trachyte, with intermediate compositions most common (Fig. 25).

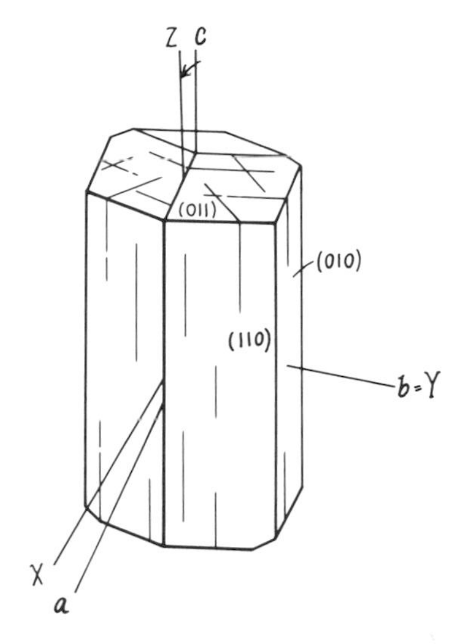

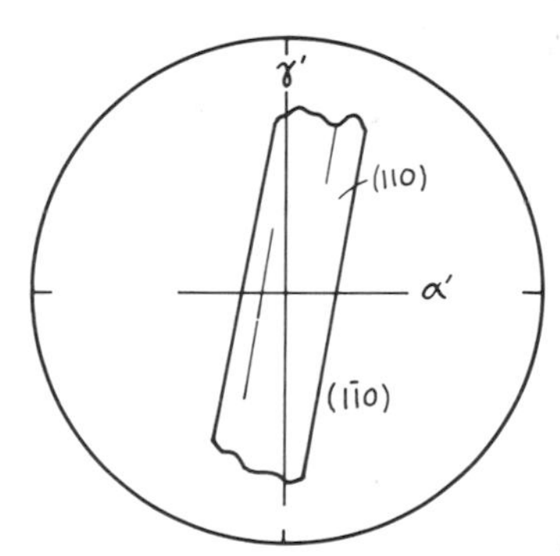

Figure 24. Oxyhornblende fragment on {110}. For low oxidation states $\gamma' \wedge \{110\} = 19°$, $\alpha' = 1.65$, and $\gamma' = 1.68$. This ranges with increasing oxidation to $\gamma' \wedge \{110\} = 0°$, $\alpha' = 1.75$, and $\gamma = 1.80$. The figure is nonsymmetric to monosymmetric Bxa (when $c \wedge Z = 0°$).

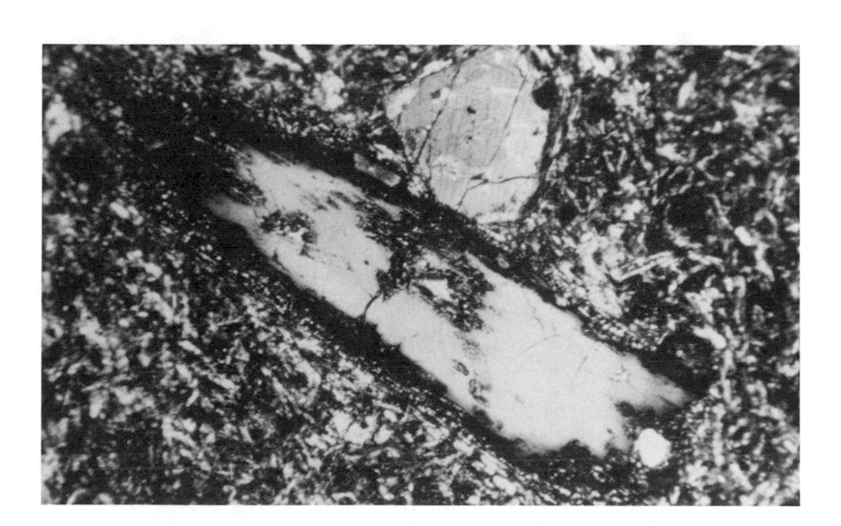

Figure 25. Oxyhornblende phenocryst in lava. The opaque rim consists of fine-grained iron oxides, and results from dehydration and oxidation during emplacement. The smaller phenocryst is clinopyroxene, and most of the groundmass consists of plagioclase microlites. Crossed nicols. Photo length is 0.55 mm.

Alteration: the commonly occurring resorption and alteration rims consist of fine-grained granular aggregates that may contain pyroxene, plagioclase, biotite, magnetite, or hematite (color photo 17); titanium-rich varieties commonly alter to chlorite and sphene.

Characteristic features: brown color; strong pleochroism; small extinction angle; high birefringence; occurrence; alteration rims are common.

Look-alikes: common hornblende is green, with lower birefringence and higher extinction angle; biotite has a single cleavage, parallel extinction, and small 2*V*.

Glaucophane-Crossite-Riebeckite Series

$Na_2(Mg,Fe^{+2})_3(Al,Fe^{+3})_2(Si_8O_{22})(OH)_2$ (Fig. 26)
Monoclinic
$\measuredangle\beta \simeq 104°$

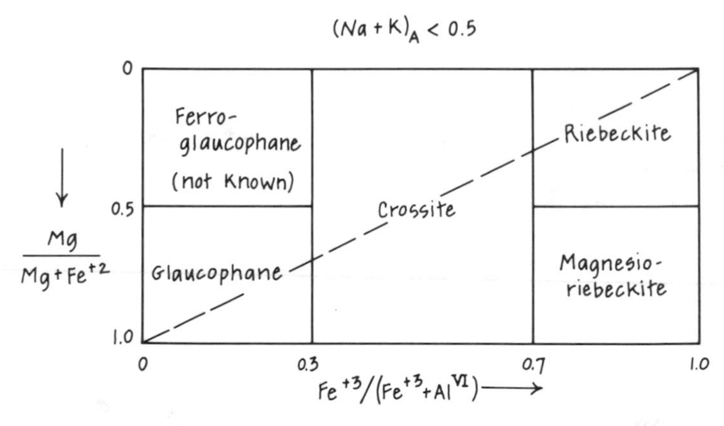

Figure 26. Compositional relationships in the glaucophane-crossite-riebeckite series (Leake, 1978). The dashed line describes the approximate compositional range in Figure 27.

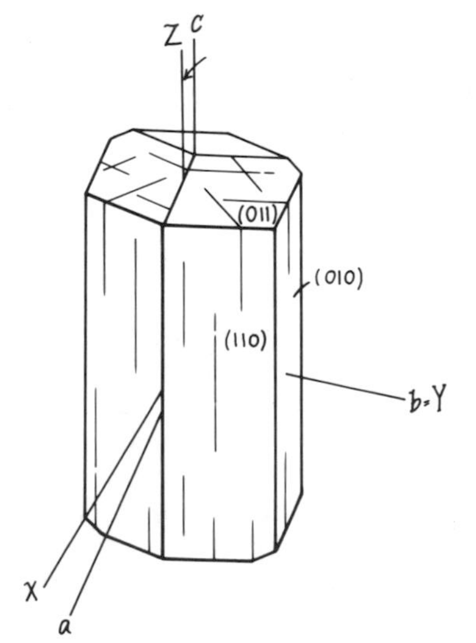

Glaucophane

$\alpha = 1.594–1.631$, $\beta = 1.614–1.648$, $\gamma = 1.619–1.649$ (Fig. 27)
$\gamma - \alpha = 0.019–0.024$ (2nd-order green)
$2V = 32°–50° (-)$
$b = Y, c \wedge Z = 5°–8°$

Pleochroism and color: moderate to strong pleochroism; X = yellow, Y = violet, Z = blue, with $Z > Y > X$ (color photo 18).

Dispersion: $v > r$ weak.

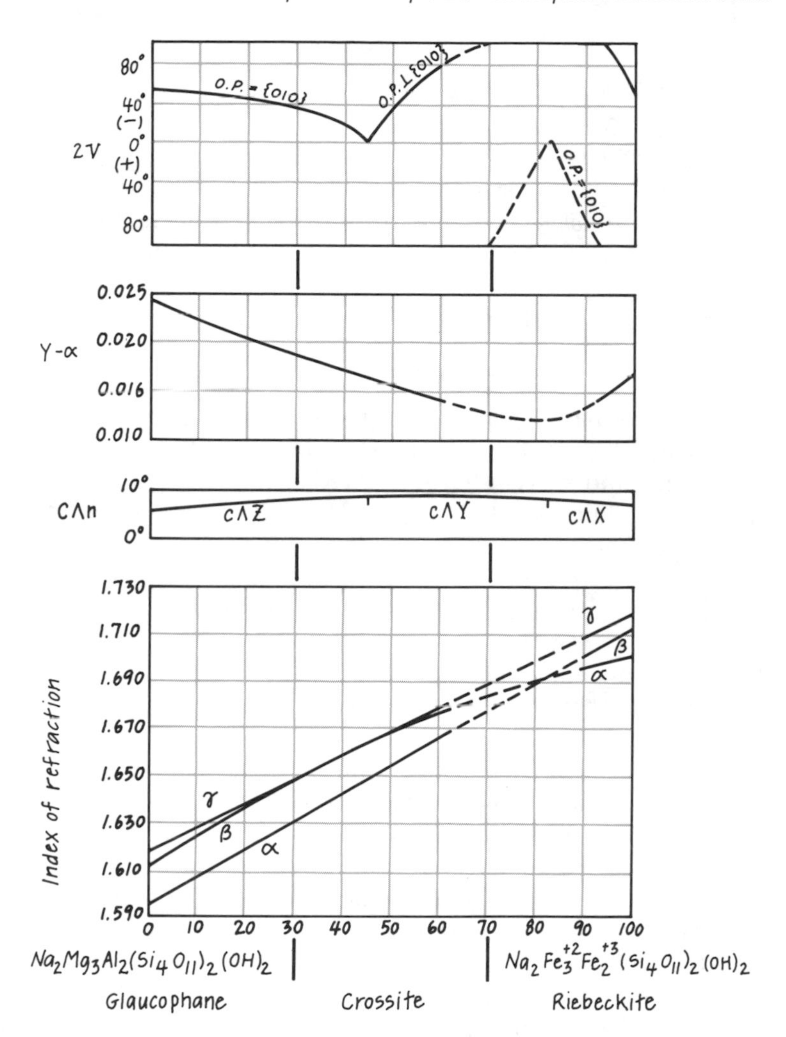

Figure 27. Optical properties in the glaucophane-crossite-riebeckite series (modified from Borg, 1967). O.P. = optic plane.

Crossite

$\alpha = 1.631–1.677$, $\beta = 1.648–1.684$, $\gamma = 1.649–1.689$
$\gamma - \alpha = 0.012–0.019$ (2nd-order bluish red)
$2V = 0°–90°\ (-)$
$b = Y$, $c \wedge Z = 8°$ or $b = Z$, $c \wedge Y = 9°$
Pleochroism and color: moderate to strong pleochroism; X = yellow, Y = blue, Z = violet, with $Z > Y > X$.
Dispersion: $r > v$ very strong.

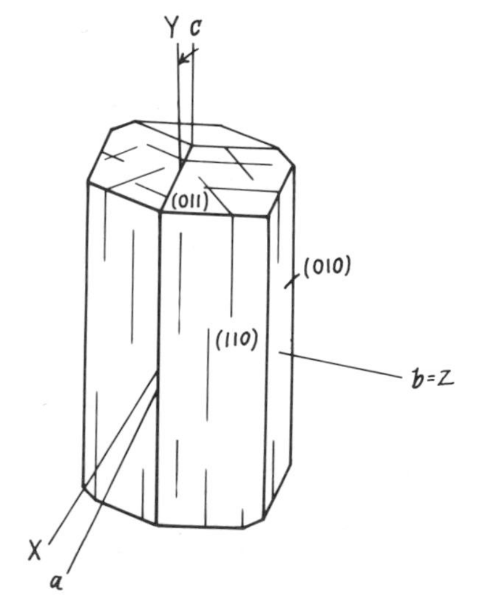

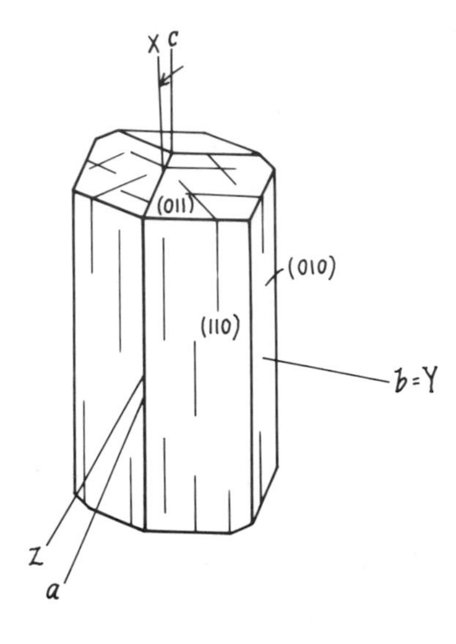

Riebeckite

$\alpha = 1.677–1.701$, $\beta = 1.684–1.714$, $\gamma = 1.689–1.720$
$\gamma - \alpha = 0.012–0.018$ (1st-order red)
$2V = 42°–90° (-)$, $0°–90° (+)$
$b = Z$, $c \wedge Y = 8°–9°$ or $b = Y$, $c \wedge X = 7°–8°$
Pleochroism and color: moderate to strong pleochroism; X = dark blue, Y = dark gray-blue, and Z = yellow-brown, with $Z > Y \simeq X$.
Dispersion: $v > r$ strong when optically (+), $r > v$ when optically (−).

Habit: glaucophane typically occurs as elongate prismatic crystals with {110} and {010} common; riebeckite may be slender prismatic and is often fibrous, felted, or asbestiform.
Cleavage: {110} perfect at 58° and 122° (Fig. 28); possible {010} and {001} partings.
Extinction in section: inclined or parallel to {110} faces and cleavages, symmetrical to {110} faces and cleavages when viewed along *c*.
Twinning: uncommon simple or polysynthetic on {100}.
Hardness: 6.
Specific gravity: 3.0–3.4.
Occurrence: glaucophane and crossite are typically found in high-pressure metamorphic rocks, commonly with epidote, pumpellyite, lawsonite, or aragonite; riebeckite is most common in soda-rich plutonic igneous rocks, often associated with aegirine, aegirine-augite, arfvedsonite, and nepheline or quartz; it is less common in metamorphic rocks, with the exception of certain bedded ironstones, where it occurs in an asbestiform variety known as crocidolite.
Alteration: glaucophane and crossite alter to actinolite; riebeckite to iron oxides, or siderite with quartz.
Characteristic features: strong pleochroism in some shade of blue; {110} faces and cleavages at 58° and 122°.
Look-alikes: arfvedsonite is always (−) and has $b = Z$; blue tourmaline is uniaxial, has parallel extinction, with strongest absorption ⊥ to *c*; chloritoid has a perfect basal cleavage; dumortierite has parallel extinction.

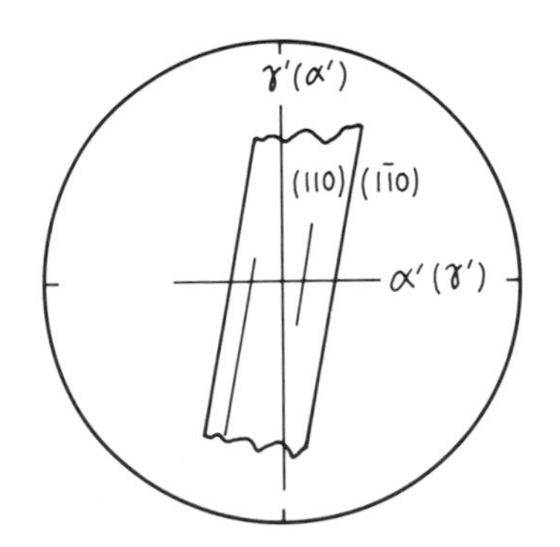

Figure 28. Fragment on {110}. Glaucophane has $\gamma' \wedge \{110\} = 9°–11°$, $\alpha' = 1.609–1.644$, $\gamma' = 1.619–1.649$. Crossite has γ' or $\alpha' \wedge \{110\} = 9°–11°$, $\alpha' = 1.644–1.684$, $\gamma' = 1.649–1.686$. Riebeckite has $\alpha' \wedge \{110\} = 4°–11°$, $\alpha' = 1.684–1.701$, $\gamma' = 1.686–1.715$. The figures are nonsymmetric.

Eckermannite-Arfvedsonite Series

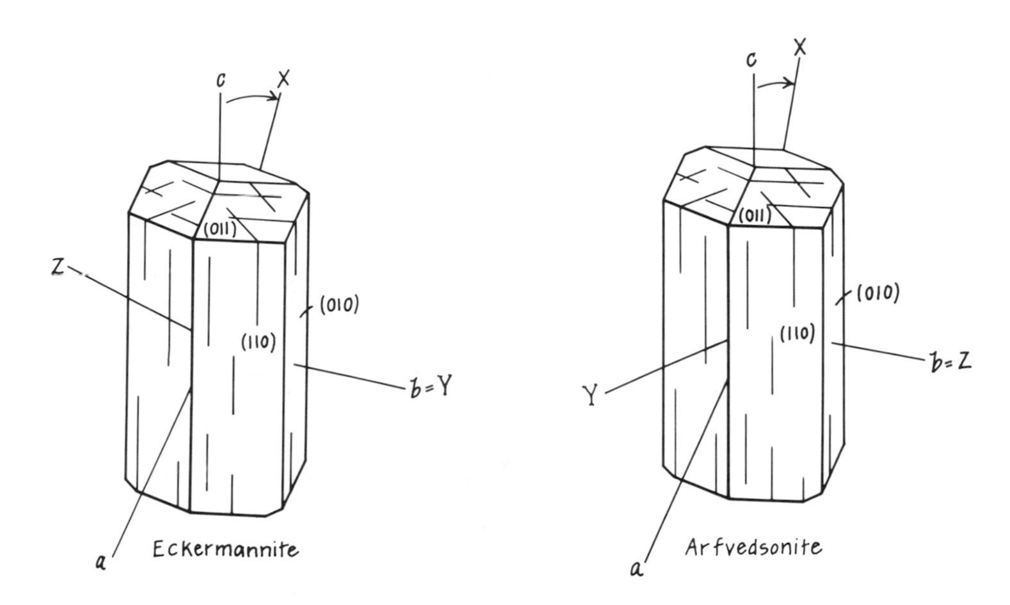

$Na(Na_{2.0-1.5}, Ca_{0-0.5})(Mg,Fe^{+2})_4(Fe^{+3},Al)(Si_8O_{22})(OH)_2$ (Fig. 29)
Monoclinic
$\measuredangle\beta \simeq 105°$
$\alpha = 1.612-1.638$, $\beta = 1.625-1.652$, $\gamma = 1.630-1.654$ (eckermannite) (Fig. 30)
$\alpha = 1.674-1.700$, $\beta = 1.679-1.709$, $\gamma = 1.686-1.710$ (arfvedsonite)

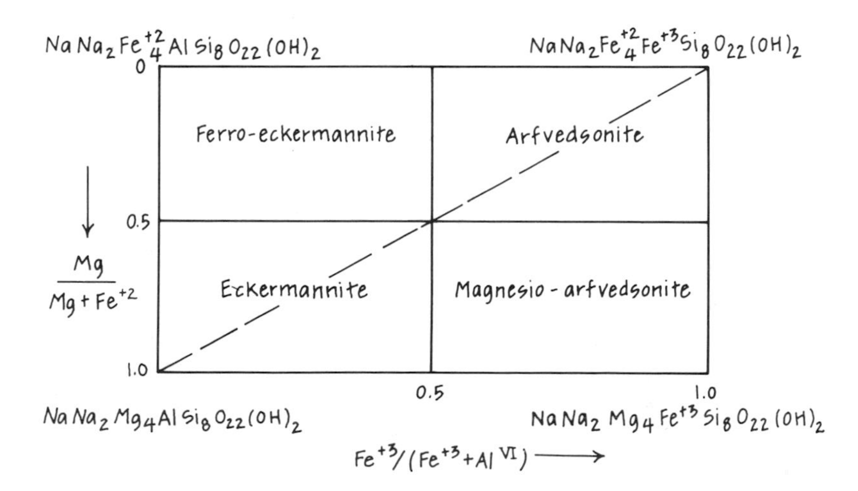

Figure 29. Compositional variation in the eckermannite-arfvedsonite series (Leake, 1978). The dashed line describes the compositional range given in Figure 30. Substitution of Ca for Na is not indicated.

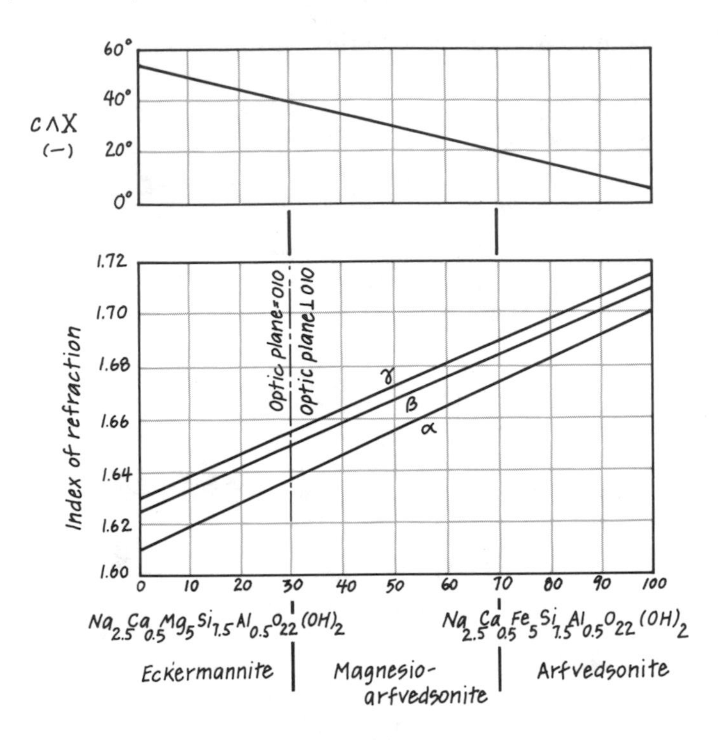

Figure 30. Variation in optical properties in the eckermannite-arfvedsonite series (after Deer, Howie, and Zussman, 1962).

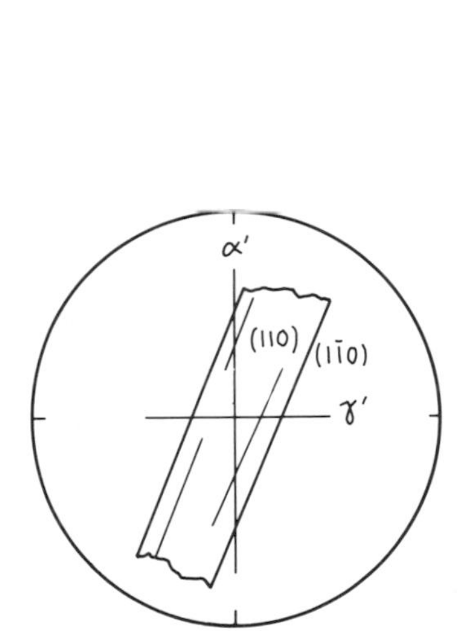

Figure 31. Fragment on {110}. Eckermannite has $\alpha' \wedge \{110\}$ = large and variable (with larger angles corresponding to larger values of $2V$), $\alpha' \simeq 1.61-1.65$, $\gamma' \simeq 1.63-1.65$. Arfvedsonite has $\alpha' \wedge \{110\} = <10°$, with $\alpha' \simeq 1.67-1.71$, $\gamma' = 1.68-1.71$. The figure is nonsymmetric, but approaches Bxo monosymmetric when $c \wedge X$ is small.

$\gamma - \alpha$ = 0.009–0.020 (2nd-order blue) (eckermannite); 0.005–0.012 (1st-order orange) (arfvedsonite)

$2V$ = 15°–80° (−) eckermannite; 0°–50° (−) arfvedsonite

$b = Y$, $c \wedge X = -30°$ to $-57°$ (eckermannite): $b = Z$, $c \wedge X = -2°$ to $-24°$ (arfvedsonite)

Pleochroism and color: eckermannite strongly pleochroic in blue-green or yellow-green with variable absorption, with $X > Y > Z$; X = deep blue-green, yellow, indigo, Y = pale blue-green, yellow-brown, gray-violet, Z = pale yellow-green, deep green, pale brownish green; arfvedsonite strongly pleochroic in blue-green, with $X > Y > Z$.

Dispersion: $r > v$ strong to very strong.

Habit: prismatic with {110} and {010} common.

Cleavage: {110} perfect at ~56° and ~124° (Fig. 31); parting may be present on {010}.

Extinction in section: inclined or parallel to elongated {110} faces or cleavage; usually length low (fast); symmetric to {110} faces and cleavage when viewed along *c;* complete extinction may not be obtained on prismatic sections due to strong directional dispersion.

Twinning: {100} simple or polysynthetic common.

Hardness: 5–6.

Specific gravity: 3.0–3.5.
Occurrence: Na-rich plutonic igneous rocks with quartz or nepheline, and/or Na-rich pyroxenes; rarely in alkali-rich volcanics and nepheline pegmatites.
Alteration: to a fine fibrous amphibole with limonite or siderite.
Characteristic features: strong pleochroism in blue-green; strong dispersion; moderate birefringence; occurrence.
Look-alikes: the glaucophane-crossite-riebeckite series commonly has a different optical orientation, sign, and pleochroism, as well as smaller extinction angles.

Analcime. See Zeolite Group (Z1).

Anatase. See Rutile Group (R2).

Andalusite. See Aluminosilicate Group (A1).

Andesine. See Feldspar Group (F1).

Andradite. See Garnet Group (G1).

Anhydrite. See Sulfate Group (S12).

Ankerite. See Carbonate Group (C2).

Anorthite. See Feldspar Group (F1).

Anorthoclase. See Feldspar Group (F1).

Anthophyllite. See Amphibole Group (A2).

Antigorite. See Serpentine Group (S3).

APATITE GROUP (A3)

Fluorapatite (Common Macroscopic Variety)

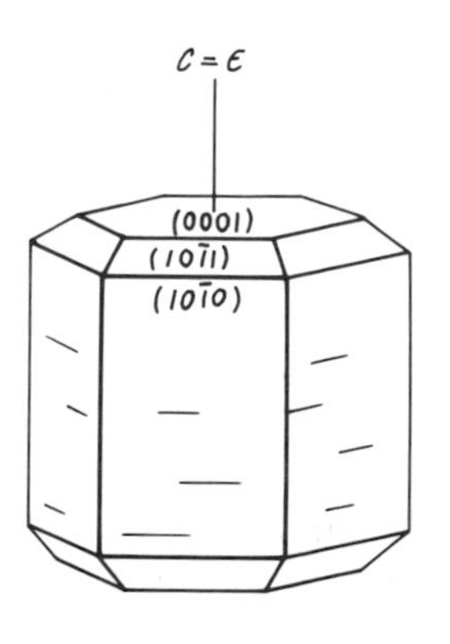

$Ca_5(PO_4)_3(F,OH,Cl)$
Hexagonal
$\omega = 1.633–1.650$, $\varepsilon = 1.629–1.646$
Less common varieties are chlorapatite ($\omega = 1.668$, $\varepsilon = 1.665$), hydroxylapatite ($\omega = 1.651$, $\varepsilon = 1.647$), carbonate fluorapatite (mean index ~1.630), carbonate hydroxylapatite (mean index ~1.520–1.610)

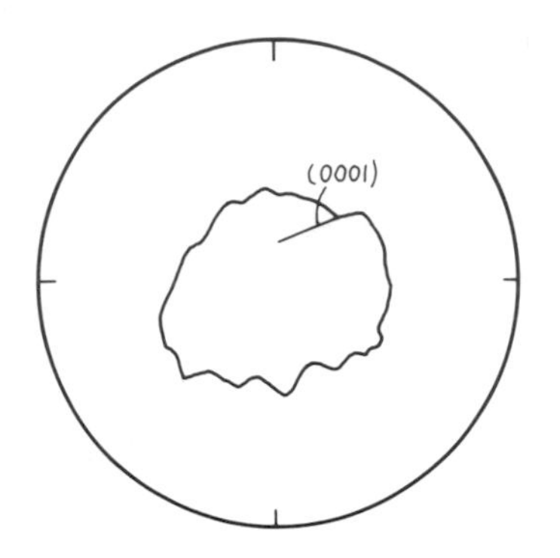

Figure 32. Apatite fragments are typically irregular in outline and orientation, as the {0001} cleavage is poor.

$\omega - \varepsilon$ = 0.003–0.005 (1st-order whitish gray), but in less common varieties may be 0.005–0.017

Uniaxial (−)

Pleochroism and color: normally colorless, but may be pale brown or gray with weak to moderate pleochroism; abs. $E > O$.

Habit: in igneous and metamorphic rocks, tiny euhedral prismatic crystals with six-sided outline; in sediments may be oolitic, botryoidal, or colloform.

Cleavage: none or poor {0001} (Fig. 32).

Extinction in section: parallel to prismatic sections, and length low (fast); continuous extinction when hexagonal outline is seen.

Twinning: rare.

Hardness: 5.

Specific gravity: 2.9–3.5.

Occurrence: a common accessory mineral in most igneous and metamorphic rocks (Fig. 33); may be present in sediments as detrital grains, cryptocrystalline aggregates (collophane), and in phosphate-rich beds.

Alteration: stable.

Characteristic features: low birefringence and high relief (color photo 19); distinctive habit in different rock types.

Look-alikes: quartz has lower relief and is uniaxial (+); nepheline has lower relief; andalusite has good prismatic cleavage and is biaxial; zoisite and melilite often show anomalous interference colors.

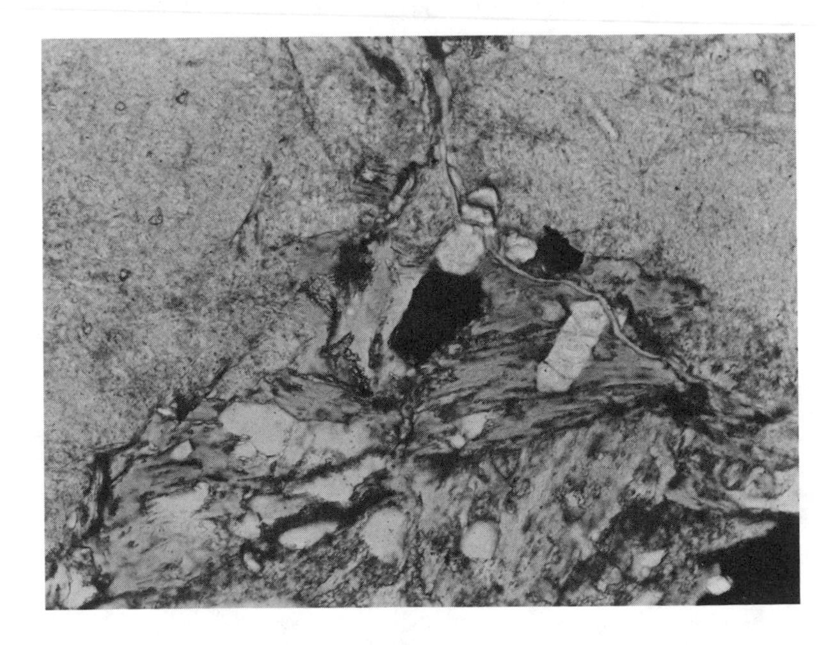

Figure 33. Apatite grains (clear) in biotite. The grain at the center exhibits a hexagonal outline, and the grain right of center has its *c* axis parallel to the section. Plane-polarized light. Photo length is 0.82 mm. (From J. Hinsch, E. Leitz, Inc.)

APOPHYLLITE (A4)

$KCa_4(Si_4O_{10})_2F{\cdot}8H_2O$
Tetragonal
$\omega = 1.531-1.536$, $\varepsilon = 1.533-1.538$
$\varepsilon - \omega = 0.000-0.003$ (1st-order gray, or strong anomalous interference colors)
Uniaxial (+) (some varieties are essentially isotropic ($n \simeq 1.542$), and others are uniaxial (−), with $\omega = 1.537-1.545$ and $\varepsilon = 1.537-1.544$)
Pleochroism and color: colorless.
Habit: anhedral granular to euhedral; may form {001} plates, or be elongated on *c* with {100} dominant.
Cleavage: {001} perfect, {110} poor (Fig. 34).
Extinction in section: parallel to cleavage traces and dominant faces; length low (fast) when {001} is dominant; parallel and length high (slow) when {100} is dominant.
Twinning: {111} rare.
Hardness: 4½–5.
Specific gravity: 2.33–2.37.
Occurrence: commonly in vugs and fissures in mafic volcanic rocks (Fig. 35); rare in silicic igneous rocks and contact metamorphic rocks; often with prehnite, pectolite, zeolites, calcite, and datolite.
Alteration: to opal, chert, clays, and calcite.

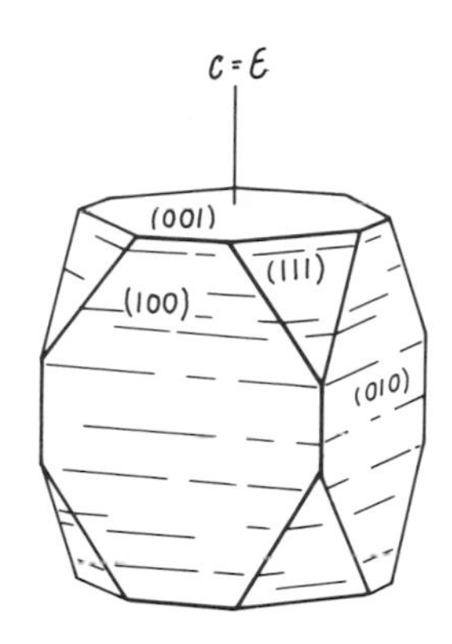

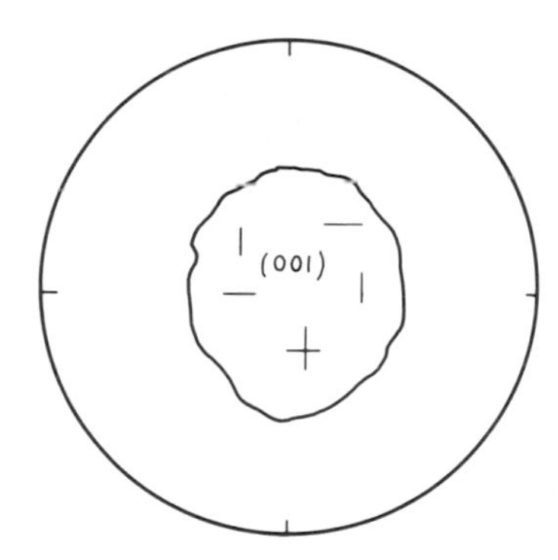

Figure 34. All apophyllite fragments lie on the perfect {001} cleavage, are in total extinction, and permit estimation of ω. Traces of the poor {110} cleavage will probably not be seen. The figure is centered O.A.

Figure 35. Apophyllite vein in altered basalt. The grains exhibit anomalous brown interference colors. Crossed nicols. Photo length is 2.22 mm.

Characteristic features: occurrence; negative relief; perfect {001} cleavage; very low birefringence.
Look-alikes: zeolites have lower indices, and different cleavage and habit.

Aragonite. See Carbonate Group (C2).

Arfvedsonite. See Amphibole Group (A2).

Augite. See Pyroxene Group (P8).

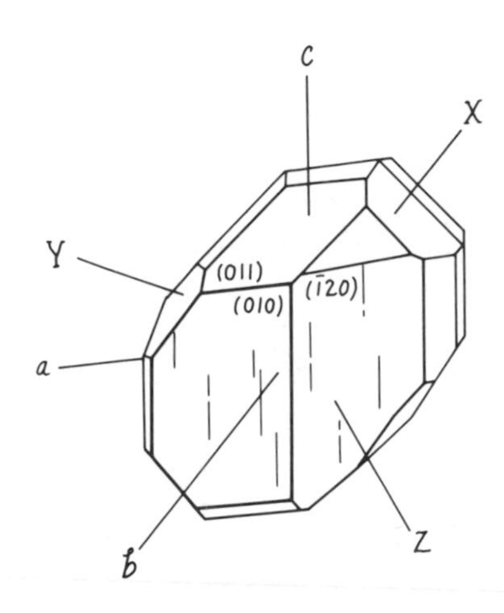

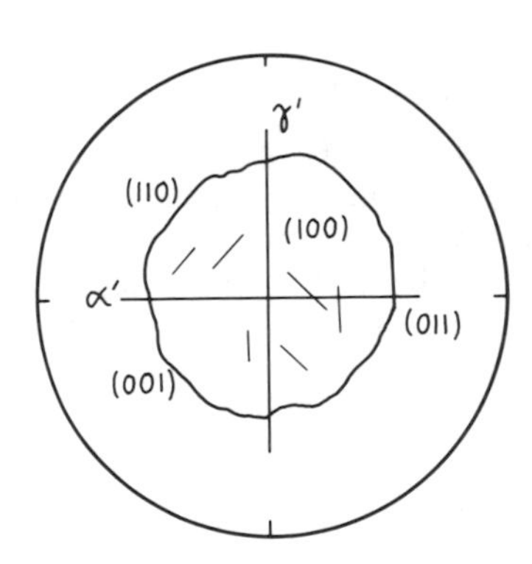

Figure 36. Axinite fragments lie on the good {100} cleavage and permit estimation of α' (1.676–1.695) and γ' (1.684–1.704). If other poor cleavages are present, $\{011\} \wedge \gamma' \simeq 4°$, $\{110\} \wedge \gamma' \simeq 42°$, $\{001\} \wedge \gamma' \simeq 50°$. The figure is nonsymmetric, but close to monosymmetric Bxo.

AXINITE (A5)

$Ca_2(Fe,Mn,Mg)Al_2BO_3(Si_4O_{12})OH$
Triclinic
$\measuredangle\alpha \simeq 88°$, $\measuredangle\beta \simeq 82°$, $\measuredangle\gamma \simeq 78°$
$\alpha = 1.674–1.693$, $\beta = 1.681–1.701$, $\gamma = 1.684–1.704$
$\gamma - \alpha = 0.009–0.011$ (1st-order yellow)
$2V = 63°–80°$ (−)
Optical orientation: varies with composition; $X \simeq \perp(\bar{1}11)$.
Pleochroism and color: clove-brown or violet-brown to colorless; $Y > X > Z$ moderate, X = pale brown, green, or yellow, Y = violet, deep yellow, blue, Z = pale violet, green, or yellow.
Dispersion: $v > r$ strong.
Habit: wedge-shaped, bladed, with {010} or {011} common.
Cleavage: {100} good, {001}, {110}, and {011} poor (Fig. 36).
Extinction in section: $(100) \wedge Z' \simeq 0°–45°$ as a function of orientation; length high (slow).
Twinning: rare, {110} polysynthetic, {100} simple.
Hardness: 6½–7.
Specific gravity: 3.26–3.36.
Occurrence: calc-silicate metamorphic rocks; sometimes in veins in igneous rocks; may be with other boron-containing minerals, as well as epidote, zoisite, diopside, grossular, idocrase, and calcite. See Figure 37.
Alteration: to chlorite and calcite.
Characteristic features: wedge-shaped crystals; possible violet-brown or clove-brown color; high relief; low birefringence; generally inclined extinction.
Look-alikes: datolite has higher birefringence; feldspars and quartz have lower indices; zoisite has lower birefringence and is (+).

Barite. See Sulfate Group (S12).

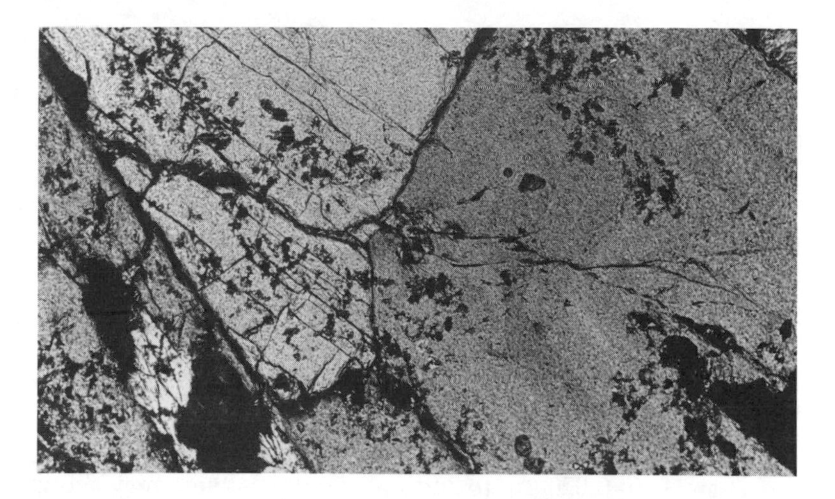

Figure 37. Axinite in two different orientations. Crossed nicols. Photo length is 2.06 mm. (From J. Hinsch, E. Leitz, Inc.)

BERYL (B1)

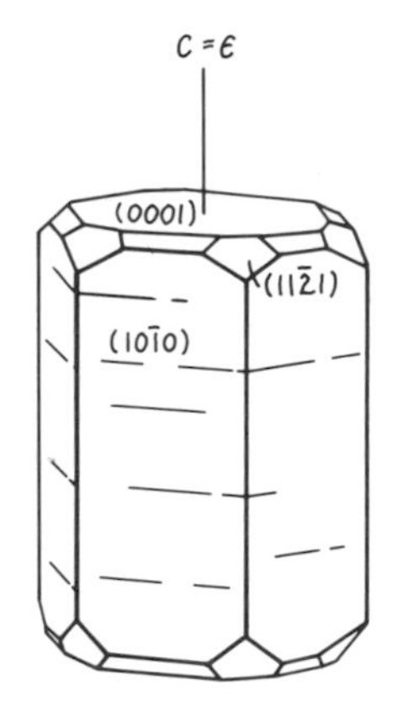

$Be_3Al_2(SiO_3)_6$
Hexagonal
$\omega = 1.567–1.594$, $\varepsilon = 1.563–1.586$
$\omega - \varepsilon = 0.004–0.008$ (1st-order white)
Uniaxial (−); may be anomalously biaxial
Pleochroism and color: typically colorless, rarely pale yellow or greenish, with abs. $E > O$.
Habit: euhedral prismatic with {10$\bar{1}$0} and {0001} common, rarely radial masses or anhedral.
Cleavage: poor {0001} (Fig. 38).
Extinction in section: parallel to prismatic faces, length low (fast); extinction is continuous or near-continuous during stage rotation when hexagonal outline is seen.
Twinning: none.
Hardness: 7½–8.
Specific gravity: ≃2.70.
Occurrence: siliceous plutonic rocks and pegmatites; less common in metamorphic rocks and in veins within limestone.
Alteration: to kaolin, sericite, or illite.
Characteristic features: low birefringence and indices; uniaxial (−); hexagonal cross-section.
Look-alikes: quartz is uniaxial (+); apatite has higher relief; nepheline has lower indices and commonly shows traces of cleavage; topaz is biaxial.

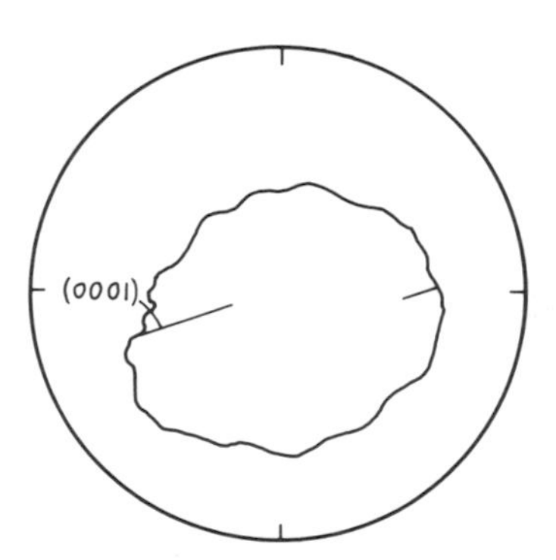

Figure 38. Most beryl grains lie in random orientation and have an irregular outline. A relatively small number will lie on the poor {0001} cleavage, exhibit total extinction, and yield only an ω index.

Biotite. See Mica Group (M2).

Boehmite. See Clays and Clay Minerals (C6).

Bowlingite. See Olivine (O1).

Bronzite. See Pyroxene Group (P8).

Brookite. See Rutile Group (R2).

BRUCITE (B2)

$Mg(OH)_2$
Hexagonal
$\omega = 1.559-1.590$, $\varepsilon = 1.579-1.600$
$\varepsilon - \omega = 0.010-0.020$ (up to 2nd-order reddish blue); may show anomalous dark brown interference color
Uniaxial (+); rarely anomalously biaxial
Pleochroism and color: colorless.
Habit: {0001} well developed, rarely fibrous.
Cleavage: {0001} perfect (Fig. 39).
Extinction in section: parallel to {0001} faces and cleavage; length low (fast).
Twinning: none.
Hardness: 2½.
Specific gravity: 2.39.
Occurrence: forms from hydration of periclase in marble; associated with magnesite and talc in serpentinite; in low-grade schist. See Figure 40.

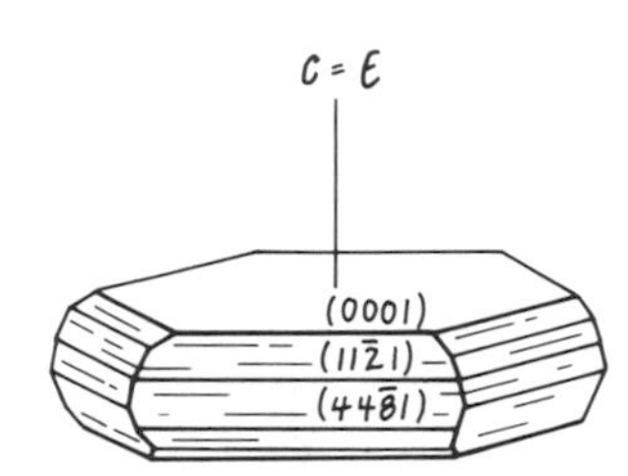

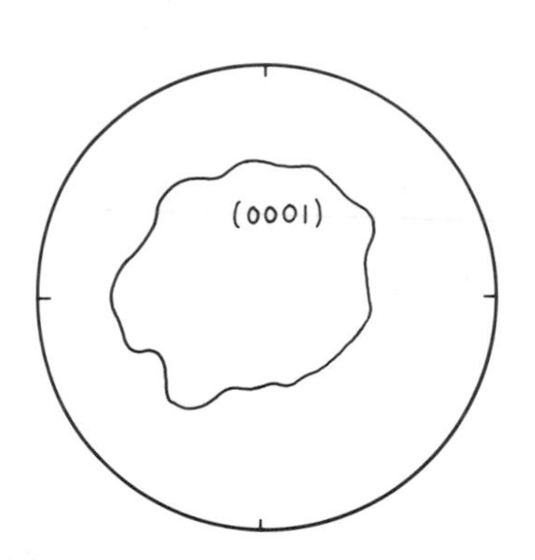

Figure 39. All brucite grains lie on {0001}, yielding the ω index and a centered O.A. figure.

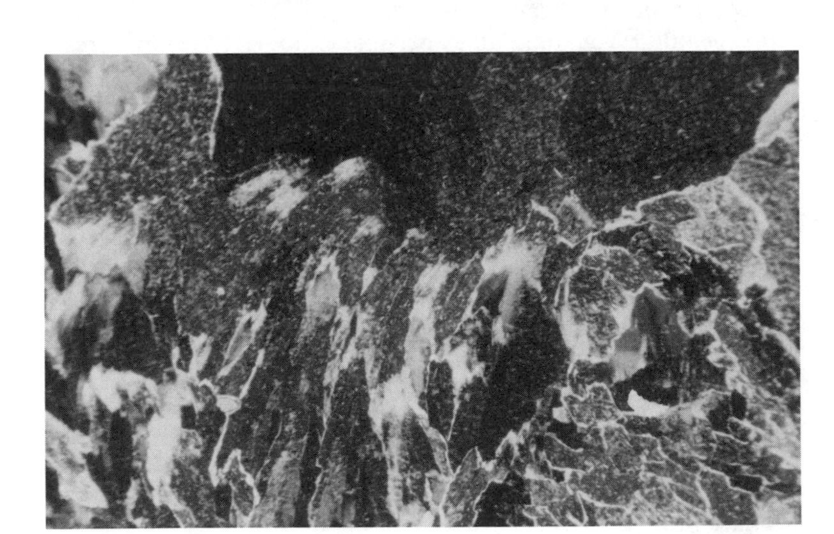

Figure 40. Aggregate of large and small grains of brucite. The uniaxial positive (+) sign distinguishes brucite from the mica group. Crossed nicols. Photo length is 2.22 mm.

Alteration: commonly to hydromagnesite ($3MgCO_3 \cdot Mg(OH)_2 \cdot 3H_2O$); less commonly to serpentine.
Characteristic features: platy habit; parallel extinction; anomalous interference colors; uniaxial (+).
Look-alikes: micas and talc have higher birefringence, are length high (slow), and are biaxial negative; chlorites and serpentines are commonly pale green, slightly pleochroic, and many are (−).

Brunsvigite. See Chlorite Group (C4).

Bytownite. See Feldspar Group (F1).

Calcite. See Carbonate Group (C2).

CANCRINITE (C1)

$(Na,Ca,K)_{6-8}(Al_6Si_6O_{24})(CO_3,SO_4,Cl)_{1-2} \cdot 1-5\ H_2O$
Hexagonal
$\omega = 1.490-1.528$, $\varepsilon = 1.488-1.503$. The indices decrease regularly from the sulfate end-member, cancrinite, to the carbonate end-member, vishnevite
$\omega - \varepsilon = 0.002-0.025$ (up to 2nd-order green)
Uniaxial (−)
Pleochroism and color: colorless.
Habit: anhedral to subhedral; may be elongate parallel to the *c* axis.
Cleavage: {10$\bar{1}$0} perfect (Fig. 41).
Extinction in section: parallel to {10$\bar{1}$0} cleavage; length low (fast).
Twinning: uncommon.

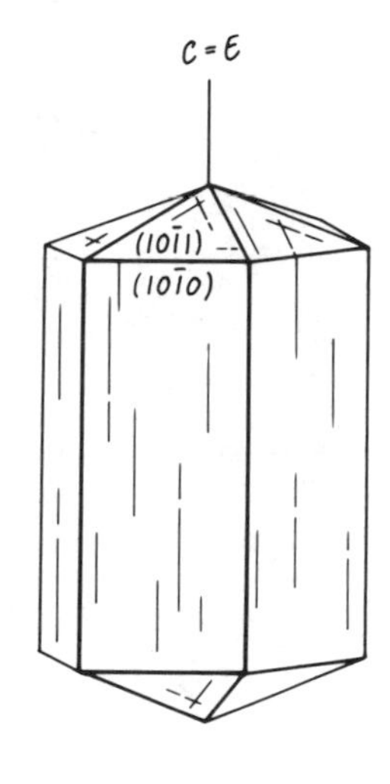

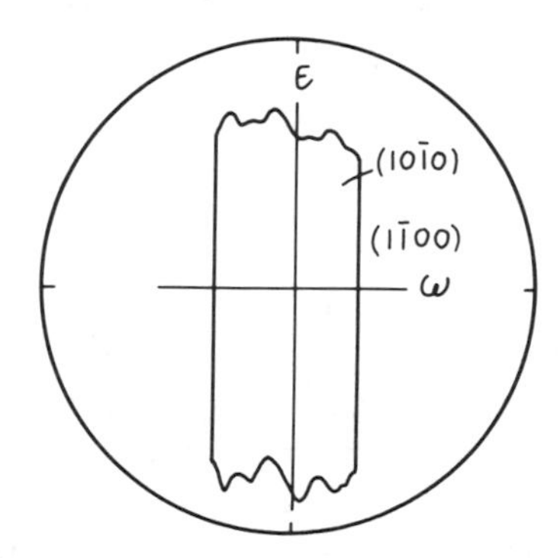

Figure 41. Cancrinite grains on {10$\bar{1}$0} have parallel extinction and yield indices of ω and ε. The figure is O.N.

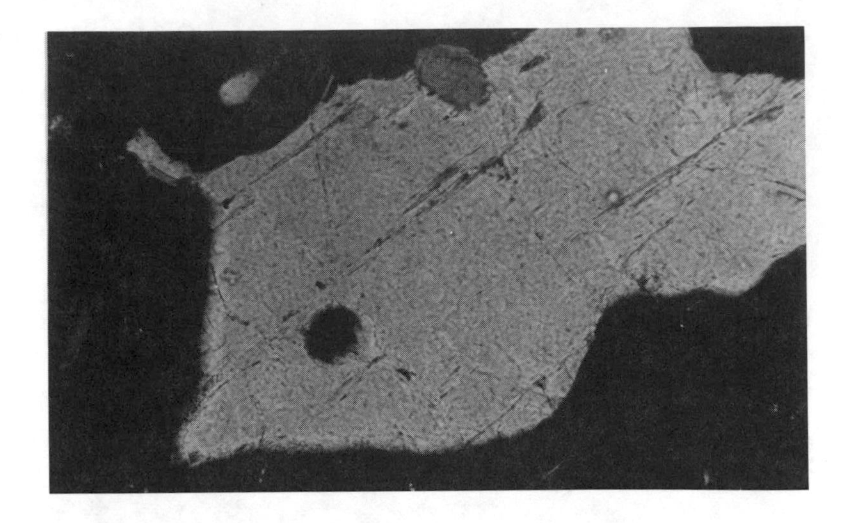

Figure 42. Cancrinite grain enclosed by nepheline (at extinction). Crossed nicols. Photo length is 0.55 mm.

Hardness: 5–6.
Specific gravity: 2.32–2.51.
Occurrence: with other silica-deficient minerals (especially nepheline and sodalite); usually in nepheline syenite; may be secondary as a replacement of nepheline or feldspar. See Figure 42.
Alteration: generally stable.
Characteristic features: negative relief; uniaxial (−); perfect $\{10\bar{1}0\}$ cleavage; associated with silica-deficient minerals.
Look-alikes: muscovite is length high (slow) parallel to cleavage traces and has higher indices; nepheline and potassium feldspars generally have lower birefringence.

THE CARBONATE GROUP (C2)

The $(CO_3)^{-2}$ ion is an essential structural unit in all members of the carbonate mineral group. The common members of the group contain *only* the CO_3 group as a complex anion, whereas less common members contain hydroxyl, halogen, sulfate, or phosphate ions in addition. The optical properties of the pure end members of the chemically simple carbonates are given in Table C2. The most commonly occurring members are calcite, dolomite, and ankerite.

The carbonate minerals are easy to recognize optically because of their extreme birefringence; interference colors consist of white of the higher order for grains of average thickness. In addition, all hexagonal carbonates possess perfect rhombohedral $\{10\bar{1}1\}$ cleavage, which results in symmetric extinction in most orientations. The orthorhombic carbonates possess various combinations of pinacoidal or prismatic cleavages which often yield

Table C2[1]
The Major Carbonate Minerals

Name	Formula	Indices		
		ω		ε
Hexagonal Carbonates (Uniaxial Negative)				
Calcite	$CaCO_3$	1.658		1.486
Dolomite	$CaMg(CO_3)_2$	1.679		1.500
Magnesite	$MgCO_3$	1.700		1.509
Ankerite	$Ca_2MgFe(CO_3)_4$	1.721		1.529
Rhodochrosite	$MnCO_3$	1.816		1.597
Smithsonite	$ZnCO_3$	1.850		1.625
Siderite	$FeCO_3$	1.875		1.633
	α	β	γ	$2V$
Orthorhombic Carbonates (Biaxial Negative)				
Strontianite	$SrCO_3$ 1.516	1.664	1.666	7°
Witherite	$BaCO_3$ 1.526	1.676	1.677	16°
Aragonite	$CaCO_3$ 1.530	1.680	1.685	18°
Cerussite	$PbCO_3$ 1.803	2.074	2.076	9°

[1]Underlined minerals are described in detail in the text.

parallel extinction against cleavage traces. The small optic angles are also distinctive for this group. Due to the occasional overlap of optical parameters, it may be necessary to identify a particular carbonate mineral by means of a flame test, staining, or other method.

Calcite

$CaCO_3$, but may vary as shown in Figure 43

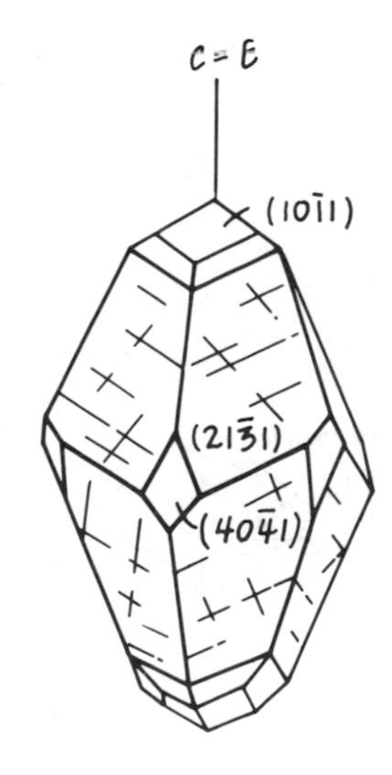

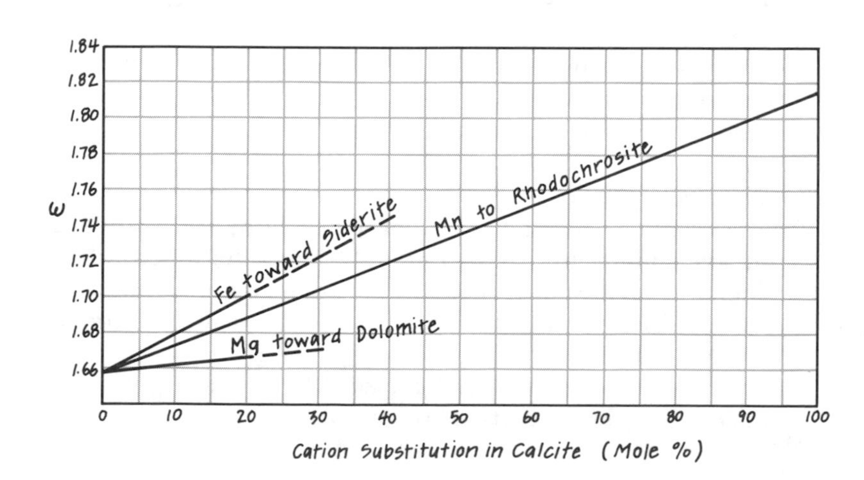

Figure 43. The change in the ω index of calcite as a result of cation substitution.

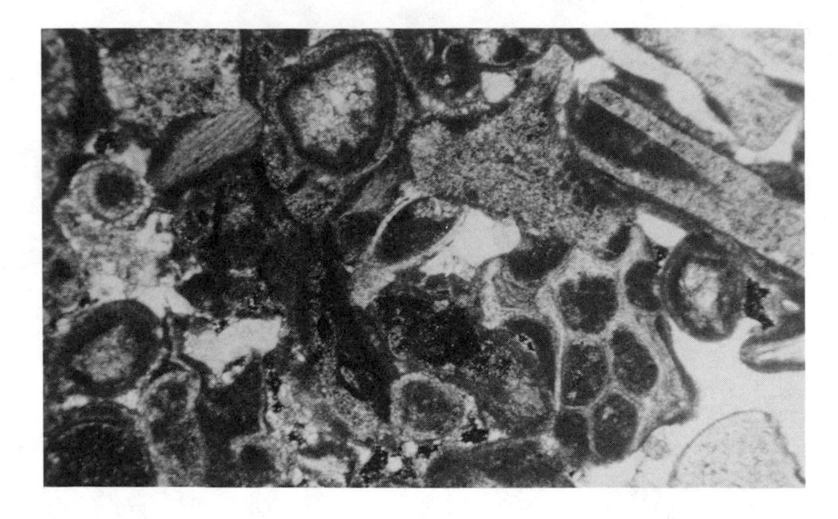

Figure 44. Calcite commonly composes the hard parts of organisms, as seen here in the Bedford Limestone. Crossed nicols. Photo length is 2.22 mm.

Hexagonal
$\omega = 1.658$, $\varepsilon = 1.486$
$\omega - \varepsilon = 0.172$; increases with chemical substitution (white of the higher order)
Uniaxial (−); may be anomalously biaxial ($2V \leq 15°$) in metamorphic rocks
Pleochroism and color: colorless or turbid.
Habit: extremely variable: when formed as single crystals, rhombohedral $\{10\bar{1}1\}$ or $\{10\bar{1}2\}$, or scalenohedral $\{21\bar{3}1\}$ faces are common; when formed as fine to coarse aggregates, it is usually anhedral; in sedimentary environments most calcite is organically precipitated as shell material, and less commonly as oolites or spherulites (Fig. 44).
Cleavage: $\{10\bar{1}1\}$ perfect (Fig. 45).
Extinction in section: usually symmetrical to faces and cleavage traces.
Twinning: very common on $\{01\bar{1}2\}$; common on $\{0001\}$; twin lamellae may be parallel to rhomb edges and the long diagonal of the rhomb, and at an obtuse angle to ε'.
Hardness: 3.
Specific gravity: ≃2.96.
Occurrence: the principal constituent of limestone and marble; a primary mineral in carbonatite and kimberlite; a secondary mineral in cavities of igneous rocks; an alteration product of feldspars, and a common vein mineral; Mg-calcite is common in recent sediments.
Alteration: easily dissolved and replaced by a variety of other minerals (of which quartz is common).

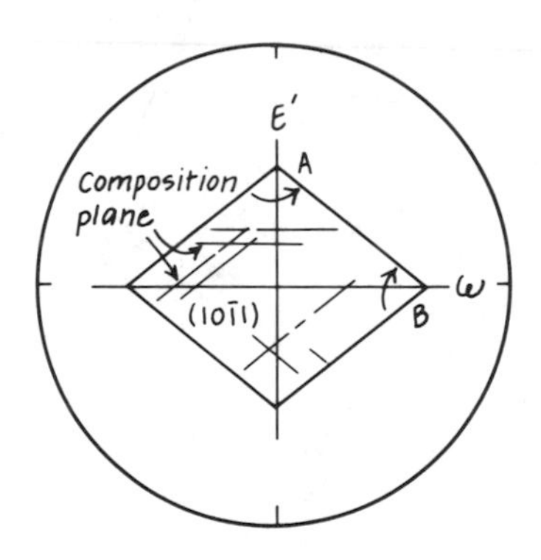

Figure 45. Almost every calcite fragment lies on $\{10\bar{1}1\}$, has symmetric extinction, and yields indices of ω and ε' (1.566). The ω index direction bisects the acute angle *B* and the ε' index direction bisects the obtuse angle *A*. Twin bands may be parallel to cleavage edges or the long diagonal. The figure is of the inclined O.A. type.

Characteristic features: extremely high birefringence; perfect {10$\bar{1}$1} cleavage; variable relief; brisk effervescence with weak acids.

Look-alikes: dolomite has higher indices, as well as twinning parallel to both the short and long diagonal of the rhomb; dolomite is usually subhedral to euhedral; siderite has much higher indices, is often brownish, and commonly is iron stained; orthorhombic carbonates often show parallel extinction to cleavage traces.

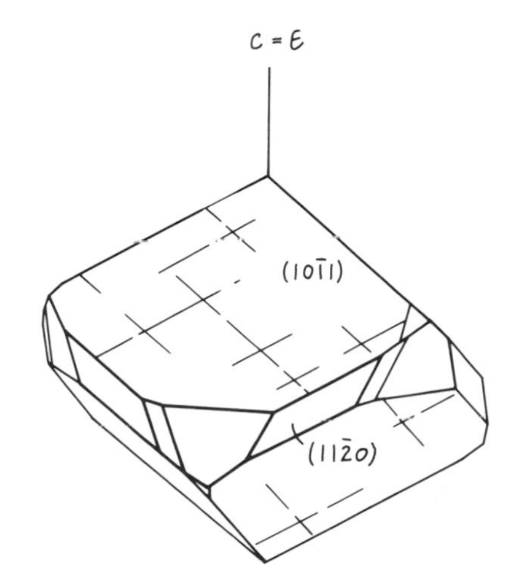

Dolomite-Ankerite Series

$CaMg(CO_3)_2$ – $Ca(Mg,Fe)(CO_3)_2$

Hexagonal

ω = 1.679–1.738, ε = 1.500–1.542 (varies with composition as shown in Figure 46)

$\omega - \varepsilon$ = 0.179–0.196 (white of the higher order); increases with Fe content

Uniaxial (−); rarely biaxial (−) due to strain

Pleochroism and color: colorless; Fe-rich varieties may become brown with alteration.

Habit: commonly anhedral aggregates; the rhombohedron {10$\bar{1}$1} (often with curved faces) is common in single crystals.

Cleavage: {10$\bar{1}$1} perfect (Fig. 47).

Extinction in section: usually symmetrical to faces and cleavage traces.

Twinning: simple on {0001}, {10$\bar{1}$0}, and {11$\bar{2}$0}; polysynthetic on {02$\bar{2}$1}; twin lamellae may lie parallel to both the short and long cleavage rhomb diagonal directions, and at an acute angle (20°–40°) to ε'.

Hardness: 3½–4.

Specific gravity: 2.86–3.10.

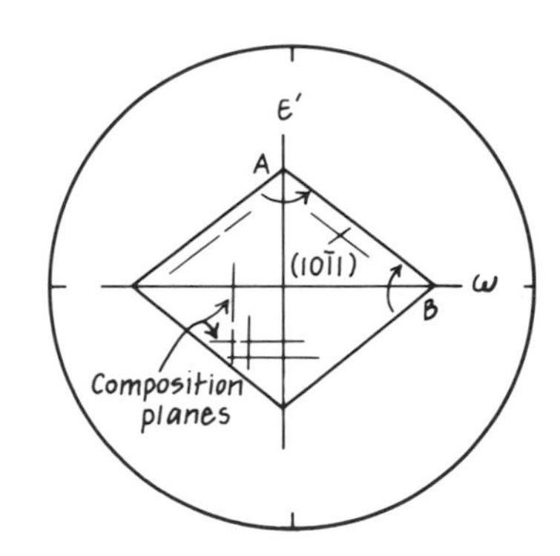

Figure 47. Almost every dolomite fragment lies on {10$\bar{1}$1}, has symmetrical extinction, and indices of ω and ε' (1.586–1.629) (increasing with iron content). The ω index direction bisects the acute angle B, and the ε' index direction bisects the obtuse angle A. Twin bands may be parallel to both the long and short diagonals. The figure is of the inclined O.A. type.

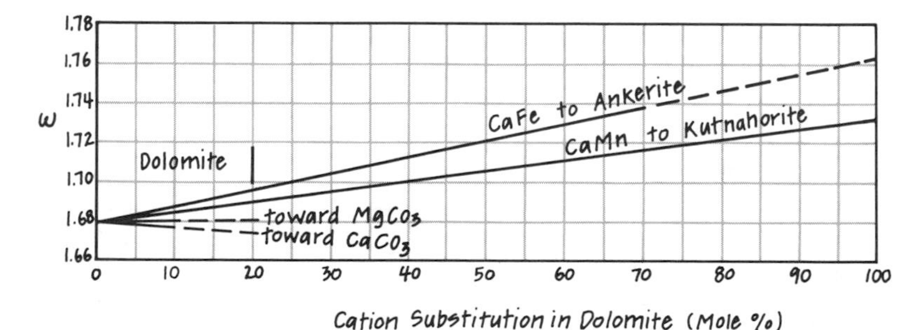

Figure 46. The change in the ω index of dolomite as a result of cation substitution.

Occurrence: dolomite is the principal constituent of the sedimentary rock dolomite, which typically forms by replacement of limestone and may contain diffuse vestiges of limestone textures; primary dolomite may be found in evaporite deposits; metamorphism of sediments may form dolomitic marble; dolomite may be present in hydrothermal veins, or as a secondary mineral in ultramafic rocks; ankerite may be present in Fe-rich sediments or hydrothermal veins with fluorite, barite, siderite, or tourmaline.

Alteration: may be dissolved and replaced by a pseudomorph; with high-grade metamorphism, may be converted to calcite and periclase, or brucite.

Characteristic features: extremely high birefringence; perfect $\{10\bar{1}1\}$ cleavage; variable relief; moderate effervescence with weak acids.

Look-alikes: calcite has lower indices and twin lamellae parallel only to the long diagonal of the rhomb; orthorhombic carbonates may have parallel extinction against cleavage traces and a small 2*V*.

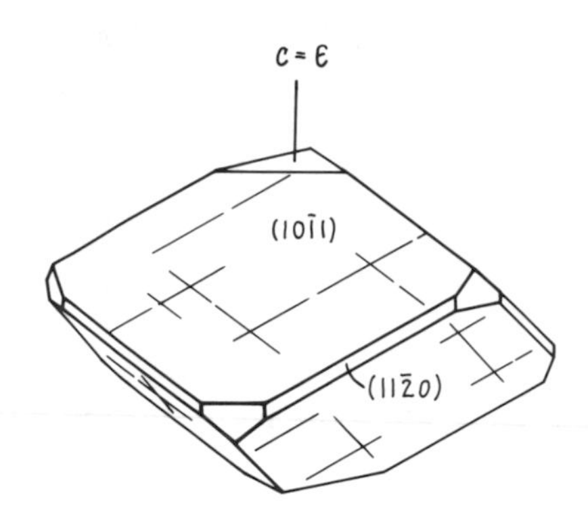

Magnesite

$MgCO_3$
Hexagonal
ω = 1.700–1.709, ε = 1.509–1.516 (varies with composition as shown in Figure 48)
$\omega - \varepsilon$ = 0.191–0.193 (white of the higher order)
Uniaxial (−)

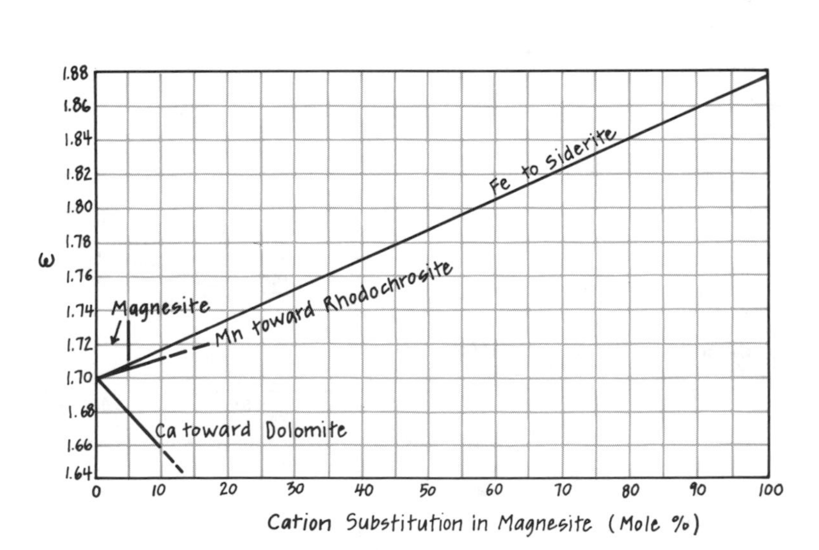

Figure 48. The change in the ω index of magnesite as a result of cation substitution.

Pleochroism and color: colorless or turbid.
Habit: mainly anhedral aggregates; rare single crystals with $\{10\bar{1}1\}$ dominant.
Cleavage: $\{10\bar{1}1\}$ perfect (Fig. 49).
Extinction in section: symmetric to rhombohedral cleavage traces.
Twinning: polysynthetic glide twinning on {0001} is rare.
Hardness: 3½–4½.
Specific gravity: ≈3.0.
Occurrence: mainly in serpentinite; may be in some metamorphic schist or contact metamorphic areas; in sediments, associated with evaporite beds.
Alteration: uncommon, but may be replaced by other minerals.
Characteristic features: extremely high birefringence; perfect $\{10\bar{1}1\}$ cleavage; variable relief; slightly effervescent with cold weak acids.
Look-alikes: calcite and dolomite have lower indices, and commonly polysynthetic twinning; siderite has higher indices and may be yellow or yellow-brown; orthorhombic carbonates often show parallel extinction and a small 2*V*.

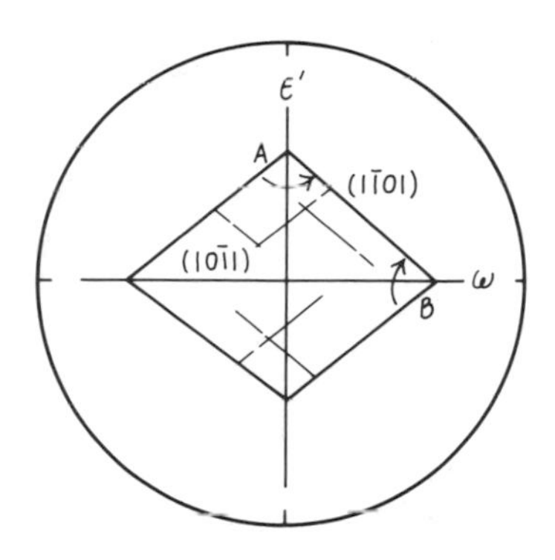

Figure 49. Almost every magnesite fragment lies on $\{10\bar{1}1\}$, has symmetric extinction, and indices of ω and ε' (1.602–1.607) (increasing with iron content). The ω index direction bisects the acute angle *B* and the ε' index direction bisects the obtuse angle *A*. The figure is of the inclined O.A. type.

Rhodochrosite

$MnCO_3$
Hexagonal
ω = 1.769–1.834, ε = 1.563–1.608 (varies with composition as shown in Figure 50)
$\omega - \varepsilon$ = 0.206–0.242 (white of the higher order) with maximum birefringence for pure $MgCO_3$

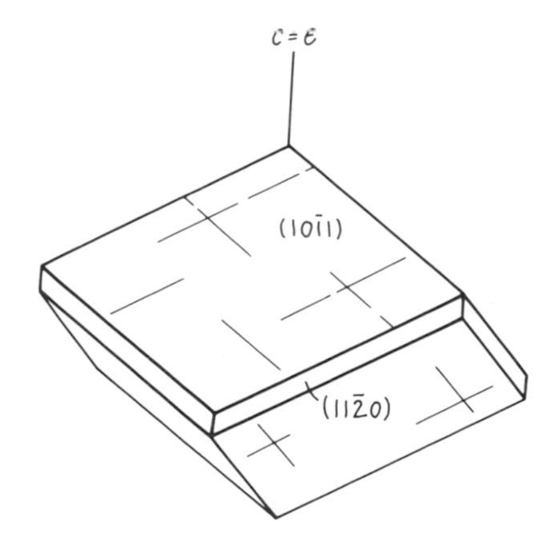

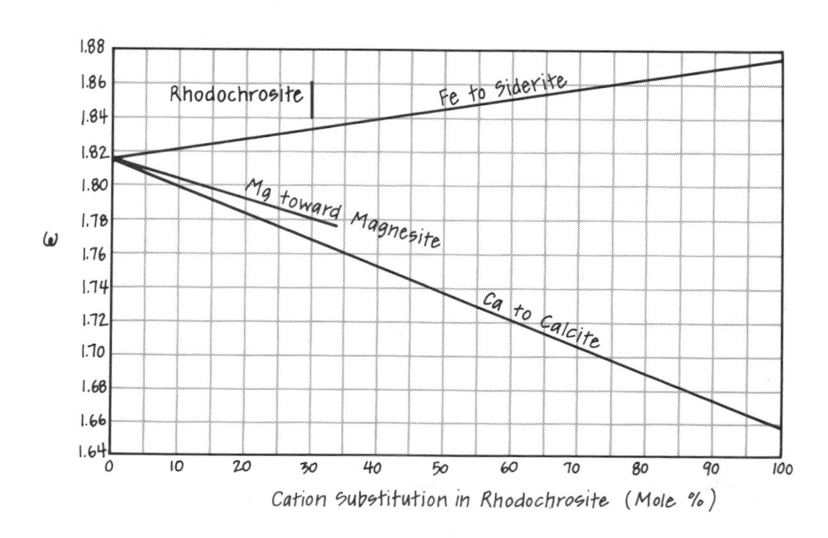

Figure 50. The change in the ω index of rhodochrosite as a result of cation substitution.

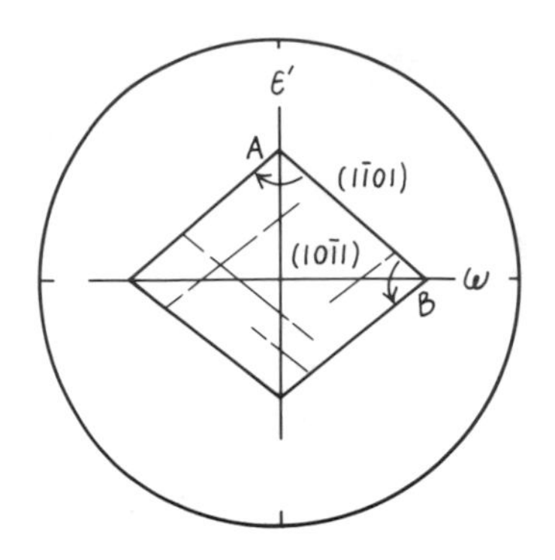

Figure 51. Almost every rhodochrosite fragment lies on {10$\bar{1}$1}, has symmetric extinction, and indices of ω and ε' (1.660–1.716) (increasing with iron and calcium content). The ω index direction bisects the acute angle *B* and the ε' index direction bisects the obtuse angle *A*. The figure is of the inclined O.A. type.

Uniaxial (−)
Pleochroism and color: colorless or pale pink, with abs. *O* > *E*.
Habit: anhedral aggregates or botryoidal masses common.
Cleavage: {10$\bar{1}$1} perfect (Fig. 51).
Extinction in section: symmetric to rhombohedral cleavage traces.
Twinning: uncommon.
Hardness: 3½–4.
Specific gravity: 3.2–3.7.
Occurrence: high-temperature metasomatic or hydrothermal veins with other Mn minerals; secondary from alteration of manganese deposits.
Alteration: to various Mn or Fe oxides, pyrolusite (MnO_2), or manganite ($MnO(OH)$).
Characteristic features: extremely high birefringence; perfect {10$\bar{1}$1} cleavage; variable relief; effervescence in warm dilute acids; pink color; occurrence.
Look-alikes: other carbonate minerals are not pink in hand specimen, have higher or lower indices, and may show significant amounts of twinning.

Siderite

$FeCO_3$
Hexagonal
ω = 1.864–1.875, ε = 1.626–1.633 (varies with composition as shown in Figure 52)
ω − ε = 0.238–0.242 (white of the higher order)
Uniaxial (−)
Pleochroism and color: colorless to pale yellow or yellow-brown.
Habit: anhedral aggregates; {10$\bar{1}$1} dominant on single crystals, and may have curved faces; radial fibers, botryoidal, earthy, or oolitic.

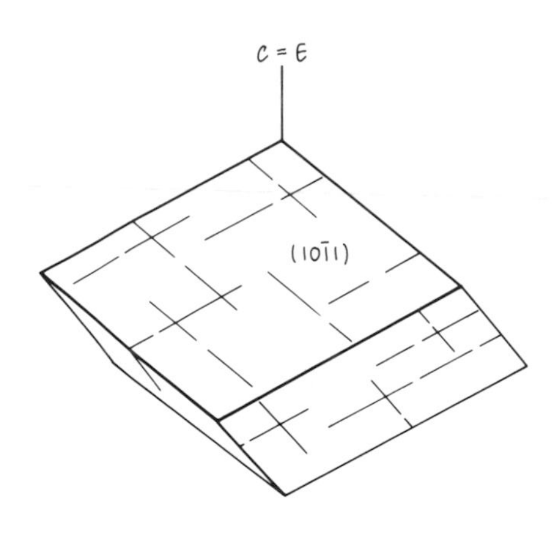

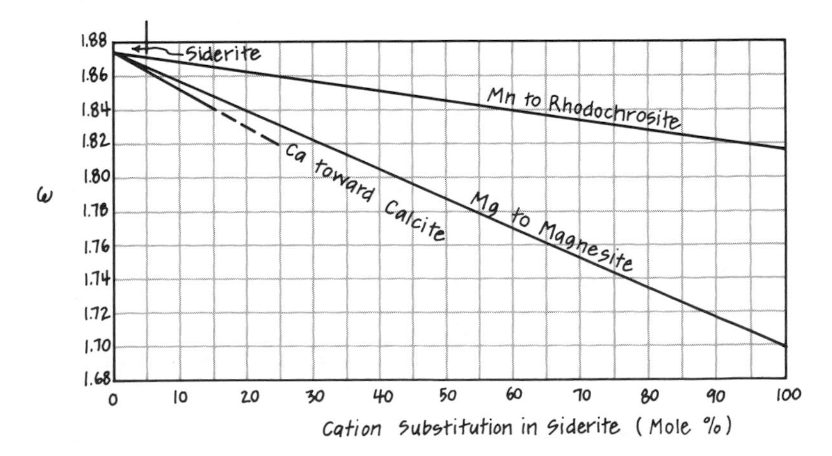

Figure 52. The change in the ω index of siderite as a result of cation substitution.

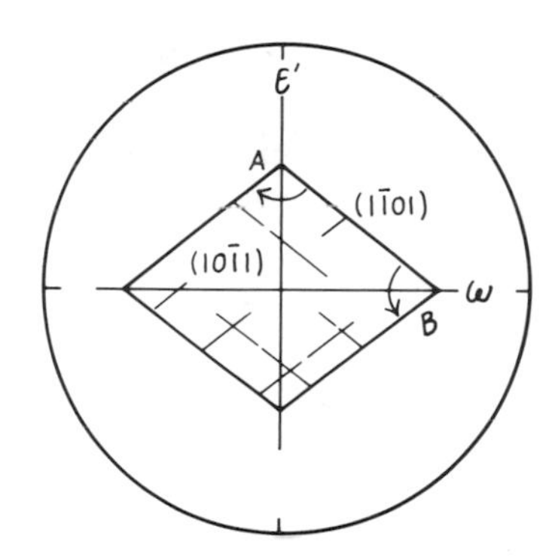

Figure 53. Almost every siderite fragment lies on {10$\bar{1}$1}, has symmetric extinction, and indices of ω and ε' (1.740–1.748) (decreasing with calcium content). The ω index direction bisects the acute angle B and the ε' index direction bisects the obtuse angle A. The figure is of the inclined O.A. type.

Cleavage: {10$\bar{1}$1} perfect (Fig. 53).
Extinction in section: symmetric to rhombohedral cleavage traces.
Twinning: occasionally polysynthetic on {01$\bar{1}$2}.
Hardness: 4–4½.
Specific gravity: 3.50–3.96.
Occurrence: in sediments; often in bog deposits; gangue mineral in hydrothermal veins.
Alteration: to iron oxides.
Characteristic features: extremely high birefringence; perfect {10$\bar{1}$1} cleavage; variable relief; slowly soluble in cold dilute HCl; dissolves with effervescence in hot acid.
Look-alikes: other carbonate minerals have lower indices, and lack associated iron oxides and yellowish color.

Aragonite

$CaCO_3$
Orthorhombic
$\alpha = 1.527–1.542$, $\beta = 1.670–1.695$, $\gamma = 1.676–1.699$
The indices of pure aragonite ($\alpha = 1.530$, $\beta = 1.680$, $\gamma = 1.685$) decrease with Sr substitution and increase with Pb substitution.
$\gamma - \alpha = 0.149–0.157$ (white of the higher order)
$2V = 18°–23°$ (−) (with higher values due to Pb substitution)
$a = Y$, $b = Z$, $c = X$
Pleochroism and color: colorless.
Dispersion: $v > r$ weak.
Habit: parallel or radial aggregates elongated on c.

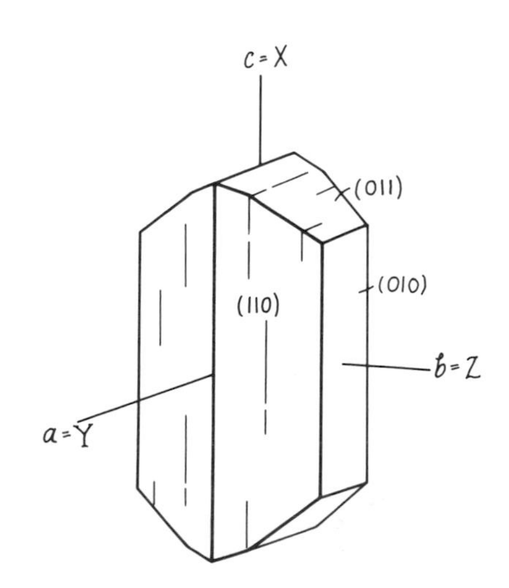

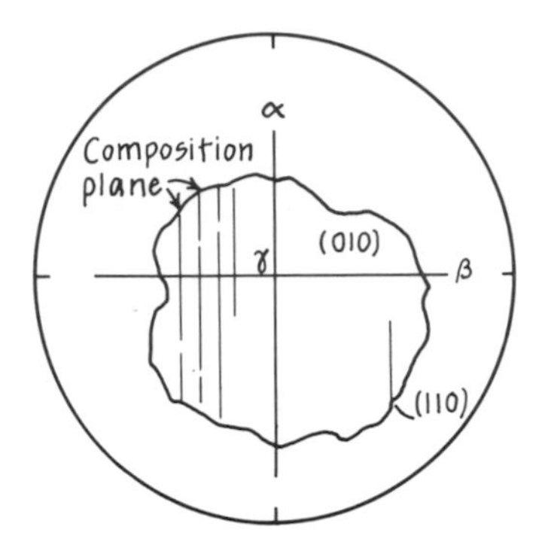

Figure 54. Aragonite fragments on {010} yield α and β indices. When {110} cleavage traces are present, extinction is parallel. Polysynthetic twins parallel to {110} are common. The figure is bisymmetric Bxo.

Cleavage: {010} poor, rarely {110} (Fig. 54).
Extinction in section: usually parallel to cleavage traces or grain edges {110} and {010}; commonly length low (fast).
Twinning: {110} polysynthetic common; often cyclic, which produces pseudohexagonal outline when viewed along the *c* axis.
Hardness: 3½–4.
Specific gravity: 2.95.
Occurrence: in sediments as oolites, needles, or hard parts of marine organisms; hot spring deposits; amygdules in mafic volcanic rocks, or in serpentinite; in metamorphic rocks it is formed in high-pressure environments with glaucophane, lawsonite, and aegirine-augite.
Alteration: to calcite.
Characteristic features: very high birefringence; parallel extinction against poor cleavage; biaxial (−) with a small $2V$; effervesces with dilute HCl.
Look-alikes: hexagonal carbonates have rhombohedral cleavage and symmetrical extinction; other orthorhombic carbonate minerals have higher indices and specific gravity; zeolites do not effervesce in acids and have negative relief.

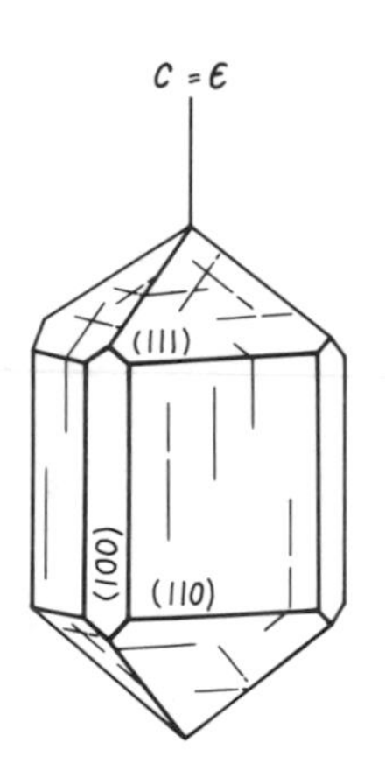

CASSITERITE (C3)

SnO_2
Tetragonal
$\omega = 1.990–2.010$, $\varepsilon = 2.093–2.100$
$\varepsilon - \omega = 0.90–0.103$ (white of the higher order)
Uniaxial (+); may be anomalously biaxial (+)
Pleochroism and color: yellow, brown, red, or almost colorless; abs. $E > O$ weak to strong; may be irregularly color zoned.
Habit: generally short prisms with {110}, {111}, and {321} dominant.
Cleavage: {100} and {110} poor, {111} parting (Fig. 55).
Extinction in section: parallel to prismatic faces and cleavages, and length high (slow); symmetric to dipyramid faces and parting.
Twinning: {101} simple or polysynthetic common; may be cyclic.
Hardness: 6–7.
Specific gravity: 6.98–7.02.
Occurrence: siliceous igneous rocks, hydrothermal veins, and pegmatites; in greisens and skarns; botryoidal (wood-tin) in low-temperature colloidal deposits.
Alteration: stable.

Characteristic features: strong coloration; very high relief and birefringence; uniaxial (+). See Figure 56.

Look-alikes: rutile is more strongly colored in red-brown, and has higher birefringence and indices; allanite has a lower birefringence; anatase is uniaxial (−); brookite is biaxial.

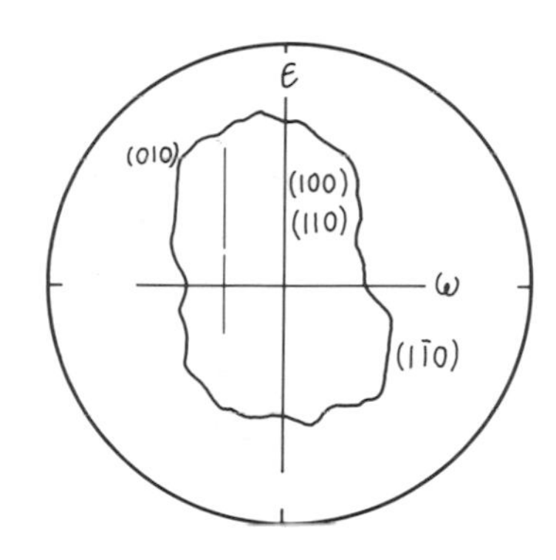

Figure 55. Cassiterite grains on {100} or {110} have parallel extinction when cleavage traces are present, yield ω and ε indices, and O.N. figures.

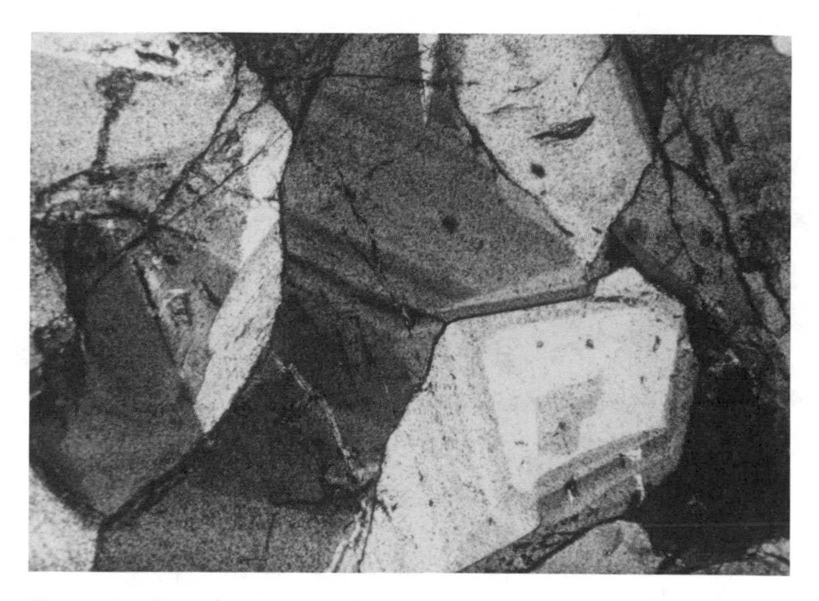

Figure 56. Zoned cassiterite. Crossed nicols. Photo length is 2.22 mm.

Celestite. See Sulfate Group (S12).

Cerussite. See Carbonate Group (C2).

Chabazite. See Zeolite Group (Z1).

Chalcedony. See Silica Group (S4).

Chalcopyrite. See Opaque Minerals (O2).

Chamosite. See Chlorite Group (C4).

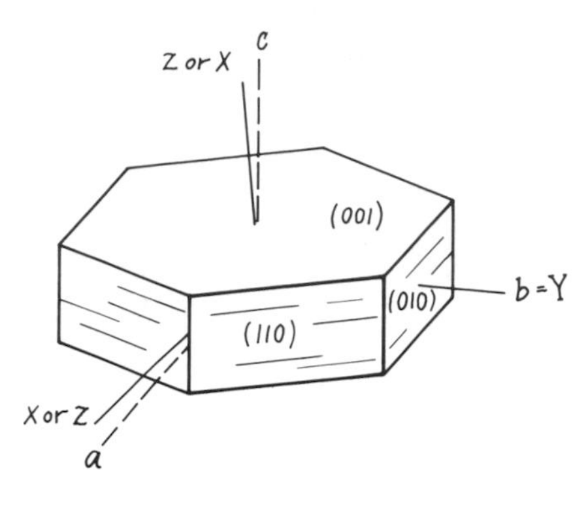

CHLORITE GROUP (C4)

Species shown in Figure 57
$(Mg,Al,Fe)_6(Si,Al)_4O_{20}(OH)_8$
Monoclinic
$\measuredangle\beta \simeq 97°$
For most chlorite, $\alpha \simeq 1.57–1.67$, $\beta \simeq 1.57–1.68$, $\gamma \simeq 1.59–1.68$; oxidized chlorites are $\alpha = 1.60–1.67$, $\beta = 1.61–1.69$, $\gamma = 1.61–1.69$
$\gamma - \alpha = 0–0.01$ (up to 1st-order yellowish white) (unoxidized), 0–0.02 (up to 2nd-order violet) (oxidized)
$2V = 0°–45° (+), 0°–30° (-)$, with small values common
$b = Y, c \wedge Z = 0°–9°$ when optically positive, or $b = Y, c \wedge X = 0°–7°$ when optically negative
Pleochroism and color: pleochroism weak to moderate; colorless to light or dark green (color photo 21), with $Z = Y > X$ or $X = Y > Z$; may show anomalous interference colors in brown, violet, or blue (see Fig. 58 and color photo 22), as well as pleochroic haloes around zircon inclusions (color photo 23).

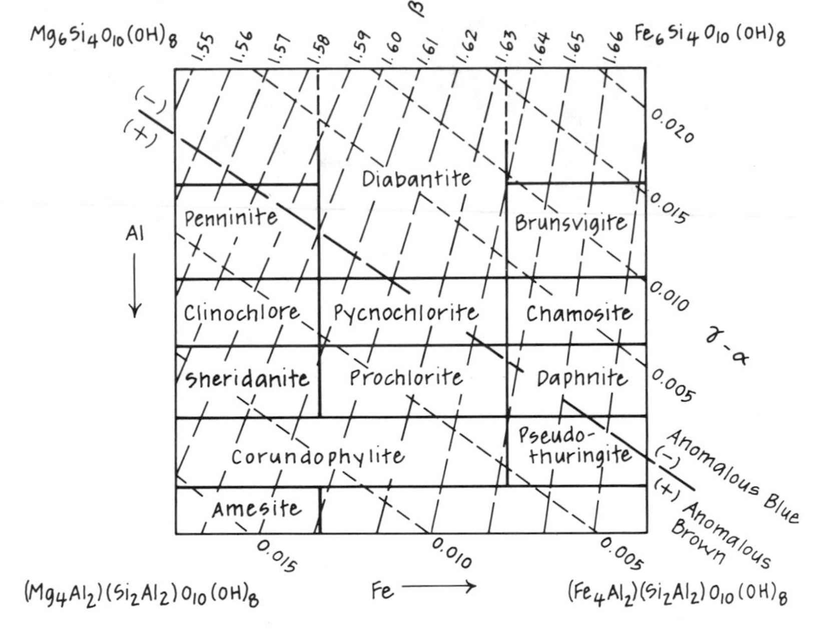

Figure 57. The optical properties of the Fe-Mg chlorites. The compositional limits of chlorite species and nomenclature are modified from patterns suggested by Hey (1954) and Phillips (1984). A number of other nomenclatures exist, the most simple of which is that of Bayliss (1975), who suggested that all Fe-rich chlorites be called chamosite and all Mg-rich chlorites be called clinochlore.

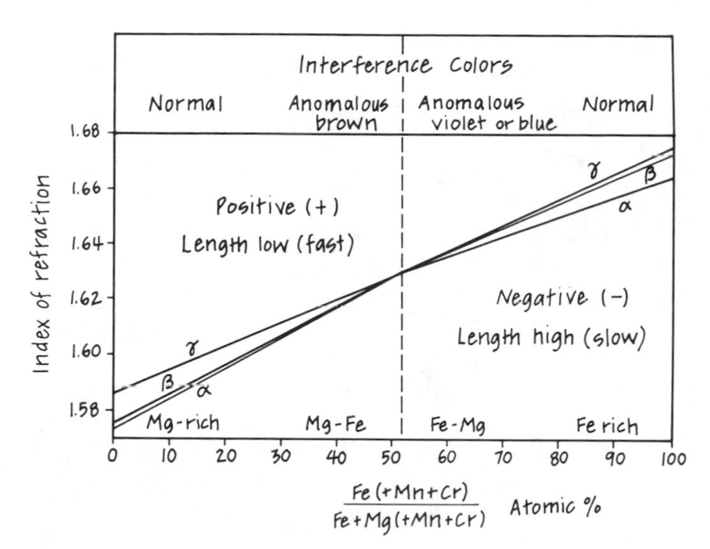

Figure 58. Correlation of optical properties and chemistry across the clinochlore-chamosite region of Figure 57 (after Albee, 1962).

Dispersion: $v > r$ strong.

Habit: {001} dominant, with {110} and {010} less so; pseudo-hexagonal plates, scaly or massive aggregates.

Cleavage: {001} perfect (Fig. 59).

Extinction in section: parallel or slightly inclined to cleavage traces and basal pinacoid form; length low (fast) for optically (+) types, and length high (slow) for optically (−) types.[1]

Twinning: {001} polysynthetic is common.

Hardness: 1–2½.

Specific gravity: 2.64–3.21.

Occurrence: fine-grained sediments; low-grade metamorphic rocks with albite, epidote, actinolite, anthophyllite, and chloritoid; hydrothermal environments; in igneous rocks as a deuteric alteration product of mafic minerals such as biotite, amphiboles, and pyroxenes.

Alteration: stable at surface conditions, but may be converted to amphiboles, pyroxenes, etc. during metamorphism.

Characteristic features: foliated; low birefringence; usually pale green and pleochroic, often with anomalous interference colors. See Figure 60.

Look-alikes: micas and many clay minerals have higher birefringence and often larger 2Vs; kaolin has lower indices; antigorite has normal interference colors; glauconite has higher birefringence.

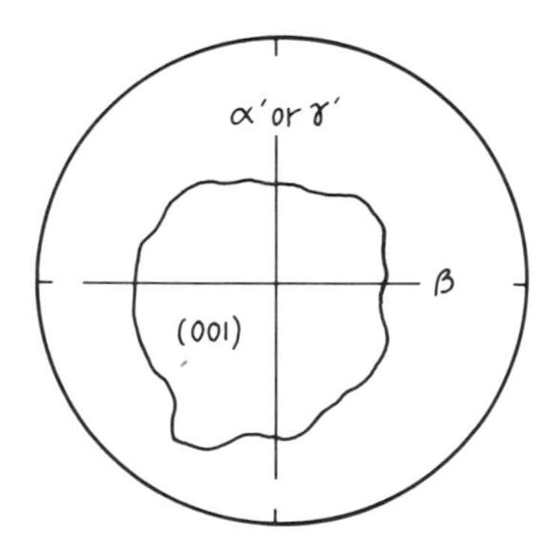

Figure 59. All chlorite grains lie on the {001} cleavage and yield β and α' or γ' indices. The values of α' or γ' are quite close to α or γ respectively. The figure is monosymmetric Bxa (and may appear bisymmetric for some compositions).

[1] Determination of the nature of elongation is often used for sign determination of grains which are too small to yield interference figures.

A

B

Figure 60. (A) Chlorite (center), containing pleochroic haloes, cross-cuts biotite. Crossed nicols. (B) The same, viewed in plane-polarized light. Chlorite (cht) is pale green, whereas biotite (bi) is strongly pleochroic in various shades of brown. Photo length is 2.06 mm. (From J. Hinsch, E. Leitz, Inc.)

CHLORITOID (C5)

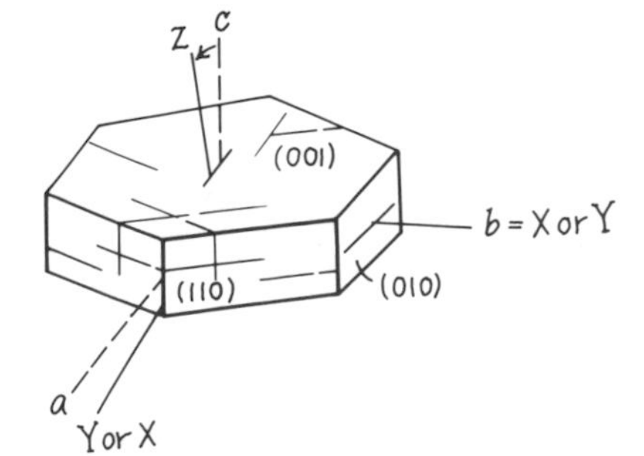

$(Fe^{+2},Mg,Mn)_2(Al,Fe^{+3})Al_3O_2(SiO_4)_2(OH)_4$
Monoclinic (triclinic less common)
$\measuredangle\beta \simeq 102°$
$\alpha = 1.713–1.728$, $\beta = 1.719–1.734$, $\gamma = 1.723–1.740$
$\gamma - \alpha = 0.010–0.012$ (1st-order yellow)
$2V$ = usually 40°–65° (+), rarely large (−)
$b = X$ or Y, $c \wedge Z = 14°–42°$

Pleochroism and color: colorless to green; pleochroic with X = green, Y = blue, and Z = colorless to yellow, with $Y > X > Z$; interference colors may be anomalous.
Dispersion: $r > v$ strong.
Habit: generally tabular with {001} dominant, {110} and {010} less so; length low (fast).
Cleavage: {001} perfect, {110} poor, {010} parting is poor (Fig. 61).
Extinction in section: inclined to basal pinacoid faces and cleavage traces; length low (fast); total extinction may not be achieved due to strong dispersion.
Twinning: common {001} polysynthetic or simple (Fig. 62).
Hardness: 6½.
Specific gravity: 3.26–3.80.
Occurrence: low-grade metamorphic rocks, usually as post-tectonic porphyroblasts (color photo 24); triclinic varieties may be found in hydrothermal veins.
Alteration: to chlorite, sericite, or kaolin.
Characteristic features: high relief; anomalous interference colors; moderate to large $2V$; strong dispersion; abundant twinning.
Look-alikes: chlorite has lower indices, lower relief, small $2V$, and may be length high (slow); biotite has small $2V$, higher birefringence, and bird's-eye maple extinction; stilpnomelane (which is incompatible with chloritoid) is biaxial (−) with a $2V$ near 0°.

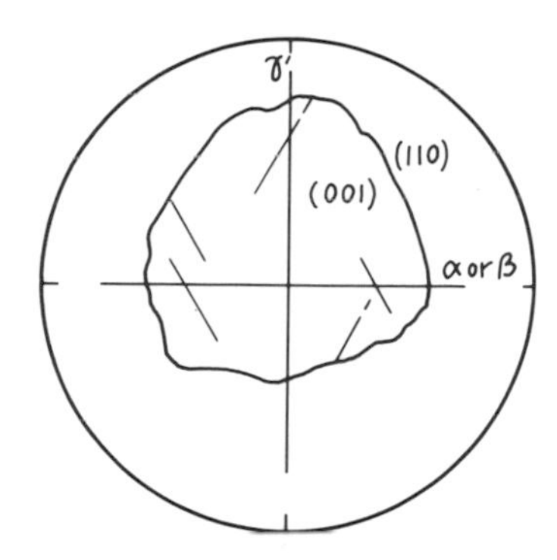

Figure 61. Almost all chloritoid grains lie on the {001} cleavage and yield α or β and an unknown value of γ' (as optical data are inadequate). Extinction of $\gamma' \wedge \{110\}$ cleavage traces $\simeq$ 30°, and the figure is monosymmetric Bxa.

Figure 62. Chloritoid, partially altered to quartz (qtz), chlorite (cht), and sericite. Light and dark bands in the chloritoid are polysynthetic twins, whose composition planes are parallel to {001}. Crossed nicols. Photo length is 2.22 mm.

Chondrodite. See Humite Group (H2).

Chromite. See Opaque Minerals (O2).

Chrysolite. See Olivine Group (O1).

Chrysotile. See Serpentine Group (S3).

CLAYS AND CLAY MINERALS (C6)

A *clay* is defined as a rock, mineral fragment, or detrital particle that has a grain size of less than 0.0039 mm (4 μm); that is, *any* solid can be considered to be a clay if its grain size is sufficiently fine. Most mudrocks are composed, however, of about 60 percent *clay minerals,* with the remainder being (mainly) quartz, feldspar, rock fragments, carbonate minerals, organic matter, and so forth. Clay minerals are a complex and loosely defined silicate mineral group composed of sheets of silicate and aluminate tetrahedra; other cations such as Mg, Fe, Cr, Li, and Mn are present in octahedral coordination with oxygen or hydroxyl; in addition, there may be exchangeable cations such as Ca, Na, K, Mg, H, or Al. As clay minerals have formed from the weathering of primary silicate minerals (see Fig. 63 and color photo 25), they are characterized by fine grain size, and thus are major constituents of most clay deposits. However, some clay deposits, such as bauxite and laterite, are mainly composed of boehmite, diaspore, and gibbsite (with subordinate iron and titanium oxides and hydroxides); these three minerals consist of hydrous alumina, and do not come under the structural and chemical definition of clay minerals.

Figure 63. Sericitic alteration of twinned plagioclase feldspar (top center). Crossed nicols. Photo length is 1.32 mm. (From J. Hinsch, E. Leitz, Inc.)

Table C6
Optical Properties of Clay Minerals and Clays

Minerals and Formula	α	β	γ	$\gamma - \alpha$	$2V$(Sign)	Remarks
			Clay Minerals			
Kaolinite $Al_2Si_2O_5(OH)_4$	1.553–1.563	1.559–1.569	1.560–1.570	0.006–0.007	20°–55°(−)	{001} flakes, $b \simeq Z$, $X \wedge [001] = 3°$
Montmorillonite $(\frac{1}{2}Ca,Na)_{0.67}(Al_{3.33}Mg_{0.67})Si_8O_{20}(OH)_4 \cdot nH_2O$	1.48–1.57	1.50–1.59	1.50–1.60	0.02–0.03	0°–30°(−)	{001} flakes, $b = Y$, $c \wedge X$ = small, $a \wedge Z$ = small
Illite $K_{1-1.5}Al_4(Si_{7-6.5}Al_{1-1.5})O_{20}(OH)_4$	1.54–1.57	1.57–1.61	1.57–1.61	0.03	0°–10°(−)	{001} flakes, $b = Z$, $c \wedge X$ = small, $a \wedge Y$ = small
Vermiculite $(Mg,Ca)[(Mg,Fe^{+2})_5(Fe^{+3},Al)](Si_5Al_3)O_{20}(OH)_4 \cdot 8H_2O$	1.52–1.56	1.54–1.58	1.54–1.58	0.02–0.03	0°–8°(−)	{001} flakes, $b = Y$, $c \wedge X = 3°–6°$, $a \wedge Z = 1°–4°$, colorless, pale brown, or green.
Sericite Usually $KAl_2(Si_3Al)O_{10}(OH)_2$	1.55–1.57	1.58–1.61	1.59–1.62	0.036–0.049	30°–47°(−)	{001} flakes, $b = Z$, $c \wedge X = 0°–5°$, $a \wedge Y = 1°–3°$
Glauconite $(K,Na)(Al,Fe^{+3},Mg)_2(Al,Si)_4O_{10}(OH)_2$	1.56–1.61	1.61–1.65	1.61–1.65	0.014–0.032	0°–20°(−)	$b = Y$, $c \wedge X \simeq 10°$, $a \simeq Z$, X = yellow green, $Y = Z$ = green
			Clays			
Boehmite γ-AlO(OH)	1.64–1.65	1.65–1.66	1.65–1.67	~0.015	~80°(+)	$a = Z$, $b = Y$, $c = X$, foliation on {010}
Diaspore α-AlO(OH)	1.682–1.706	1.705–1.725	1.730–1.752	0.046–0.052	84°–86°(+)	$a = Z$, $b = Y$, $c = X$, foliation on {010} or fibrous
Gibbsite $Al(OH)_3$	1.568–1.580	1.568–1.580	1.587–1.600	~0.019	0°–40°(+)	$b = X$, $c \wedge Z = -21°$, tabular on {001} with pseudohexagonal outline

Due to the extremely fine grain size of clays and clay minerals, identification by optical methods is inaccurate or impossible. Other techniques (such as X-ray diffraction, differential thermal analysis, thermogravimetric analysis, electron microscopy, or chemical analysis) are required. Only brief descriptions are given in Table C6. Note that the optical properties vary as a function of chemical substitution.

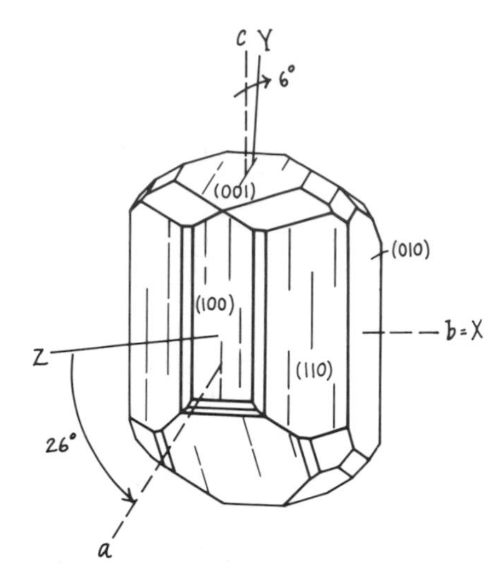

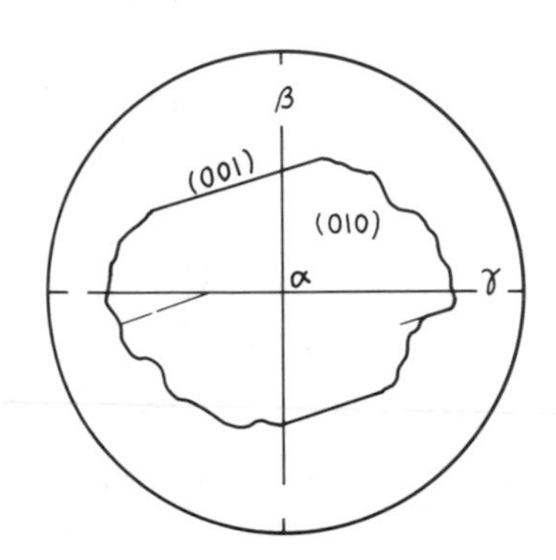

Figure 64. Colemanite grains on the {010} cleavage yield β and γ indices. $\gamma \wedge \{001\} = 26°$ and the figure is bisymmetric Bxo.

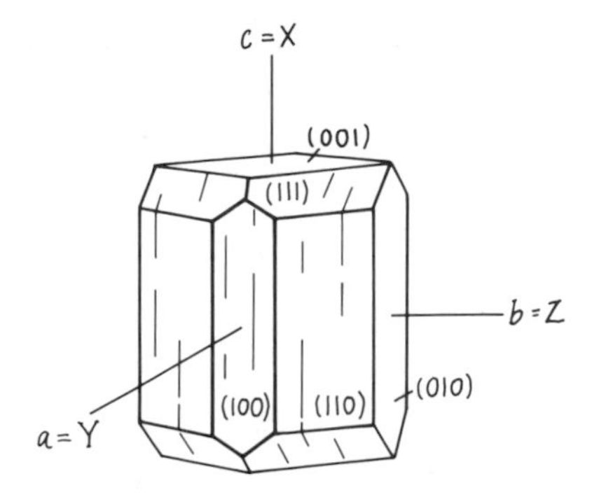

Clinochlore. See Chlorite Group (C4).

Clinohumite. See Humite Group (H2).

Clinozoisite. See Epidote Group (E1).

COLEMANITE (C7)

$Ca_2B_6O_{11}\cdot5H_2O$
Monoclinic
$\measuredangle\beta \simeq 110°$
$\alpha = 1.586, \beta = 1.592, \gamma = 1.614$
$\gamma - \alpha = 0.028$ (2nd-order green)
$2V = 55° (+)$
$b = X, c \wedge Z = 84°$
Pleochroism and color: colorless.
Dispersion: $v > r$ weak.
Habit: stubby prismatic, with dominant {110} form common; {001} and other complex forms are variable in size.
Cleavage: {010} perfect, {001} poor (Fig. 64).
Extinction in section: parallel or slightly inclined to {110} faces when viewing elongate grains; parallel when viewed parallel to {010} cleavage traces and length high (slow).
Twinning: none.
Hardness: 4½.
Specific gravity: 2.42.
Occurrence: appears to have formed from alteration of ulexite ($NaCaB_5O_9\cdot8H_2O$) and borax in boron-rich playas; associated are borate minerals, as well as gypsum, calcite, and celestite.
Alteration: to calcite.
Characteristic features: occurrence and associated borate minerals; stubby crystals and positive relief in thin section.
Look-alikes: associated borate minerals usually show negative relief in thin section, and commonly a fibrous or acicular habit; gypsum has low negative relief in thin section and is often twinned.

CORDIERITE (C8)

$(Mg,Fe)_2Al_3(Si_5Al)O_{18}$
Orthorhombic (pseudohexagonal)
$\alpha = 1.522–1.560, \beta = 1.524–1.574, \gamma = 1.527–1.578$
$\gamma - \alpha = 0.005–0.018$ (up to 1st-order red, but usually gray to white)

$2V$ = usually 40°–90° (−), but may be 75°–90° (+)
$a = Y, b = Z, c = X$

Pleochroism and color: usually colorless in section; Fe-rich varieties may be X = colorless and Z = pale violet; Mg-rich varieties in thick section may have strong pleochroism with $Z > Y > X$, and X = colorless to pale yellow, Y = pale blue, and Z = blue; inclusions of zircon may have yellow pleochroic haloes.

Dispersion: $v > r$ weak.

Habit: euhedral poikiloblastic to euhedral pseudohexagonal forms.

Cleavage: {010} good, {100} and {001} poor, may have {001} parting (Fig. 65).

Extinction in section: generally parallel to cleavage traces and larger faces.

Twinning: commonly polysynthetic or cyclic on {110} or {130}; less commonly, simple {110} or {130}; polysynthetic twins may resemble those in plagioclase (color photo 26); cyclic twins may produce a star-shaped pattern with three, six, or twelve individuals, and a pseudohexagonal outline; some cordierite is untwinned. See Figure 66.

Hardness: 7–7½.

Specific gravity: 2.53–2.78.

Occurrence: common as ovoid poikiloblastic grains in thermally metamorphosed (high-temperature, low-pressure) pelitic rocks associated with andalusite and biotite; in high-grade hornfels it may be associated with anthophyllite, sillimanite,

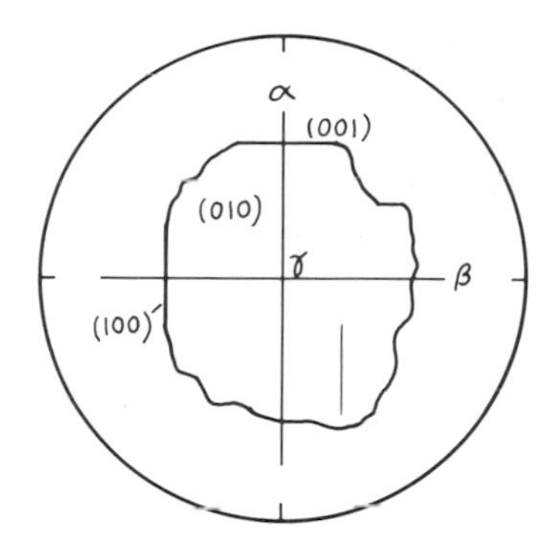

Figure 65. Cordierite grains on the {010} cleavage yield β and α indices. Extinction against the poor {100} or {001} cleavage is parallel. The figure is commonly bisymmetric Bxo, but may be bisymmetric Bxa.

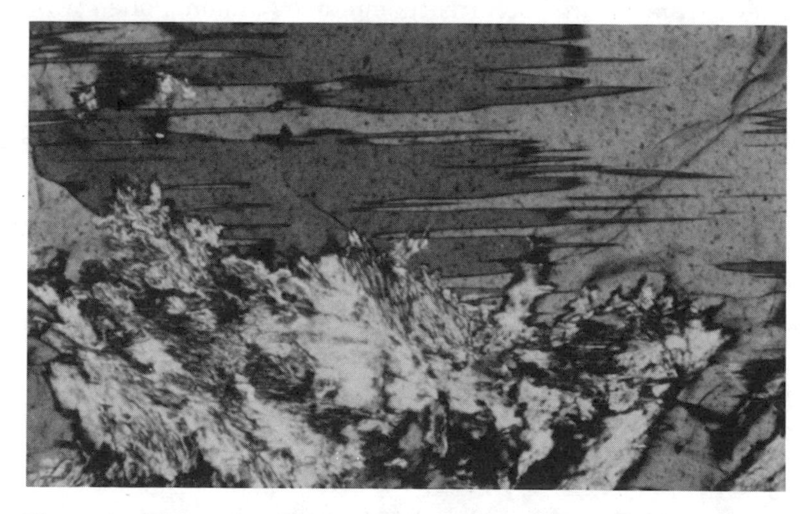

Figure 66. Twinned cordierite (top) partially altered to sericite. Crossed nicols. Photo length is 0.91 mm. (From C. Shultz, Slippery Rock University)

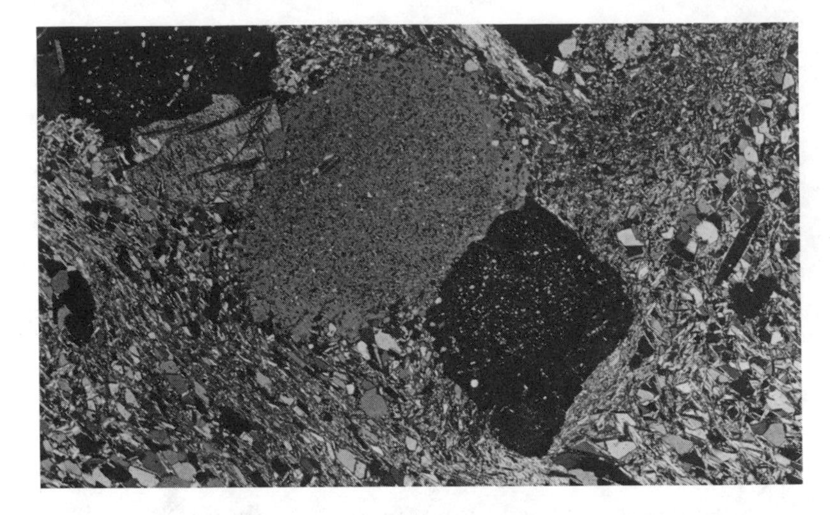

Figure 67. Several porphyroblasts of cordierite containing abundant inclusions, within a mica-quartz matrix. Crossed nicols. Photo length is 5.16 mm. (From J. Hinsch, E. Leitz, Inc.)

K-feldspar, muscovite, corundum, or spinel; less commonly, found in regionally metamorphosed rocks (color photo 27); garnet is usually absent and anthophyllite characteristically present; a high-temperature hexagonal polymorph (indialite) has been found in fused sediments; in igneous rocks, may be present in partly assimilated inclusions, in some granitic pegmatites, or rarely as phenocrysts in lavas; rare in the light fraction of sediments.

Alteration: to a fine-grained greenish aggregate of chlorite and muscovite or biotite (commonly called pinite); also present may be zoisite, serpentine, tourmaline, or garnet.

Characteristic features: occurrence; multiple twinning; often with pale blue-violet color; abundant inclusions of fine-grained opaque materials (Fig. 67); pleochroic haloes about zircon inclusions. See color photo 26.

Look-alikes: commonly mistaken for quartz, feldspar, or nepheline; quartz is uniaxial (+); feldspar has two perfect cleavages; nepheline is uniaxial (−).

Corundophyllite. See Chlorite Group (C4).

CORUNDUM (C9)

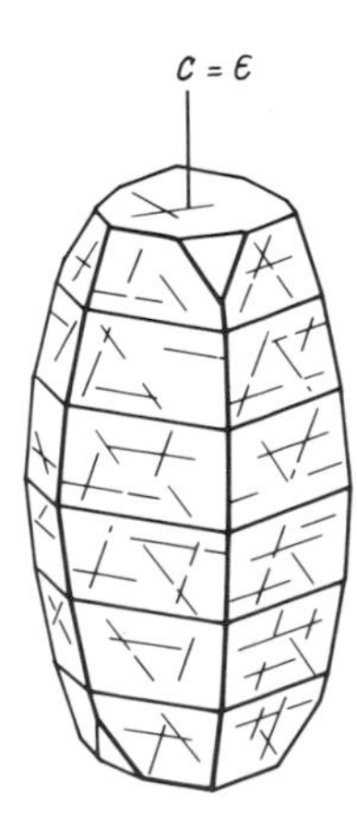

Al_2O_3
Hexagonal
$\omega = 1.767-1.772$, $\varepsilon = 1.759-1.763$
$\omega - \varepsilon = 0.007-0.010$ (up to 1st-order yellow)

Uniaxial (−), commonly anomalous biaxial

Pleochroism and color: commonly colorless, but may be pale yellow, blue, pink or green (due to tiny rutile or hematite inclusions); color may be patchy or zoned, and is deeper in detrital grains; abs. $O > E$ usually weak.

Habit: commonly euhedral with prisms and steep dipyramids, yielding barrel-shaped crystals; also anhedral, granular.

Cleavage: none, but well developed {0001} and $\{10\bar{1}1\}$ partings are common (Fig. 68).

Extinction in section: parallel or symmetric to principal faces and partings.

Twinning: polysynthetic on $\{10\bar{1}1\}$ very common; contact twinning on $\{10\bar{1}1\}$ or {0001} less common.

Hardness: 9.

Specific gravity: 3.98–4.10.

Occurrence: Si-poor and Al-rich igneous and metamorphic rocks, which include syenite, monzonite, nepheline syenite, and associated pegmatites; in dikes within mafic and ultramafic rocks; in xenoliths with spinel; in Al-rich metapelite; locally abundant as a heavy mineral in sediments.

Alteration: to Al-rich minerals such as margarite (Ca mica), or muscovite; less commonly to diaspore, gibbsite, zoisite, andalusite, kyanite, spinel, or chloritoid.

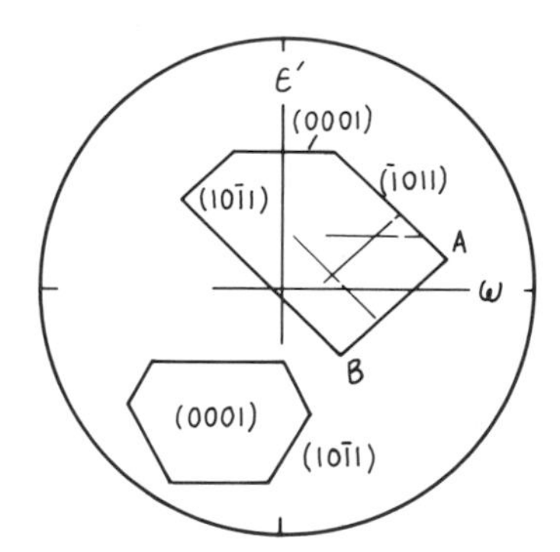

Figure 68. Corundum grains on the $\{10\bar{1}1\}$ parting permit estimation of ω and ε' (1.761–1.765). Extinction is symmetric about other rhombohedral partings. The ω vibration direction bisects the obtuse angle A (94°) and ε' bisects the acute angle B (86°). The ε' vibration is normal to traces of the {0001} parting. The figure is inclined O.A. Grains on the {0001} parting have total extinction, permit estimation of ω, and yield an O.A. figure. In many instances the $\{10\bar{1}1\}$ and {0001} partings may be absent.

Characteristic features: extreme relief and hardness; weak birefringence; uniaxial (−); parting and twinning; irregular or zoned colors; incompatible with quartz.
Look-alikes: apatite has lower indices and birefringence; the rare mineral sapphirine is biaxial and untwinned.

Cristobalite. See Silica Group (S4).

Crossite. See Amphibole Group (A2).

Cummingtonite. See Amphibole Group (A2).

Daphnite. See Chlorite Group (C4).

Diabantite. See Chlorite Group (C4).

Diaspore. See Clays and Clay Minerals (C6).

Diopside. See Pyroxene Group (P8).

Dipyre. See Scapolite Group (S2).

Dolomite. See Carbonate Group (C2).

DUMORTIERITE (D1)

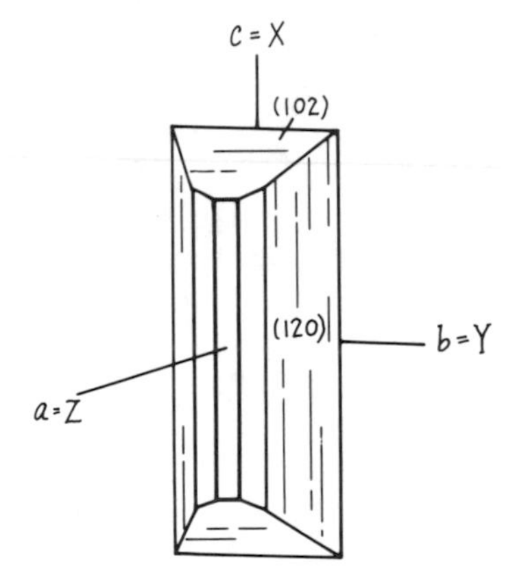

$(Al,Fe^{+3})_7O_3(BO_3)(SiO_4)_3$
Orthorhombic
$\alpha = 1.655–1.686$, $\beta = 1.675–1.722$, $\gamma = 1.684–1.723$
$\gamma - \alpha = 0.011–0.027$ (up to 2nd-order green)
$2V = 15°–52°$ (−)
$a = Z, b = Y, c = X$
Pleochroism and color: colorless to strongly pleochroic, commonly in blue and violet, but also in brown or pink, with $X > Y > Z$.
Dispersion: $v > r$ strong, rarely $r > v$.
Habit: fibrous, bladed, or acicular, elongated on *c;* euhedral crystals are rare.
Cleavage: {100} good, {110} poor (Fig. 69).
Extinction in section: commonly parallel to major faces and cleavage; elongate grains are length low (fast).
Twinning: cyclic on {110}; may yield trillings or pseudohexagonal forms.
Hardness: 7–8½.

Specific gravity: 3.26–3.41.
Occurrence: commonly associated with other Al-rich minerals in hydrothermal or metasomatic environments; found in granite pegmatites, aplite, quartz veins, quartzite, gneiss, and altered igneous rocks.
Alteration: to fine-grained muscovite.
Characteristic features: deep blue or violet color; parallel extinction; length low (fast).
Look-alikes: tourmaline has maximum absorption normal to direction of elongation; sillimanite is length high (slow); Na-rich amphiboles have characteristic {110} cleavage; sodalite is isotropic; piemontite is pleochroic in deep red.

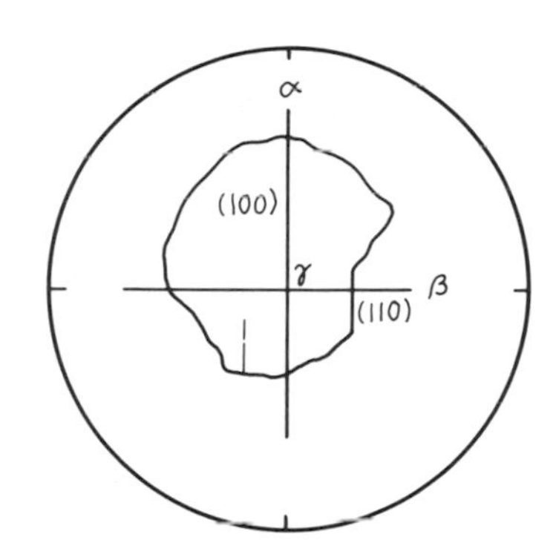

Figure 69. Dumortierite fragments on {100} permit estimation of α and β, have parallel extinction, and have maximum absorption parallel to poor {110} cleavage (when present). The figure is bisymmetric Bxo.

Eckermannite. See Amphibole Group (A2).

Edenite. See Amphibole Group (A2).

Enstatite. See Pyroxene Group (P8).

EPIDOTE GROUP (E1)

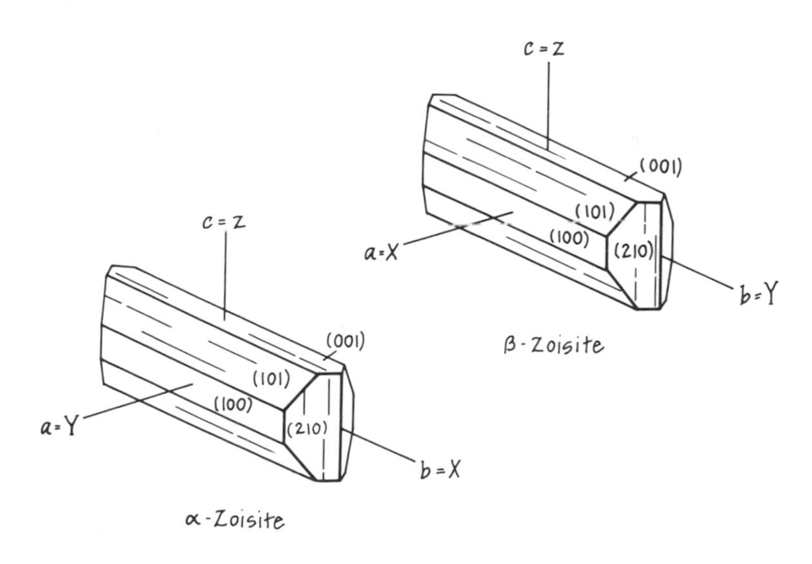

Zoisite

$Ca_2Al_3O[Si_2O_7][SiO_4](OH)$. Iron-free zoisite is called α-zoisite. Ferrian or β-zoisite may have up to 10 percent Fe^{+3} in octahedral (Al^{+3}) sites.

Orthorhombic

$\alpha = 1.685–1.707$, $\beta = 1.688–1.711$, $\gamma = 1.697–1.725$

$\gamma - \alpha = 0.005–0.020$ (commonly gray to white, but up to 1st-order deep red for β-zoisite)

$2V = 0°–60°$ (+)

$a = Y, b = X, c = Z$ (α-zoisite)

$a = X, b = Y, c = Z$ (β-zoisite)

Pleochroism and color: most commonly colorless; with small amounts of Mn^{+3} (variety *thulite*), X = deep pink, Y = pale pink or colorless, and Z = yellow; $X > Y > Z$; may be color zoned.

Dispersion: $v > r$ strong (α-zoisite); $r > v$ strong (β-zoisite); dispersion commonly results in blue anomalous interference colors (color photo 28).

Habit: columnar, bladed, or fibrous parallel to *b;* also granular.

Cleavage: {100} perfect, {001} poor (Fig. 70).

Extinction in section: commonly parallel to direction of elongation and cleavage traces; α-zoisite is length low (fast); β-zoisite may be either length low (fast) or length high (slow) as a function of grain orientation.

Twinning: none.

Hardness: 6.

Specific gravity: 3.15–3.36.

Occurrence: common in medium-grade metamorphic rocks with albite, calcite, quartz, garnet, biotite, hornblende, or idocrase (Fig. 71); rare as a primary mineral in igneous rocks; may be

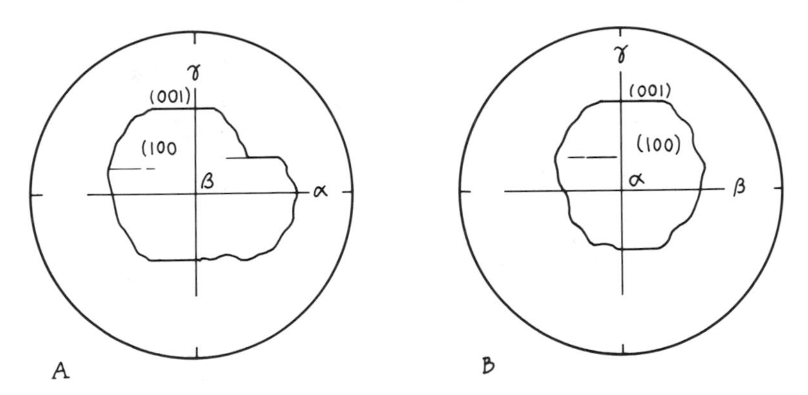

Figure 70. (A) α-zoisite, when lying on {100}, permits estimation of α and γ. If traces of the poor {001} cleavage are present, extinction is parallel. The figure is bisymmetric O.N. (B) β-zoisite, when lying on {100}, permits estimation of β and γ. If traces of the poor {001} cleavage are present, extinction is parallel. The figure is bisymmetric Bxo.

Figure 71. (A) Zoisite aggregate with minor quartz. Much of the zoisite exhibits anomalous blue interference colors. Crossed nicols. (B) The same, in plane-polarized light. Note the high relief of zoisite relative to quartz. Photo length is 2.06 mm. (From J. Hinsch, E. Leitz, Inc.)

secondary as a result of deuteric alteration of plagioclase (in the fine-grained aggregate of zoisite, prehnite, chlorite, sericite, and zeolite known as saussurite).

Alteration: resistent to weathering.

Characteristic features: high relief; weak birefringence; parallel extinction; anomalous interference colors common; strong dispersion.

Look-alikes: clinozoisite and epidote often display inclined extinction when viewed along *b;* epidote is (−) and has higher birefringence; melilite and apatite are (−); idocrase, usually (−), is uniaxial and has {110} cleavage.

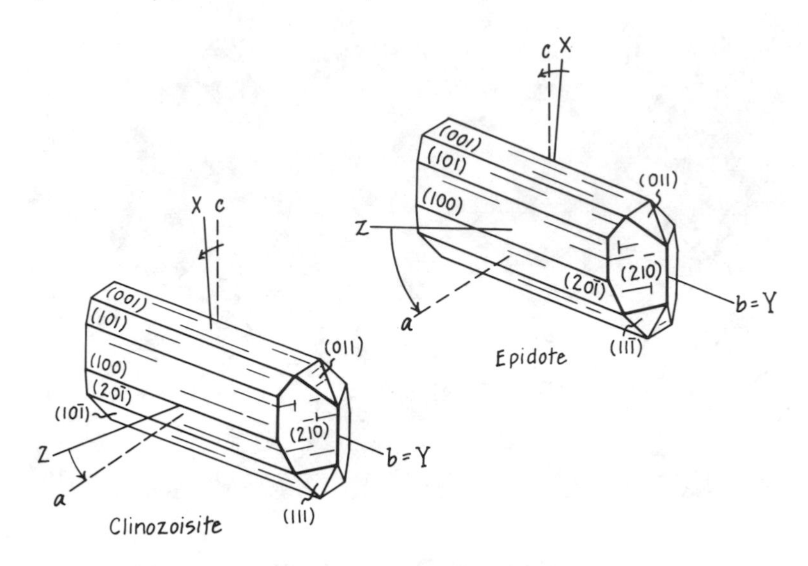

Clinozoisite-Epidote Series

$Ca_2(Al,Fe^{+3})Al_2O[Si_2O_7][SiO_4](OH)$
Monoclinic
$\measuredangle\beta \simeq 115°25'$
$\alpha = 1.697–1.714$, $\beta = 1.699–1.722$, $\gamma = 1.702–1.729$ (clinozoisite)
$\alpha = 1.714–1.728$, $\beta = 1.722–1.758$, $\gamma = 1.729–1.776$ (epidote)
$\gamma - \alpha = 0.005–0.015$ (up to 1st-order orange in clinozoisite)
$\gamma - \alpha = 0.015–0.048$ (up to 3rd-order yellow in epidote)
$2V = 14°–90°$ (+) (clinozoisite); $64°–90°$ (−) (epidote): clinozoisite is distinguished on the basis of the (+) optic sign, which changes to (−) at about 15 percent substitution of Fe^{+3} for Al.
$b = Y$, $X \wedge c = 0°–90°$ (clinozoisite); $b = Y$, $X \wedge c = 0°$ to $5°$ (epidote)
Pleochroism and color: clinozoisite is colorless; increasing iron content in epidote results in increasing coloration with X = colorless, pale yellow, pale green, Y = yellow-green, brownish green, Z = colorless, pale yellow-green, and $Y > Z > X$. Small amounts of Mn^{+3} produce reddish colors and strong pleochroism from deep pink or red to yellow (as transition is made to piemontite). The 1st-order white interference color in both clinozoisite and epidote is commonly replaced by anomalous pale blue or yellow-brown.
Dispersion: $v > r$ strong (clinozoisite); $r > v$ strong (epidote).
Habit: columnar or acicular aggregates elongated on b; also fine to coarse granular aggregates.

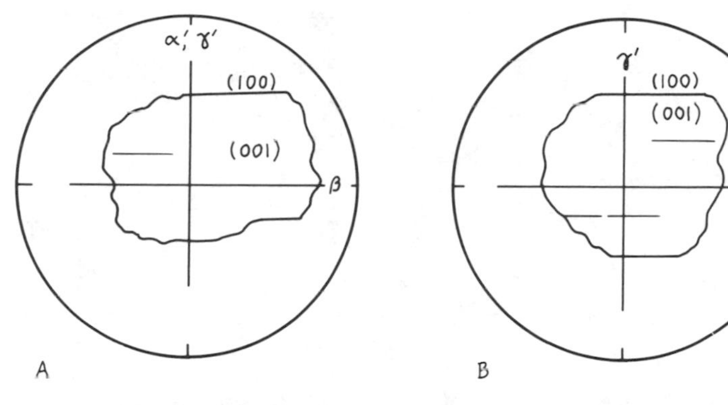

Figure 72. (A) Clinozoisite, when lying on {001}, permits estimation of β and α' or γ' (1.698–1.726). Extinction is parallel to traces of good {100} cleavage. Grains are length high (slow) for Al-rich compositions, and length low (fast) for Fe-rich compositions. The figure ranges from monosymmetric Bxa, through O.A., to monosymmetric Bxo with increasing iron content. (B) Epidote, when lying on {001}, permits estimation of β and γ' (1.72–1.76). Extinction is parallel to traces of good {100} cleavage. Grains are length low (fast). The figure is monosymmetric Bxa (due to change in optic sign).

Cleavage: perfect {001}, good {100} (Fig. 72).

Extinction in section: generally parallel to cleavage traces and larger faces when viewing elongate grains; length high (slow) or length low (fast).

Twinning: {100} polysynthetic common.

Hardness: 6-6½.

Specific gravity: 3.21–3.49 (increasing with Fe content).

Occurrence: common in low- to medium-grade metamorphic rocks (particularly argillaceous carbonate rocks), with chlorite, actinolite, albite, and/or hornblende (Fig. 73); in contact zones, epidote and clinozoisite may be associated with idocrase, grossular, anorthite, diopside, and/or calcite; rare as a primary mineral in igneous rocks; deuteric or hydrothermal alteration of Ca-rich plagioclase may form saussurite (a fine-grained aggregate that may contain prehnite, chlorite, sericite, zeolites, and one or more epidote-group minerals—see color photo 29); common as a detrital mineral.

Alteration: resistant to weathering.

Characteristic features: anomalous interference colors; elongate grains may be length high or low; high relief; generally extinction parallel to a perfect cleavage; occurrence (Fig. 74).

Figure 73. Elongate epidote (center) in chlorite-actinolite-quartz schist. Crossed nicols. Photo length is 0.55 mm.

Figure 74. Aggregate of epidote and quartz. Crossed nicols. Photo length is 0.55 mm.

Look-alikes: clinozoisite and epidote are distinguished from each other by optic sign; zoisite always has parallel extinction when viewed normal to the *b* axis and commonly displays very distinct anomalous interference colors; amphiboles and pyroxenes have two perfect cleavages; idocrase and melilite may have anomalous interference colors, but both are uniaxial or nearly so.

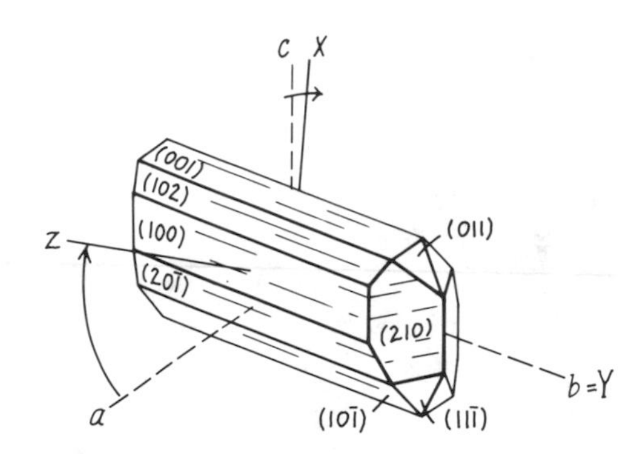

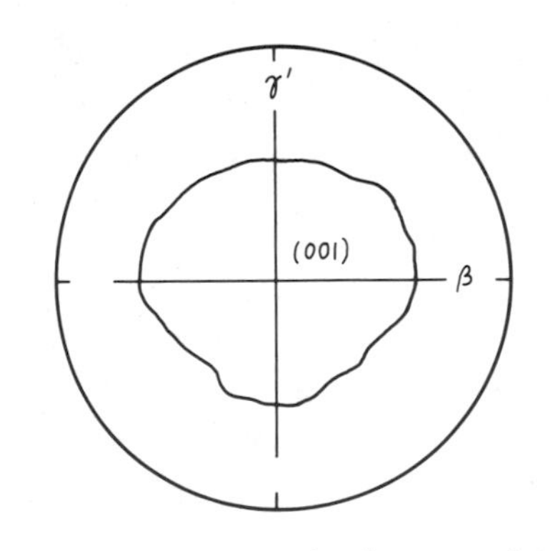

Figure 75. A piemontite fragment lying on {001} permits estimation of β and γ' (1.74–1.82). The figure is monosymmetric Bxo.

Piemontite

$Ca_2(Mn,Fe^{+3},Al)_2AlO[Si_2O_7][SiO_4](OH)$
Monoclinic
$\measuredangle\beta = 115°42'$
$\alpha = 1.725–1.794$, $\beta = 1.730–1.807$, $\gamma = 1.750–1.832$
$\gamma - \alpha = 0.025–0.082$ (up to 4th-order, but often masked by intense mineral color)
$2V = 50°–86°$ (+)
$b = Y$, $c \wedge X = -2°$ to $-9°$
Pleochroism and color: deep colors and strong pleochroism, with X = pale yellow, pink, Y = pale violet, red-violet, and Z = deep red, brownish red, with $Z > Y > X$.
Dispersion: $r > v$ strong, rarely $v > r$.
Habit: columnar, bladed, or acicular parallel to *b*.
Cleavage: {001} perfect (Fig. 75).
Extinction in section: commonly parallel to cleavage and major faces when viewing elongate grains; length high (slow) or length low (fast).
Twinning: polysynthetic {100} uncommon.
Hardness: 6–6½.
Specific gravity: 3.40–3.52.

Occurrence: mainly in low-grade metamorphic rocks; with Mn-rich minerals in hydrothermal or metasomatic deposits; in vesicles or fissures within volcanic rocks.
Alteration: resistant to weathering.
Characteristic features: strongly pleochroic from red to yellow; high indices; parallel extinction common.
Look-alikes: Mn-rich epidote is (−); dumortierite is (−) and has a small $2V$.

Allanite (Orthite)

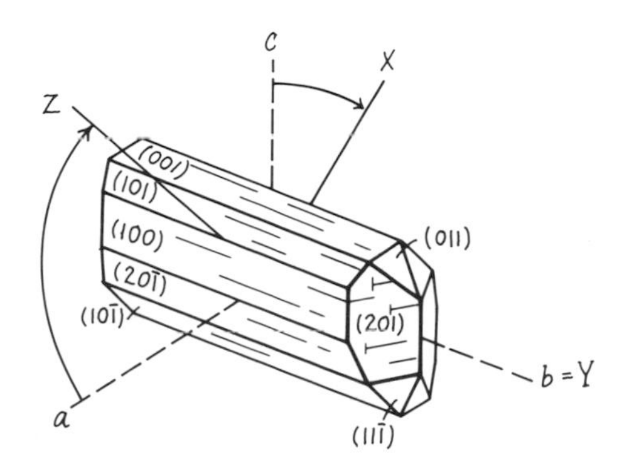

$(Ca,Ce)_2(Fe^{+2},Fe^{+3})Al_2O[Si_2O_7][SiO_4](OH)$
Monoclinic
$\measuredangle\beta = 115°$
$\alpha = 1.690–1.791$, $\beta = 1.700–1.815$, $\gamma = 1.706–1.828$; some metamict allanites are isotropic with $n = 1.54–1.72$
$\gamma - \alpha = 0.013–0.036$ (up to 2nd-order red)
$2V = 40°–90°$ (−), rarely $57°–90°$ (+)
$b = Y$, $c \wedge X = -1°$ to $-42°$
Pleochroism and color: usually brownish yellow, brown, or less commonly green; distinctly pleochroic, with $Z > Y > X$.
Dispersion: variable, usually $r > v$ strong, rarely $v > r$.
Habit: elongate parallel to b, {100} tablets, or anhedral.
Cleavage: {001} moderate to poor, {100} and {110} poor (Fig. 76).
Extinction in section: commonly parallel to cleavage and larger faces when viewing elongate grains; length high (slow) or low (fast).
Twinning: {100} uncommon.
Hardness: 5–6½.
Specific gravity: 3.4–4.2, but may be as low as 2.7 when metamict.
Occurrence: an accessory mineral in granite, granodiorite, monzonite, and nepheline syenite; may be in larger amounts in limestone skarns and pegmatites (with other rare-earth or radioactive minerals); a rare accessory in some schist and gneiss; when enclosed in biotite, chlorite, or hornblende it is usually surrounded by a pleochroic halo; often rimmed by epidote. See Figure 77.
Alteration: to limonite, silica, and alumina; metamict to isotropic state.
Characteristic features: high relief; poor cleavage; irregular brownish color and pleochroism; often surrounded by pleochroic halo.
Look-alikes: brown amphiboles and pyroxenes have perfect prismatic cleavage and inclined extinction; when metamict, allanite cannot be distinguished optically from other dark-brown isotropic minerals.

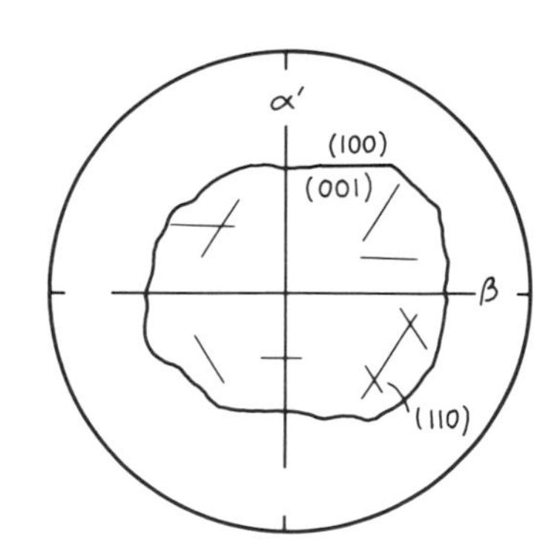

Figure 76. Allanite fragments are commonly in random orientation with irregular outline. When lying on {001}, estimation can be made of β and some value of α'. Extinction is parallel to traces of poor {100} cleavage, and symmetric to poor {110} cleavage. The figure is commonly monosymmetric Bxo, but for certain compositions may be bisymmetric Bxo.

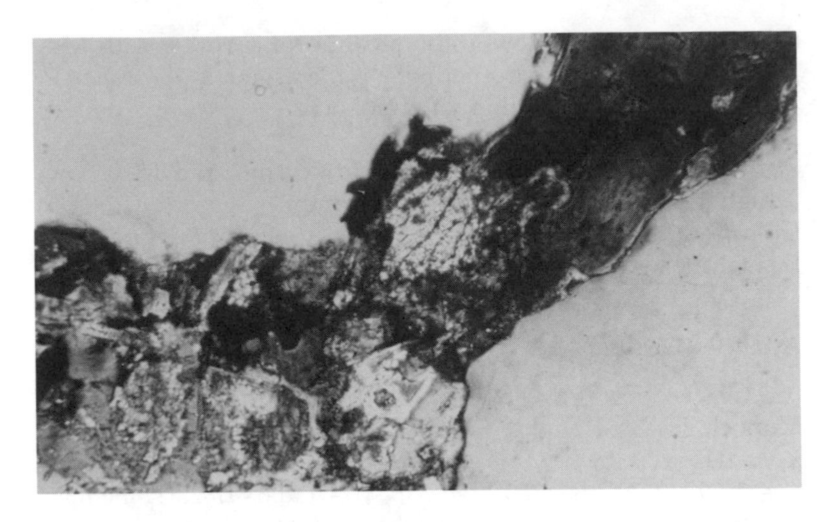

Figure 77. Allanite grain (center) within altered biotite (dark). The surrounding colorless grains are quartz. Crossed nicols. Photo length is 0.55 mm.

Eulite. See Pyroxene Group (P8).

Fayalite. See Olivine Group (O1).

THE FELDSPAR GROUP (F1)

Feldspars are the most abundant minerals in the earth's crust, and unfortunately (from the standpoint of identification) one of the most complex. Precise determination of feldspar composition and structural state require use of both the electron microprobe and X-ray diffraction. Optical methods, however, are sufficiently accurate for most mineralogical and petrological purposes.

Composition

The common rock-forming feldspars are composed of various mixtures of three end-member components—$NaAlSi_3O_8$, $KAlSi_3O_8$, and $CaAl_2Si_2O_8$. The darkened area in Figure 78 shows feldspar compositional limits. Compositions (such as A) outside of the darkened area, when crystallized, will yield two feldspars (such as B and C), whereas compositions (such as B or C) within the darkened area, when crystallized, will yield a single feldspar that corresponds to the original composition.

Complete solid solution exists at high temperatures between $CaAl_2Si_2O_8$ and $NaAlSi_3O_8$; this mineral series is known as *plagioclase*. Complete solid solution between $KAlSi_3O_8$ and

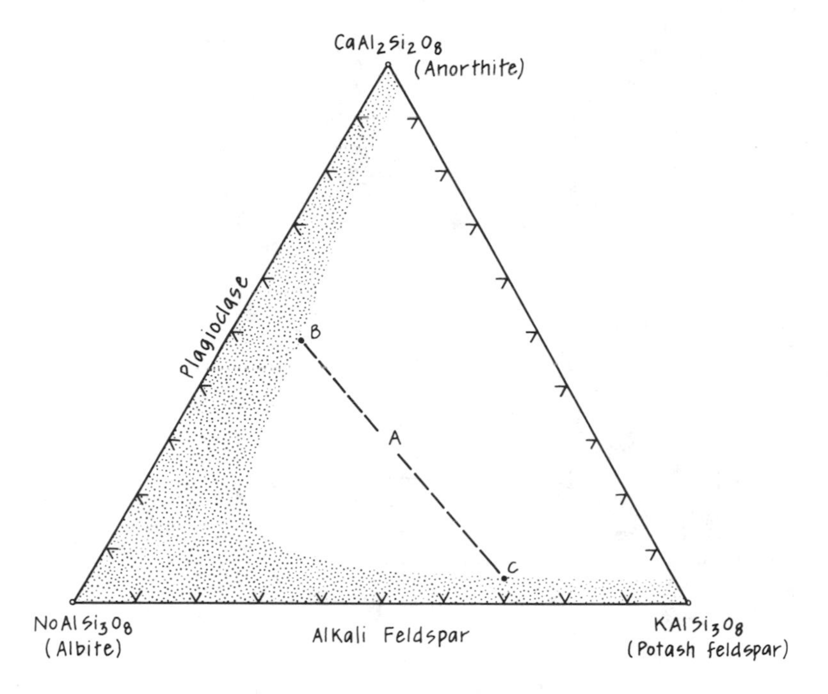

Figure 78. Compositional limits of the common rock-forming feldspars (under anhydrous conditions) are shown in darkened area. A composition such as A, outside of the darkened area, crystallizes to form two feldspars such as B and C.

$NaAlSi_3O_8$ comprises the *alkali feldspar* series.[1] Notice that in addition to the two principal components, the plagioclase series can contain some $KAlSi_3O_8$, and the alkali feldspar series can contain some $CaAl_2Si_2O_8$.

The limits of solid solution vary as a function of the temperature at which the feldspars have reached equilibrium. Feldspars equilibrated at high temperatures (as in volcanic rocks) show maximum solid solution. Any composition within the darkened portion of Figure 78 is possible. At lower temperatures of equilibration (as in plutonic or metamorphic environments), not only are the darkened compositional bands narrower than shown in Figure 78, but also compositional gaps are present within the two solid-solution series.

The extent of solid solution within the alkali feldspar series as a function of temperature is shown in Figure 79. The darkened region of the diagram shows the range of temperatures and compositions that yield a single feldspar. Point A represents a single

[1] The Na-rich end member, albite, is in both the plagioclase and alkali feldspar series. For purposes of rock classification, albite containing 0–5 percent of the $CaAl_2Si_2O_8$ component is considered to be an alkali feldspar; plagioclase contains 5–100 percent of the $CaAl_2Si_2O_8$ component.

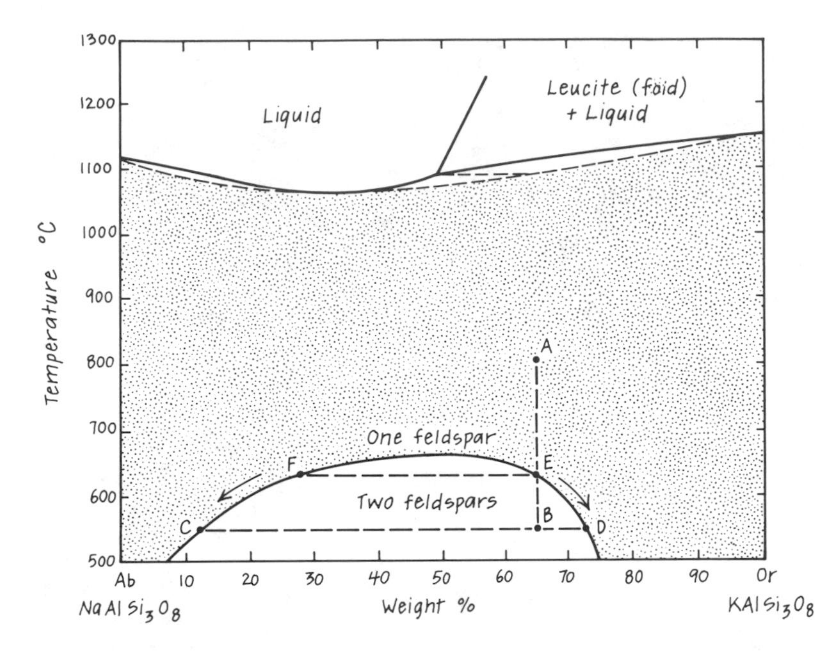

Figure 79. Phase diagram showing the limits of stability of the alkali feldspars (under anhydrous conditions) in terms of composition and temperature (after Tuttle and Bowen, 1958). A single alkali feldspar such as A, when cooled to the solvus at E, exsolves a second feldspar of composition F. Continued cooling to B causes additional exsolution, and change of composition along the solvus. At a temperature of 550°C, the composition B consists of two feldspars of compositions C and D. The two feldspars commonly form an intergrowth known as perthite.

Figure 80. Perthitic feldspar in granite. Perthite usually consists of sodium-rich plagioclase exsolved from microcline. Crossed nicols. Photo length is 0.82 mm.

feldspar at 800°C; the composition of this feldspar, as determined on the horizontal coordinate, is 35 percent $NaAlSi_3O_8$ and 65 percent $KAlSi_3O_8$. At lower temperatures the diagram shows a curved line called a *solvus;* the solvus defines the upper limits of a region of immiscibility. Assuming equilibrium conditions, any composition within the solvus consists of two feldspars. For example, if metamorphic recrystallization occurred at B, two feldspars (C and D) would form. The compositions of the two feldspars are found at the intersections of a horizontal tie line (through B) with the solvus (see also pp. 71–72 and 155–156 of Ehlers and Blatt, 1982).

A single feldspar, such as A, when cooled, would encounter the subsolvus region at E. With very rapid cooling, the feldspar may persist metastably to room temperature as a single phase. With slower cooling, the single feldspar begins to unmix at the solvus. At temperature (E), the feldspar exsolves a second feldspar of different composition (F). The second feldspar may form discrete crystals external to the original host; or it may remain within the host as irregular blebs or bands (Fig. 80). Continued cooling (to B) results in additional exsolution. During this process the composition and proportions of both the host and exsolved material change; host E changes to D, and the exsolved feldspar changes to C. Continued cooling to room temperatures results in continued exsolution and compositional change of the two feldspars along the solvus.

The megascopic texture that may result from unmixing of alkali feldspars is called *perthite*.[2] If the exsolution texture is visible only on the microscopic scale it is called *microperthite* (Fig. 80). Submicroscopic exsolution (detected by X-ray diffraction) is called *cryptoperthite;* cryptoperthites, as seen in the microscope, appear as more-or-less homogeneous grains.

Three immiscibility regions exist in the plagioclase feldspar series, the most important of which is between 2 percent and 16 percent anorthite (An_{2-16}).[3] Although plagioclase exsolution is submicroscopic, it is indicated by optical scattering, which yields an iridescent play of colors in some hand specimens (such a mineral is known as peristerite). Similar optical scattering is also common in labradorite.

Structural and Compositional Varieties

The *alkali feldspars* are composed of five solid-solution series. The existence of each series is mainly dependent upon temperature of crystallization; however, slow cooling (often in the presence of a vapor phase) may result in the conversion of high-temperature varieties into lower temperature types.

Figure 81 shows the alkali feldspar varieties as a function of temperature of formation and composition. At the K-rich side of the diagram, the sequence of minerals with decreasing temperature is high sanidine, sanidine, orthoclase, microcline, and

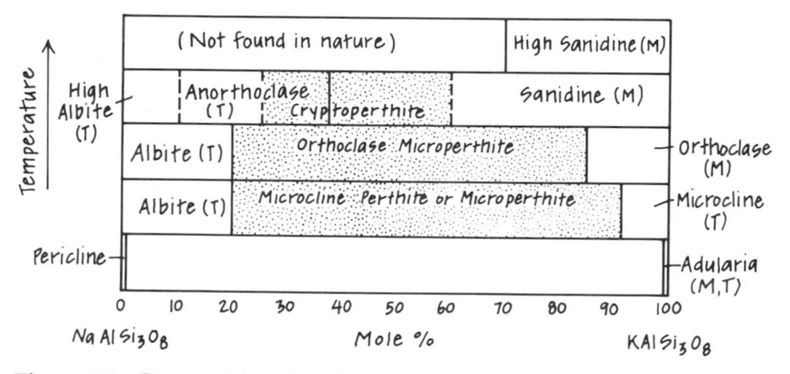

Figure 81. Compositional and structural varieties of alkali feldspars as a function of temperature of equilibration. Darkened area shows the approximate region of unmixing. M = monoclinic, T = triclinic.

[2]Some perthites have been formed by partial replacement of one feldspar by another.

[3]Standard abbreviations used for feldspars are Ab = albite, Or = orthoclase, and An = anorthite.

adularia. High sanidine (uncommon) and sanidine (common) are found in volcanic rocks; orthoclase may be in volcanic or plutonic rocks; microcline is limited to plutonic and metamorphic environments, and adularia (uncommon) is limited to hydrothermal veins. The Na-rich side of the diagram shows anorthoclase and high albite, albite, and pericline. High albite and anorthoclase may be present in soda-rich lavas, but are uncommon. Albite is common in plutonic and metamorphic rocks, and pericline is rare (in hydrothermal veins).

The shaded central portion of Figure 81 shows regions of exsolution. Intermediate compositions of the sanidine–high albite series form cryptoperthites. Microperthites are common in the orthoclase-albite series, as are perthites and microperthites in the microcline-albite series. The rare minerals adularia and pericline crystallize as pure end members and display no exsolution effects.

In the *plagioclase* series, complete solution exists between albite (Ab), $NaAlSi_3O_8$, and anorthite (An), $CaAl_2Si_2O_8$ at near-liquidus temperatures. Composition is commonly stated in terms of anorthite percentage. Members of the series are named as follows: albite, An_{0-10}; oligoclase, An_{10-30}; andesine, An_{30-50}; labradorite, An_{50-70}; bytownite, An_{70-90}; anorthite, An_{90-100}. Both high- and low-temperature varieties are known respectively in volcanic and plutonic environments. Although three solvi are present in this series, exsolution is submicroscopic.

Crystallography, Habit, and Twinning

Sanidine and orthoclase are monoclinic. The other rock-forming feldspars are triclinic. Adularia, generally classified as monoclinic (see orthoclase description), has been demonstrated to possess optical domains identical with and intermediate to both sanidine and microcline. Although most feldspars are triclinic, the angle (α) between the *b* and *c* crystallographic axes is within a few degrees of being 90°; this results in triclinic single crystals appearing to be monoclinic. Principal optical directions, however, range widely from this pseudomonoclinic orientation.

Most feldspars have one of three *crystal habits* (Fig. 82): (1) stubby prisms with predominant {110} and {010} faces; (2) tabular crystals with {010} predominant; (3) laths elongate on *a*. Feldspar cleavage is essentially independent of composition and symmetry, with {001} perfect, {010} nearly perfect, and {110} poor at 59°–61°.

Twinning in feldspar is very common (Table F1A). Twins are described in terms of the *twin axis* (the axis about which one in-

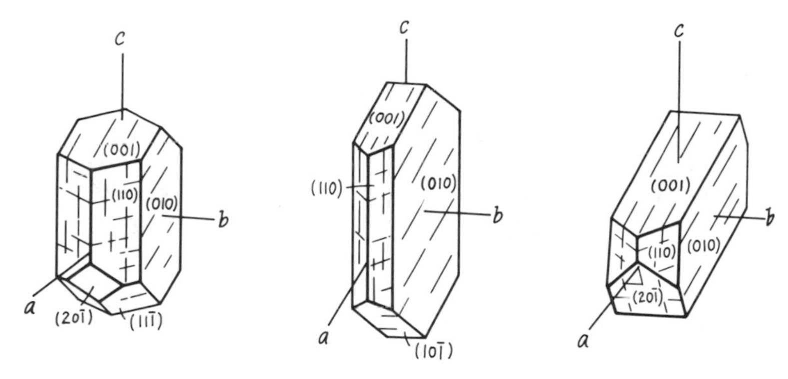

Figure 82. The three common habits of feldspar. Traces of the {001} and {010} cleavages are indicated.

Table F1A
Feldspar Twin Laws

Name	Twin Axis	Composition Plane	Remarks
Albite	⊥ (010)	(010)	Normal twin, very common, polysynthetic, only in triclinic feldspars.
Manebach	⊥ (001)	(001)	Normal twin, uncommon, usually simple.
Baveno	⊥ {021}	{021}	Normal twin, uncommon, usually simple.
Carlsbad	*c*	(010)	Parallel twin, very common, usually simple, contact or penetration.
Pericline	*b*	Rhombic section[1]	Parallel twin, very common, polysynthetic, only in triclinic feldspars.
Acline	*b*	(001)	Parallel twin, uncommon, polysynthetic, only in triclinic feldspars.
Albite–Carlsbad	⊥ *c* in (010)	(010)	Complex twin, common, usually polysynthetic, only in triclinic feldspars.

[1]The rhombic section is a generally nonrational plane that contains the *b* axis; its angle of inclination is a function of feldspar compositions. The rhombic section is so named because its intersection with (110) and (1$\bar{1}$0) faces forms the sides of a rhombus.

dividual of a twin crystal may be rotated (usually 180°) in order to coincide with the other individual) and the *composition* plane (the surface upon which the twinned individuals are joined). It should be noted that monoclinic species have only simple twins (two individuals), whereas triclinic species may have polysynthetic (repeated) twins or simple twins (color photo 30). Feldspar twinning develops during growth, deformation, or inversion from monoclinic to triclinic symmetry.

Optical Identification of Feldspars

When determining feldspars in either thin section or mineral separates, do not assume that only a single feldspar is present. Many rocks contain both plagioclase and alkali feldspar. In some

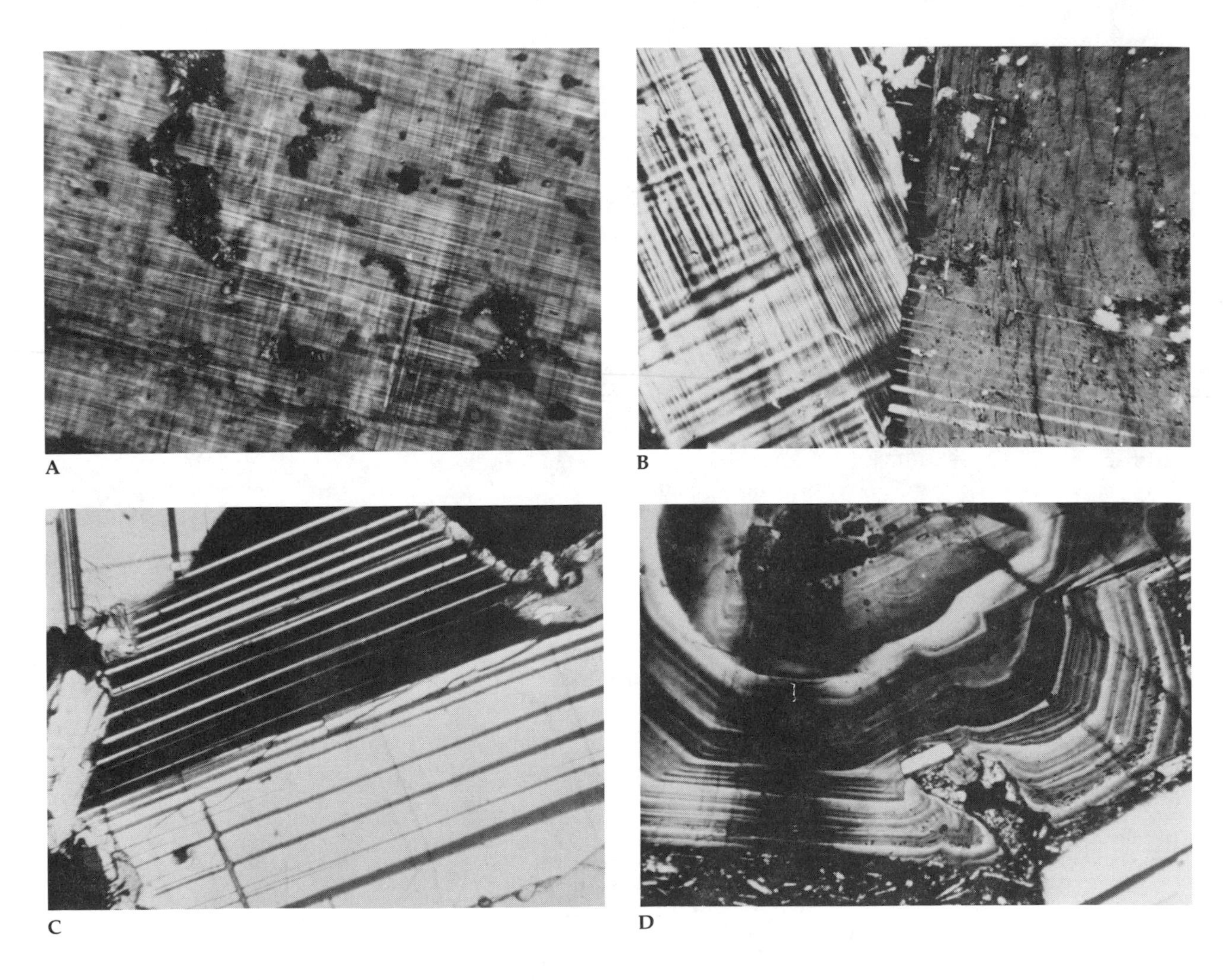

instances, two alkali feldspars may be present and in others two or more plagioclases. Volcanic rocks often contain plagioclase phenocrysts that are considerably more calcic than matrix plagioclase. Many feldspars are compositionally zoned as a result of disequilibrium during growth; this may require separate identification of both cores and rims.

Feldspars are distinguished from other minerals on the basis of relatively low indices of refraction (1.518–1.590), low birefringence (0.005–0.013), two cleavages at or near 90°, almost ubiquitous twinning, and lack of color. Feldspar varieties and textures are shown in Figure 83 and color figures 29–36.

After determining that a feldspar is present, the next step is to distinguish between plagioclase and alkali feldspar. This is most easily done on the basis of index of refraction.

E

F

Figure 83. (A) Very narrow double polysynthetic twinning in anorthoclase. Crossed nicols. Photo length is about 2.24 mm. (B) Adjacent grains of microcline (left) and plagioclase (right). Microcline typically has two directions of polysynthetic twinning (albite and pericline laws), whereas plagioclase typically has one direction (albite law). Crossed nicols. Photo length is 0.82 mm. (C) A small number of plagioclase grains have both albite and Carlsbad twinning. The Carlsbad composition plane is located between the predominantly dark and light portions of the grain. Crossed nicols. Photo length is 0.82 mm. (D) A compositionally zoned plagioclase phenocryst in a lava. The photo shows how inclusions are trapped by a growing crystal. Crossed nicols. Photo length is 0.82 mm. (E) Myrmekitic intergrowth of plagioclase and quartz. The branching quartz (in extinction left of center) has a single optical orientation within the plagioclase. Crossed nicols. Photo length is 0.82 mm. (From J. Hinsch, E. Leitz, Inc.) (F) Graphic intergrowth of triangular quartz within alkali feldspar (in extinction). Crossed nicols. Photo length is 0.32 mm.

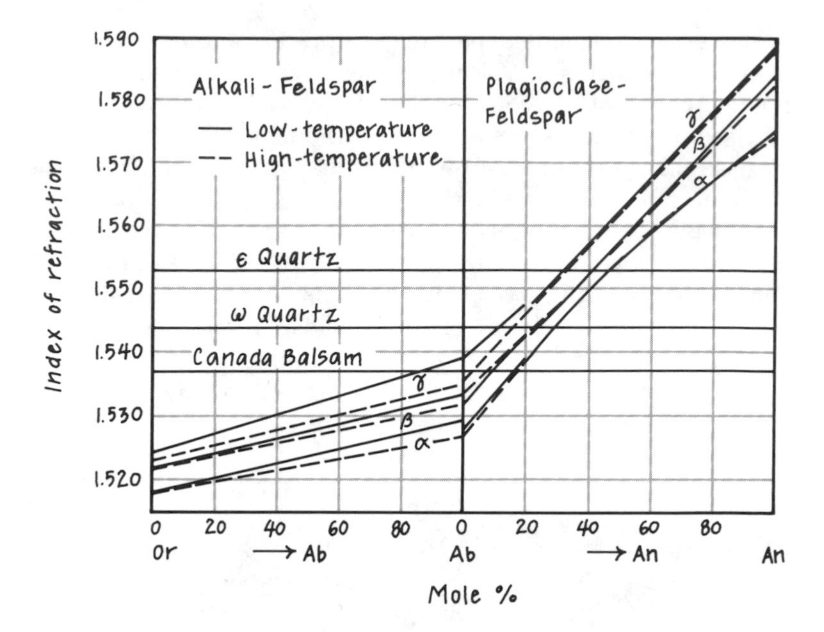

Figure 84. Variation in indices of refraction of alkali feldspar and plagioclase with composition (Smith, 1958, Fig. 3; Troeger, 1971, p. 123).

Indices of refraction for the various feldspars (Fig. 84) increase from K-feldspar, through albite, to anorthite. When working with fragments and immersion liquids, make a mount of n = 1.530. The smaller index of albite (either α or α' on a cleavage surface) is very slightly less than 1.530; if albite is present the grains will show almost no relief. If the low index of the grain is distinctly lower in index than the immersion liquid, an alkali feldspar is present; if the low index of the grain is equal or higher than the immersion liquid a plagioclase is present.

If feldspar determination is made in thin section, index determination can be made against either the mounting medium or quartz (if present); Canada balsam (n = 1.537) was used as a mounting medium in older sections, and epoxy (n = 1.533–1.540) is used in newer sections. Sections with Canada balsam may have a slight yellowish cast, whereas those with epoxy are colorless.

Find a feldspar grain at the edge of the section and compare its low index direction (α, α', or β) against the cementing material. A perfect match indicates that the feldspar is a Na-rich oligoclase. If the feldspar has a higher index than the cement, the feldspar is a plagioclase that is more calcic than Na-rich oligoclase. If the feldspar has very low relief and an index slightly lower than the cement, the feldspar is albite. A distinctly lower grain index indicates an alkali feldspar.

If quartz (uniaxial (+)) is present, it is easy to compare the ω index (1.544) against the adjacent feldspar index. With an accessory, determine the low (fast) vibration direction in quartz; then rotate the stage until this direction is parallel to the privileged direction of the lower polarizer, and perform the Becke test with uncrossed nicols. A perfect match is obtained for oligoclase. A higher index for the feldspar indicates a more calcic plagioclase, a lower index either albite or alkali feldspar.

Be sure to check the index of refraction of the feldspar on a number of grains. If nonuniform results are obtained, it is likely that more than one feldspar is present. In this case there is usually some difference in size, twinning, shape, zoning, or alteration between the feldspar species present. Thus, the different feldspars can be recognized rapidly without tediously checking the index of each grain.

Alkali Feldspars

The composition of alkali feldspars can be determined by precise measurement of refractive indices. The particular mineral species is determined by a combination of $2V$, type of twinning, occurrence, and perhaps extinction angles.

Figure 85 shows the variation in $2V$ as a function of composition and structural state. Sanidine has a small $2V$; orthoclase is moderate, and microcline is large. Sanidine and orthoclase (being monoclinic) cannot display the cross-hatched twins (albite

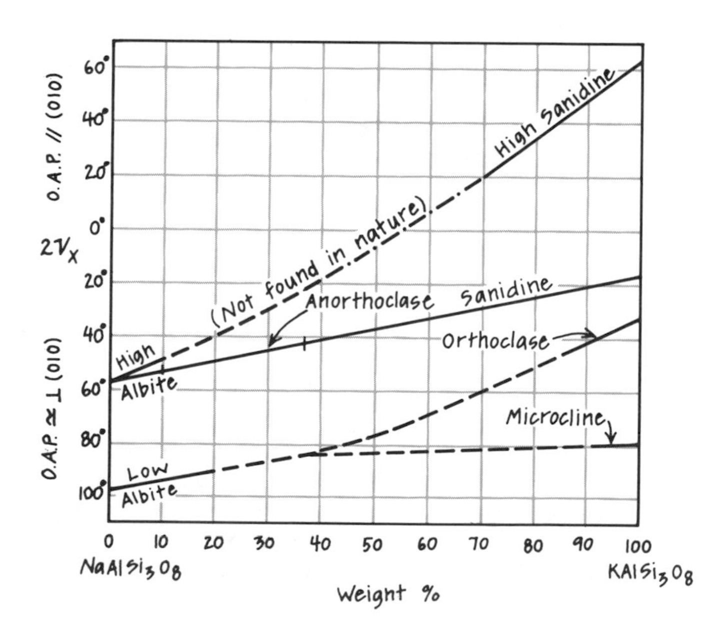

Figure 85. Variation in $2V$ of the alkali feldspars (Tuttle, 1952; MacKenzie and Smith, 1956). The compositional range of each mineral species is dependent upon temperature.

and pericline combination) that are very common in microcline and anorthoclase. Sanidine and anorthoclase are limited to high-temperature volcanic rocks whereas microcline is found in plutonic or metamorphic environments. High albite is (−) and limited to volcanic rocks, whereas low albite is (+) and is limited to lower temperature environments; albite twins are common.

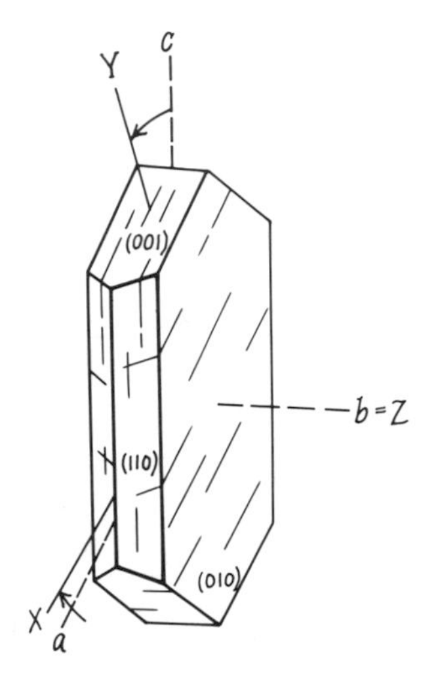

Sanidine

$(K_{1-0.77},Na_{0-0.33})AlSi_3O_8$
Monoclinic
$\measuredangle\beta \simeq 116°$
$\alpha = 1.518–1.524$, $\beta = 1.522–1.529$, $\gamma = 1.522–1.530$
$\gamma - \alpha = 0.005–0.006$ (1st-order gray-white)
$2V = 18°–24°$ (−)
$b = Z$, $c \wedge Y \simeq 21°$
Pleochroism and color: colorless.
Dispersion: $r > v$ weak.
Habit: euhedral phenocrysts, {010} tablets common, elongate parallel to *a*, or subhedral microlites.
Cleavage: {001} perfect, {010} nearly perfect, {110} poor (Fig. 86).
Extinction in section: when habit is tabular, extinction is parallel to elongate {010} edges, length low (fast); when grains are elongate on *a*, extinction is parallel or inclined, length low (fast).
Twinning: Carlsbad twins common; no polysynthetic twins.
Hardness: 6.
Specific gravity: 2.56–2.62.

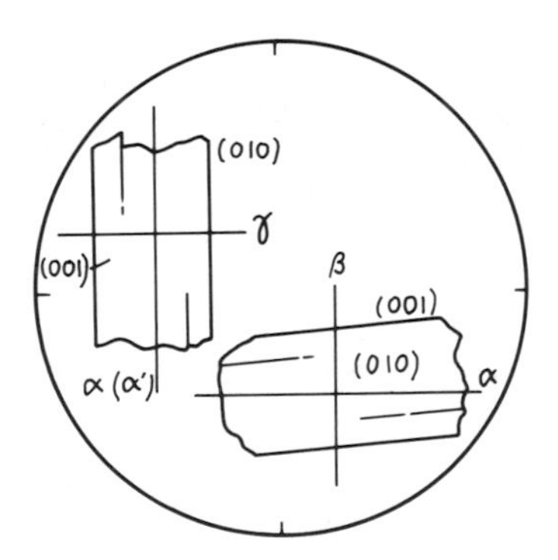

Figure 86. Sanidine fragment on {001} permits estimation of α (as $\alpha' \simeq \alpha$) and γ. Extinction is parallel and length low (fast) against {010}. The figure is almost bisymmetric O.N. A fragment on {010} permits estimation of α and β. The extinction is inclined, with $X \wedge \{001\} = 5°$, and the grain is length low (fast). The figure is bisymmetric Bxo.

Occurrence: as phenocrysts or matrix crystals in felsic volcanic rocks (color photo 31); as spherulites in rhyolite and obsidian.
Alteration: to kaolin or sericite.
Characteristic features: small 2*V* (−); occurrence; simple twins.
Look-alikes: other feldspars have moderate to large 2*V*s, polysynthetic twins, and may be found in nonvolcanic environments.

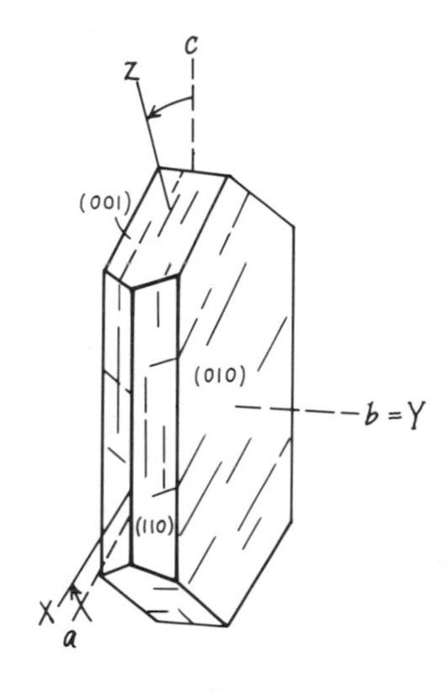

High Sanidine

$(K_{1-0.70},Na_{0-0.30})AlSi_3O_8$
Monoclinic
$\measuredangle\beta \simeq 116°$
$\alpha = 1.518–1.521$, $\beta = 1.523–1.525$, $\gamma = 1.524–1.526$
$\gamma - \alpha = 0.006$ (1st-order gray-white)
$2V = 15°–63°\ (-)$
$b = Y, c \wedge Z \simeq 20°$
Pleochroism and color: colorless.
Dispersion: $v > r$ weak.
Habit: euhedral phenocrysts, {010} tablets common.
Cleavage: {001} perfect, {010} nearly perfect, {110} poor (Fig. 87).
Extinction in section: when habit is tabular, extinction is parallel to elongate {010} edges, length high (slow) or low (fast).
Twinning: Carlsbad twins common; no polysynthetic twins.
Hardness: $\simeq 6$.
Specific gravity: 2.56–2.62.
Occurrence: uncommon, as phenocrysts or matrix crystals in felsic volcanic rocks.

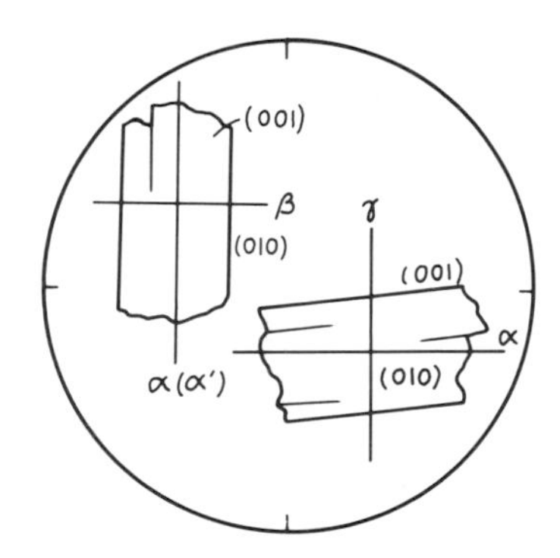

Figure 87. High sanidine fragment on {001} permits estimation of α (as $\alpha' \simeq \alpha$) and β. Extinction is parallel and length low (fast) against {010}. The figure is almost bisymmetric Bxo. A fragment on {010} permits estimation of α and γ. Extinction is inclined, with $X \wedge \{001\} = 6°$, and the grain is length low (fast). The figure is bisymmetric O.N.

Alteration: to kaolin or sericite.
Characteristic features: small to moderate $2V$ $(-)$; occurrence; simple twins.
Look-alikes: most feldspars have moderate to large $2V$s, polysynthetic twins, and may be found in nonvolcanic environments; sanidine has $b = Z$.

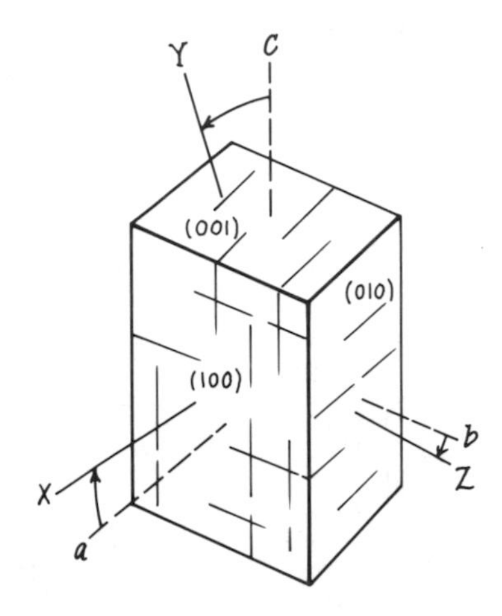

Anorthoclase

$(K_{0.37-0.10}, Na_{0.63-0.90})AlSi_3O_8$
Triclinic
$\measuredangle\alpha \simeq 92°$, $\measuredangle\beta \simeq 116°$, $\measuredangle\gamma \simeq 90°$
$\alpha = 1.524-1.526$, $\beta = 1.529-1.532$, $\gamma = 1.530-1.534$
$\gamma - \alpha = 0.006-0.007$ (1st-order gray-white)
$2V = 42°-52°$ $(-)$
$b \wedge Z \simeq 5°$, $c \wedge Y \simeq 20°$, optic plane almost $\perp$ to (010)
Pleochroism and color: colorless.
Dispersion: $r > v$ weak.
Habit: commonly rhomb-shaped crystals with pinacoids predominant.
Cleavage: {001} perfect, {010} nearly perfect, {110} poor (Fig. 88).
Extinction in section: commonly inclined, but may be almost parallel to side and basal pinacoid faces and cleavage traces.
Twinning: very fine cross-hatched (albite and pericline) twins are common; Carlsbad twins are possible.
Hardness: 6.
Specific gravity: 2.56–2.62.

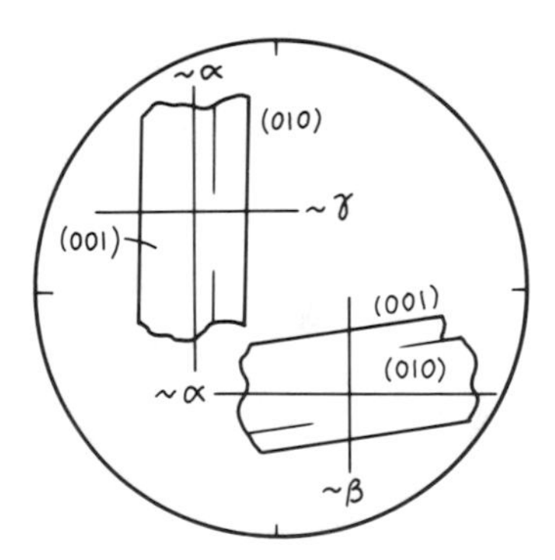

Figure 88. Anorthoclase fragment on {001} permits estimation of γ (as $\gamma' \simeq \gamma$) and α (as $\alpha \simeq \alpha'$). Extinction of $X' \wedge \{010\} = 0°-3°$. The grain is length low (fast) and the figure is nonsymmetric O.N. that appears to be bisymmetric. A fragment on {010} permits estimation of indices almost identical with α and β. Extinction of $X \wedge \{001\} = 8°-9°$. The grain is length low (fast) and the figure appears to be bisymmetric Bxo.

Occurrence: uncommon, as phenocrysts or matrix crystals in sodic volcanic rocks (color photo 32).
Alteration: to kaolin or sericite.
Characteristic features: occurrence; very fine cross-hatched twinning; moderate 2*V*.
Look-alikes: orthoclase and sanidine have only simple twins; sanidine has a small 2*V*; microcline has coarser cross-hatched twins and a larger extinction angle on the basal surface.

Orthoclase

$(K_{1-0.85},Na_{0-0.15})AlSi_3O_8$
Monoclinic
$\measuredangle\beta \simeq 116°$
$\alpha = 1.518–1.520$, $\beta = 1.522–1.524$, $\gamma = 1.522–1.525$
$\gamma - \alpha = 0.005$ (1st-order gray-white)
$2V = 35°–50°\ (-)$
$b = Z$, $c \wedge Y = 13°–21°$
Pleochroism and color: colorless.
Dispersion: $r > v$ moderate, parallel.
Habit: commonly anhedral; as phenocrysts may be blocky prismatic (with {110}, {010}, and {001}), or elongate on *a*; as microlites may be elongate on *a*; radiating fibers in spherulites.
Cleavage: {001} perfect, {010} nearly perfect, {110} poor (Fig. 89).
Extinction in section: commonly parallel to {010} faces and cleavage traces; commonly parallel or slightly inclined to {001} faces and cleavage traces.
Twinning: Carlsbad twins common, Manebach (001) and Baveno (021) twins less common, but may be untwinned; no polysynthetic twins.
Hardness: 6–6½.
Specific gravity: 2.55–2.63.
Occurrence: common in felsic plutonic rocks, but may be present in volcanic rocks; found in high-grade metamorphic rocks (often with sillimanite), but microcline is more common; in sediments, relatively common as detrital grains, and may be present as authigenic overgrowths.
Alteration: to kaolin or sericite.
Characteristic features: moderate 2*V* (−); low relief and birefringence; cleavages; simple twins.
Look-alikes: quartz is uniaxial (+), untwinned, unaltered, and has no cleavage; nepheline is uniaxial (−), has a lesser negative relief, and is commonly more altered; sanidine has a small 2*V*; albite, usually polysynthetically twinned, has a positive sign, lower relief, and slightly inclined extinction parallel to {010}; other feldspars have a larger 2*V*, polysynthetic twins, and/or

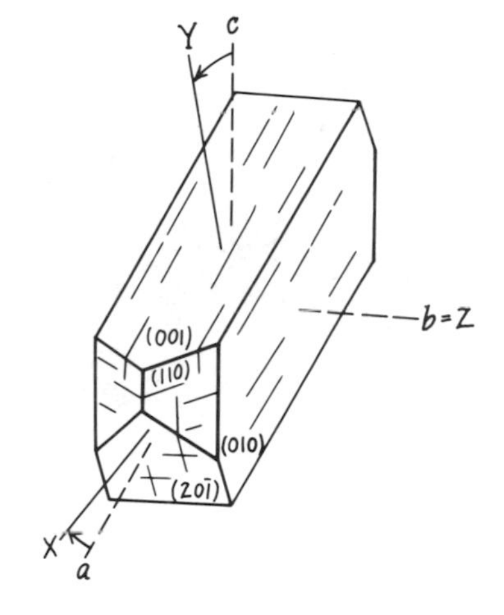

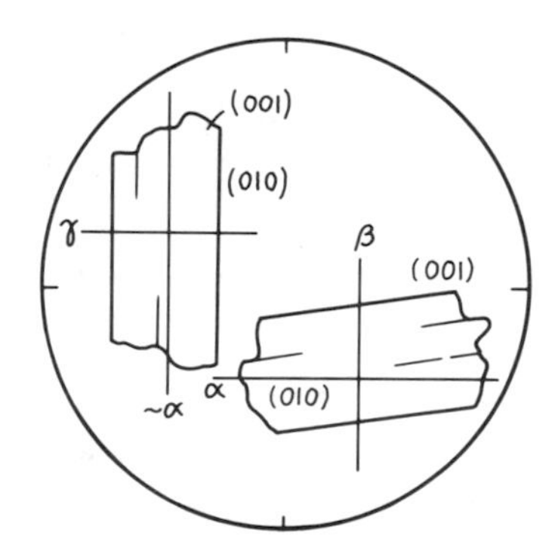

Figure 89. Orthoclase fragment on {001} permits estimation of γ and α (as $\alpha \simeq \alpha'$). Extinction is parallel to {010} and grain is length low (fast). The figure is almost bisymmetric O.N. A fragment on {010} permits estimation of α and β. Extinction of $X \wedge \{001\} = 5°–13°$, and the grain is length low (fast). The figure is bisymmetric Bxo.

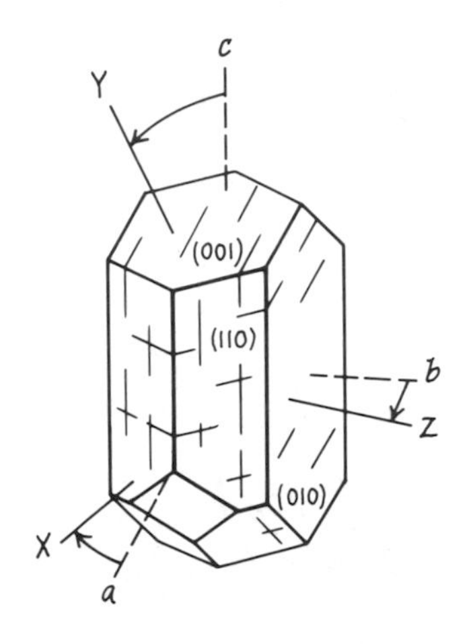

higher indices; adularia is restricted to hydrothermal veins, has a distinctive habit of dominant {110} and minor {001} and {101} faces, and may be identical to orthoclase in optical characteristics.

Microcline

$(K_{1-0.92},Na_{0-0.08})AlSi_3O_8$
Triclinic
$\measuredangle\alpha \simeq 91°$, $\measuredangle\beta \simeq 116°$, $\measuredangle\gamma \simeq 87°30'$
$\alpha = 1.514–1.516$, $\beta = 1.518–1.519$, $\gamma = 1.521–1.522$
$\gamma - \alpha = 0.007$ (1st-order white)
$2V = 66°–68°$ (−)
Optic plane near {001}, $c \wedge Y \simeq 18°$ (22°–14°), $a \wedge X \simeq -8°$ (−4° to −12°), b is near Z
Pleochroism and color: colorless.
Dispersion: $r > v$ moderate, parallel.
Habit: commonly anhedral; may be blocky prismatic, or elongate on a.
Cleavage: {001} perfect, {010} nearly perfect, {110} poor (Fig. 90).
Extinction in section: inclined to dominant faces and cleavage traces.
Twinning: lenticular cross-hatched (albite and pericline) twins very common (color photos 30 and 33); pericline twins are at ~82° from {001}; simple twins are uncommon.
Hardness: 6–6½.
Specific gravity: 2.56–2.63.
Occurrence: common in metamorphic and siliceous plutonic igneous rocks, pegmatites, and as detrital grains and authigenic overgrowths.
Alteration: to kaolin and sericite (color photo 25).
Characteristic features: lenticular cross-hatched twins; large $2V$; (−) sign; occurrence; low relief and birefringence; inclined extinction; cleavages.
Look-alikes: anorthoclase has very thin cross-hatched twinning and occurs in volcanic environments; plagioclase has higher indices, either (+) or (−) sign, and the angle between pericline twins and {001} is never greater than 40°; orthoclase does not have polysynthetic twinning, and extinction on {001} against {010} is parallel.

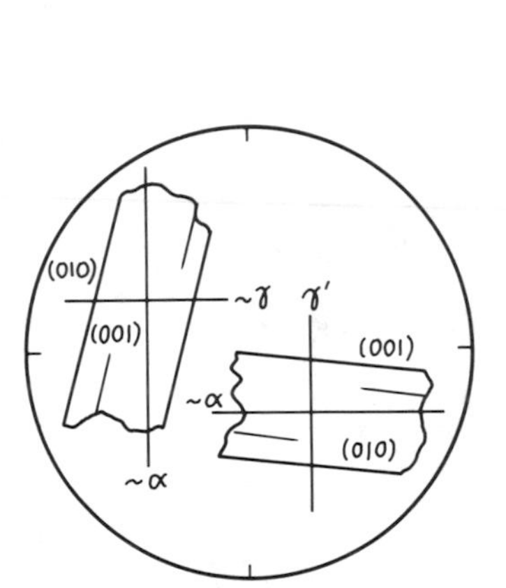

Figure 90. Microcline fragment on {001} permits estimation of α (as $\alpha \simeq \alpha'$) and γ (as $\gamma \simeq \gamma'$). Extinction of $\alpha \wedge \{010\} = 15°$. The grain is length low (fast) and the figure is almost bisymmetric O.N. A fragment on {010} permits estimation of α (as $\alpha \simeq \alpha'$) and γ' (1.519). Extinction of $\alpha \wedge \{001\} = 6°$, and the grain is length low (fast). The figure is almost monosymmetric Bxo.

Plagioclase

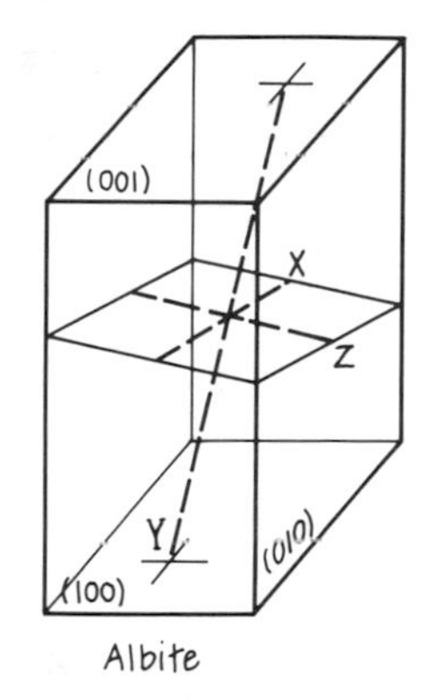

Albite

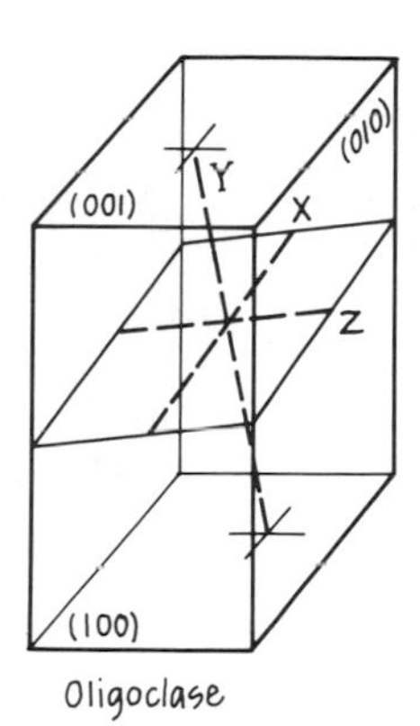

Oligoclase

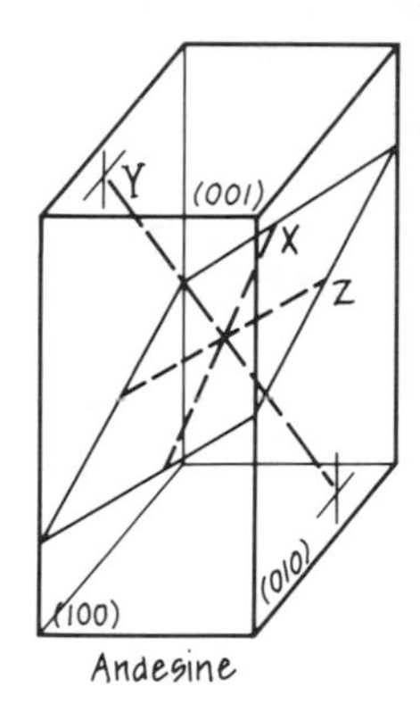

Andesine

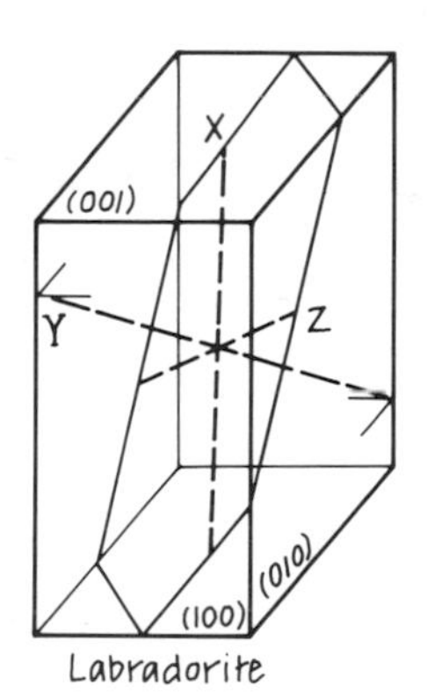

Labradorite

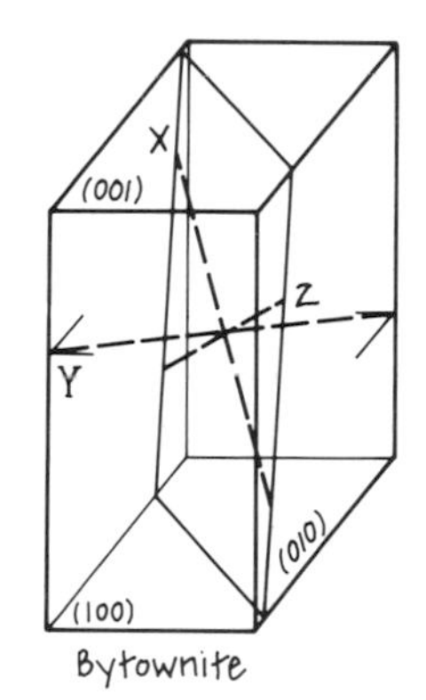

Bytownite

(NaSi,CaAl)$AlSi_2O_8$, with up to 8 percent potassium substitution
Triclinic
$\measuredangle\alpha \simeq 93°30'$, $\measuredangle\beta \simeq 116°$, $\measuredangle\gamma \simeq 87°–91°$
α, β, γ (see Table F1B and Figure 91)
$\gamma - \alpha = 0.008–0.013$ (1st-order white to yellow)
$2V$ = large (+) or (−) (see Table F1B and Figure 92)
Optic orientation: changes with composition (see diagrams above, from Emmons, 1943).
Pleochroism and color: colorless, but may be clouded by alteration.
Dispersion: see Table F1B.
Habit: euhedral to anhedral; in {010} tablets; laths or microlites elongated on *a*.
Cleavage: {001} perfect, {010} nearly perfect, {110} poor.
Extinction in section: commonly inclined to cleavage traces, twin planes, and larger crystal faces.
Twinning: polysynthetic albite twinning is very common (color photo 34); this may occur with pericline and Carlsbad twins,

Table F1B
Optical Properties of the Low-Temperature Plagioclase Series

Mineral	α	β	γ	$\gamma - \alpha$	$2V$ (Sign)	Dispersion
Albite (An_0)	1.527	1.531	1.538	0.010	77°(+)	$v > r$ weak
Oligoclase (An_{20})	1.538	1.542	1.546	0.008	87°(−)	$r > v$ weak
Andesine (An_{40})	1.549	1.553	1.557	0.008	83°(+)	$r > v$ weak
Labradorite (An_{60})	1.560	1.564	1.568	0.008	80°(+)	$r > v$ weak
Bytownite (An_{80})	1.569	1.574	1.579	0.010	85°(−)	$v > r$ weak
Anorthite (An_{100})	1.577	1.585	1.590	0.013	78°(−)	$v > r$ weak

Figure 91. The indices of refraction of the plagioclase series (J. R. Smith, 1958).

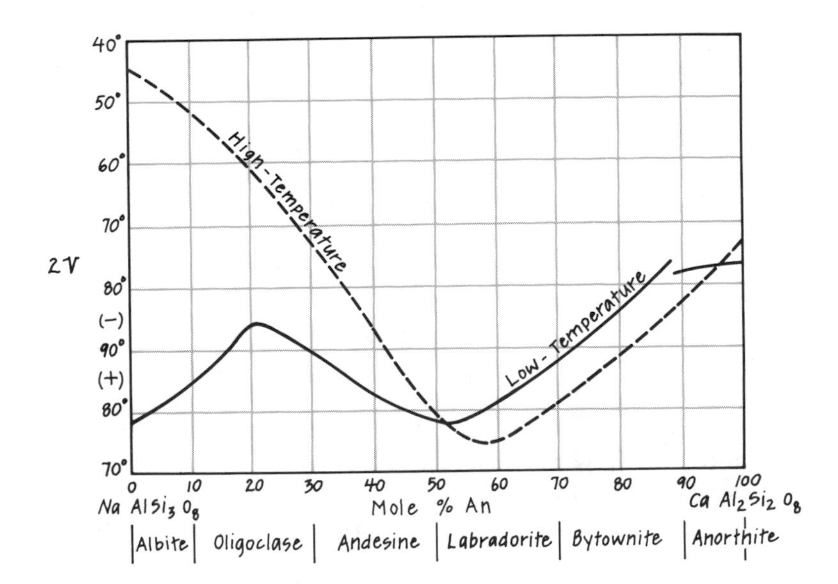

Figure 92. The range in $2V$ for the plagioclase series. The solid line applies to natural low-temperature plagioclase. The dashed line applies to high-temperature (heated) plagioclase (J. R. Smith, 1958).

and other less common types. Several methods of plagioclase identification in *thin section* utilize the angular relationships among twins, optical, and crystallographic directions.

Hardness: 6–6½.

Specific gravity: 2.63 (Ab) to 2.76 (An).

Occurrence: volcanic rocks commonly contain high-temperature plagioclase, with composition ranging from albite through labradorite; more calcic types are uncommon (color photo 35). Plutonic rocks may have any composition of plagioclase, but anorthite is uncommon. Metamorphic rocks commonly contain plagioclase of any composition, with calcium content commonly increasing as a function of grade; anorthite is common in calc-silicate rocks and marble. Detrital rocks often contain (somewhat altered) plagioclase, as well as authigenic sodic plagioclase.

Alteration: to kaolin and sericite (color photo 36); calcic plagioclase alters more readily than sodic plagioclase; metamorphism may convert plagioclase to epidote and other minerals.

Characteristic features: polysynthetic twinning; low birefringence; low relief; cleavages.

Look-alikes: cordierite is commonly characterized by abundant inclusions, pleochroic haloes, sector twins, pale color, and poor cleavage; quartz has no cleavage or twinning, and is uniaxial (+); nepheline is uniaxial (−) and has negative relief; other feldspars have lower indices.

The Michel–Lévy Method

This method relates plagioclase composition to the maximum extinction angle between the low (fast) index direction and {010}. A number of extinction angles (at least eight to ten) must be measured in order to be reasonably certain that the maximum value has been obtained. The method requires the finding of grains with distinct albite twinning; also, {010} must be normal to the section. Such grains have the following characteristics: (1) composition planes between adjacent twin laminae are sharp and narrow, (2) when the composition plane {010} is rotated into a NS or EW orientation, adjacent laminae have identical interference colors.

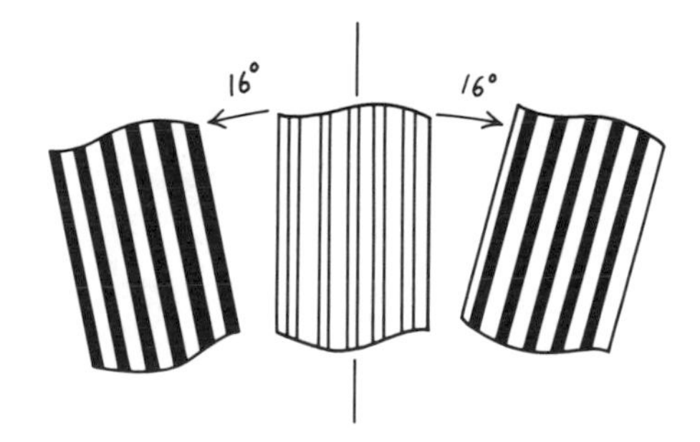

Figure 93. The Michel–Lévy method of determining the extinction angles in albite twins. The {010} composition plane is normal to the section. When the composition plane is oriented NS, the albite twins have identical interference colors. Clockwise and counterclockwise rotations bring alternate sets of twins into extinction.

The extinction angle is then determined (Fig. 93). The grain is rotated clockwise from an orientation in which {010} is NS, until one set of laminae becomes extinct. The number of degrees of rotation is recorded (16° in Fig. 93). The grain is then rotated counterclockwise from the NS orientation until the alternate set

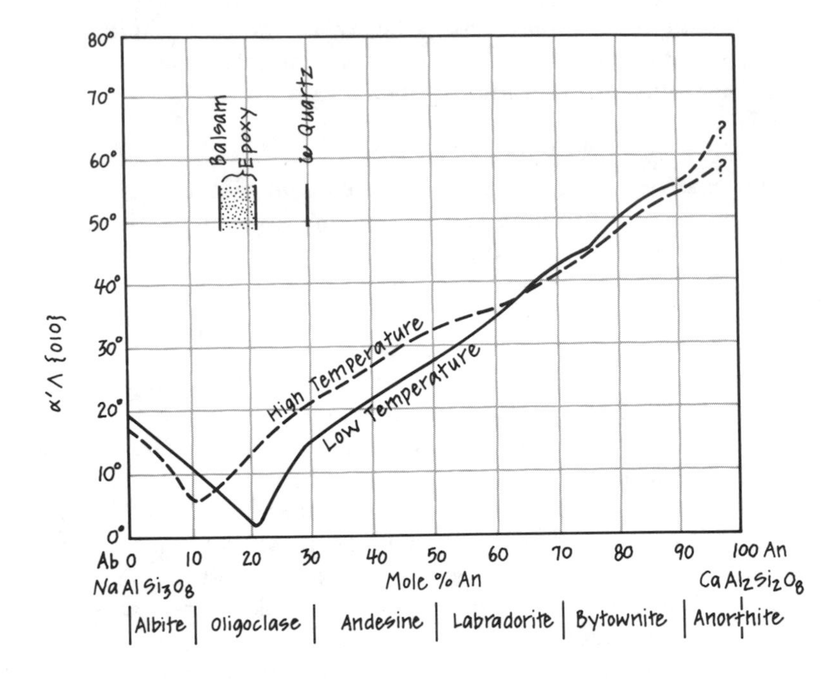

Figure 94. The values of $\alpha' \wedge \{010\}$ to be used with the Michel–Lévy method of plagioclase determination (Shelley, 1985). The vertical lines indicated as ω quartz, balsam, and epoxy are included so as to facilitate a rough comparison of plagioclase compositions with the indices of these media. For example, a plagioclase more calcic than An_{30} has higher indices than the ω index of quartz; a plagioclase more sodic than An_{30} has lower indices than the ω index of quartz. See Figure 91.

of twins becomes extinct. The rotation angle is recorded (again 16° in Fig. 93). If the two rotational angles correspond within 4° or 5°, their *averaged* value is taken as representing a proper extinction angle measurement: the {010} plane has been established as being essentially normal to the section. Grains that yield greater differences in rotational angles are rejected (as {010} is established as being at an oblique angle to the section). Only the single grain with the *maximum* extinction angle is used to determine plagioclase composition. Do *not* average the extinction angle values obtained.

The method depends upon the measurement of the maximum extinction angle between {010} and the *low* (fast) vibration direction. The character of the vibration directions is easily determined with the aid of an accessory plate. Note in the determinative diagram (Fig. 94), that in the usual case (compositions An_0–An_{75}) the low (fast) vibration direction makes an angle of less than 45° with {010}; that is, it is the extinction angle represented by the first set of twin laminae to become extinct on either side of the NS orientation. If the composition is An_{75}, both sets of laminae become extinct at 45° from the NS position. In the unusual case of a plagioclase whose composition is more calcic than An_{75}, the low (fast) vibration direction makes an angle greater than 45° with {010}, and is thus the larger of the two extinction angles that can be measured against {010}. Although plagioclase of composition An_{75}–An_{100} is unusual, it is good procedure to verify the character of the vibration direction in order to be certain of the determination.

Finally notice that for extinction angles of less than 20°, Figure 94 indicates two possible compositions. The distinction between the two possibilities is made by comparing the low index of refraction of the plagioclase against that of the mounting material or quartz.

The Albite-Carlsbad Method

This method provides a determination of plagioclase composition using a single grain. The trick here is to find the right grain. The grain must be oriented with {010} normal to the section; in addition it must be divided by a Carlsbad twin, with albite twins on both sides of the Carlsbad composition plane. The twinning in such a grain can be recognized by noting that the albite twins on one side of the grain have different extinction angles and amounts of illumination than those on the other.

Rotate the grain until the composition planes are in the diagonal (45°) position. If the orientation is correct (with {010} vertical), the individual albite twin laminae on either side of the Carlsbad composition plane are uniformly illuminated. The left side of the grain will normally have a different (but uniform) interference color than the right side; the line of demarcation marks the position of the Carlsbad composition plane. When the composition plane is oriented NS (Fig. 95), both albite and Carlsbad twins disappear or are almost invisible. Now (by clockwise and counterclockwise rotations) from the NS orientation determine the extinction angle of the albite twins on one side of the Carlsbad composition plane (in the manner of the Michel-Lévy method). If the clockwise and counterclockwise rotations do not correspond to within 4° or 5°, discard the grain and find another. If the rotational values agree, they are averaged. Then obtain a similar averaged extinction angle of the albite twins on the other side of the Carlsbad composition plane. Normally, different values for the two averaged extinction angles will be obtained.

The smaller extinction angle is found on the dashed lines of Figure 96, and the larger angle is found on the solid lines. The intersection point of the appropriate dashed and solid lines indicates the plagioclase composition (on the abscissa). If more than one possible composition is indicated, the distinction is made by estimating the index of refraction of the plagioclase, as described earlier in the Michel–Lévy method.

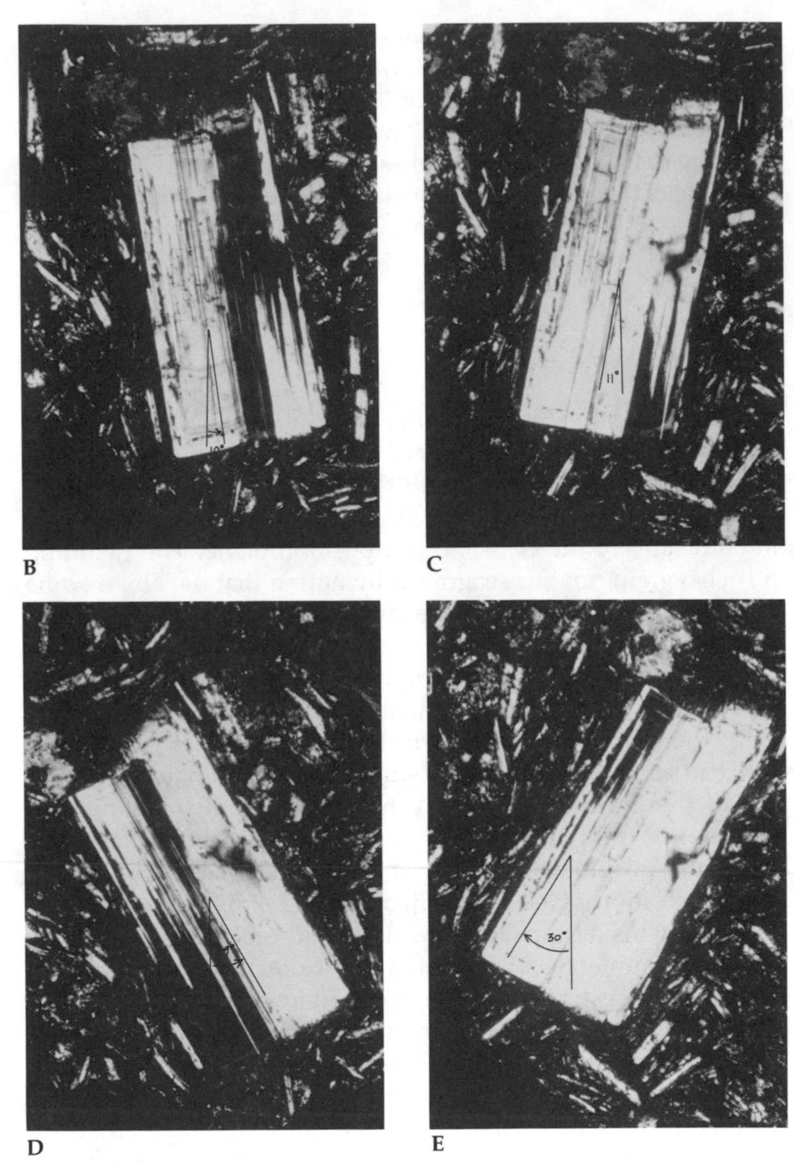

Figure 95. The albite-Carlsbad method. Part (A) shows the composition planes oriented NS. (B) and (C) show measurement of the small extinction angle on the right half of the grain. (D) and (E) show measurement of the extinction angle on the left half of the grain. The averaged values of the smaller and larger extinction angles (10.5° and 29.5°) yield a composition of An_{53} on Figure 96. See text for explanation. Crossed nicols. Photo length is 0.91 mm. (From C. Shultz, Slippery Rock University)

Determination of the Extinction Angle in a Section Cut Normal to a

In this method, plagioclase composition can be obtained from a single suitably oriented grain. Scan the section until a plagioclase grain is found that has two (almost perpendicular) cleavage directions normal to the section; this orientation indicates that *a* is normal to the section (as both {001} and {010} cleavages are parallel to *a*). Such a grain is recognized by the sharpness of the

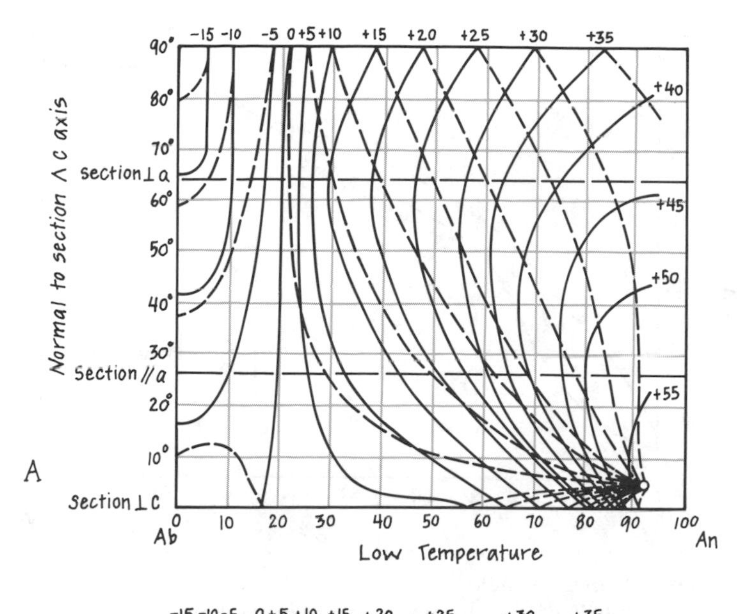

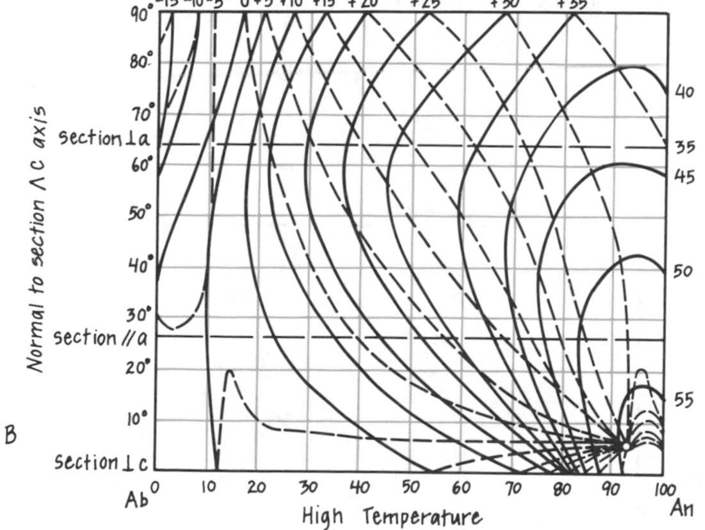

Figure 96. Extinction angles of albite twins on the two sides of a Carlsbad twin for low and high temperature plagioclase (after Tobi and Kroll, 1975). Use the solid curves for the larger angle, and the dashed curves for the smaller, to obtain the plagioclase composition (on the abscissa). The ordinate gives the angle between the twin's *c* axis and *a* normal to the section.

cleavage traces. Changing the microscope focus does not result in apparent lateral displacement of either of the cleavage traces. In addition, if albite twins are present, their interference colors are uniform when the composition plane is oriented NS. Determine the extinction angle between the low (fast) vibration direction and the trace of the {010} cleavage. Note that {010} is also the composition plane of the albite twins. If albite twins are not present, determine the angles between the low (fast) vibration direction and both cleavage traces; as $\alpha' \wedge \{010\}$ is normally less

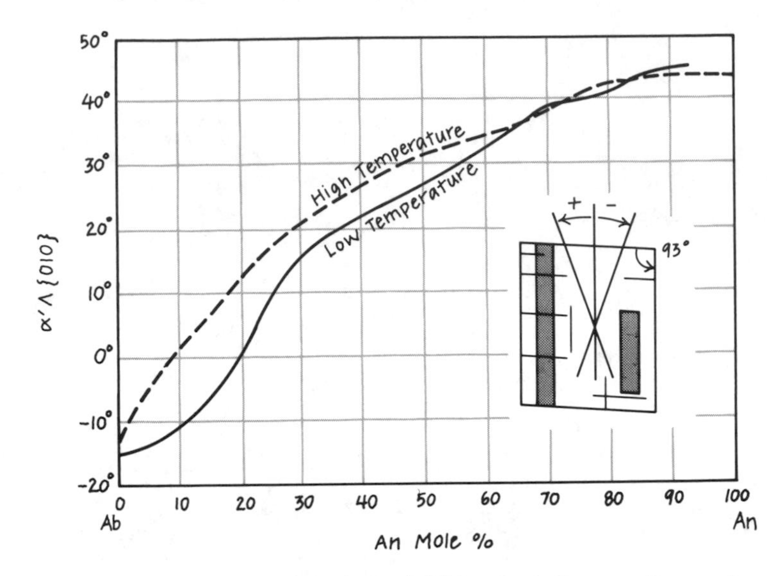

Figure 97. Extinction angles of $\alpha' \wedge \{010\}$ in plagioclase sections cut normal to *a* (Shelley, 1985, from the data of Burri, Parker, and Wenk, 1967, p. 307).

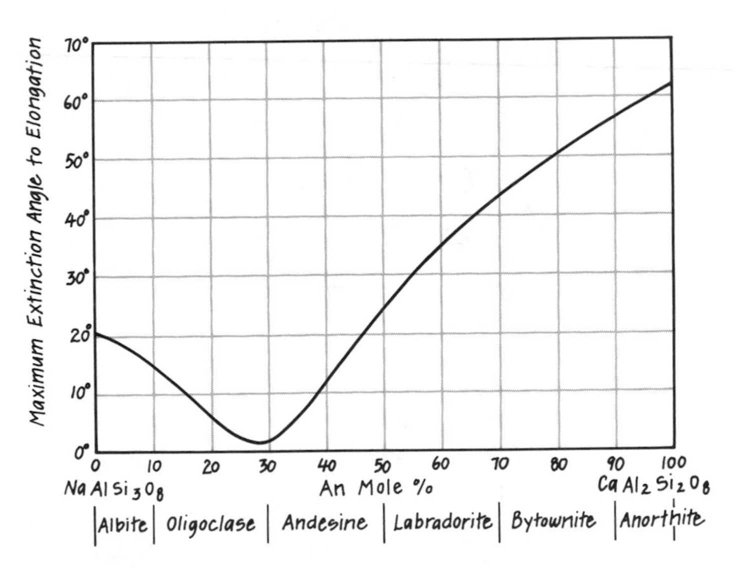

Figure 98. Plagioclase composition related to the maximum extinction angle ($\alpha \wedge$ length) of microlites (Heinrich, 1965).

than 45°, the smaller angle is the one that is used in Figure 97 to determine the plagioclase composition. If the extinction angle is less than 15°, two choices of composition are possible. Distinction can be made by estimating the index of refraction of plagioclase against the mounting material or the ω value of quartz.

The Microlite Method

Plagioclase microlites are commonly elongate on the *a* axis (see Fig. 82C). Measure the extinction angle of the low (fast) vibration against the direction of elongation for at least 10 to 12 grains. The *maximum* extinction angle is used with Figure 98 to obtain the plagioclase composition. For extinction angles less than 20°, two choices are possible, and it is necessary to compare the index of refraction of the plagioclase against the mounting material or the ω value of quartz.

Immersion Methods

Plagioclase identification of *grains in immersion liquids* uses a variety of techniques involving optical, crystallographic, and twin relationships.

As plagioclase has perfect {001} and nearly perfect {010} cleavage, most grains are found on these surfaces (Fig. 99). Grains on {001} may display albite twins parallel to the length of the grain, and grains on {010} may display pericline twins, that are usually at a distinct angle to the grain length.

The angle that the pericline twins make with {001} varies as a function of plagioclase composition (Fig. 100). When these

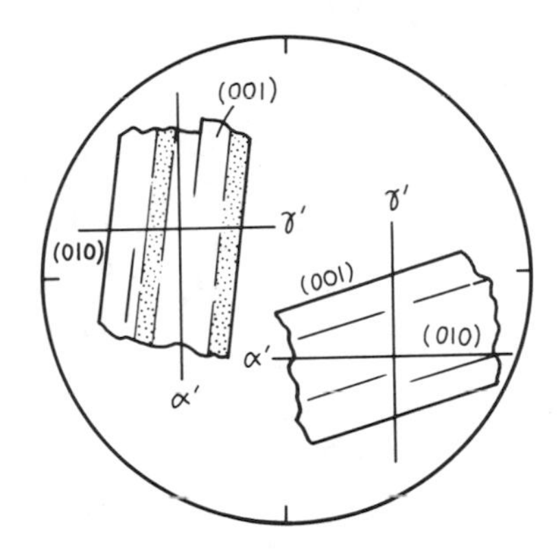

Figure 99. Plagioclase grains on {001} and {010} cleavages yield a variety of index measurements (Fig. 101), extinction angles (Fig. 100), and nonsymmetric interference figures. Grains on {001} often show albite twins parallel to {010}, whereas fragments on {010} commonly appear untwinned.

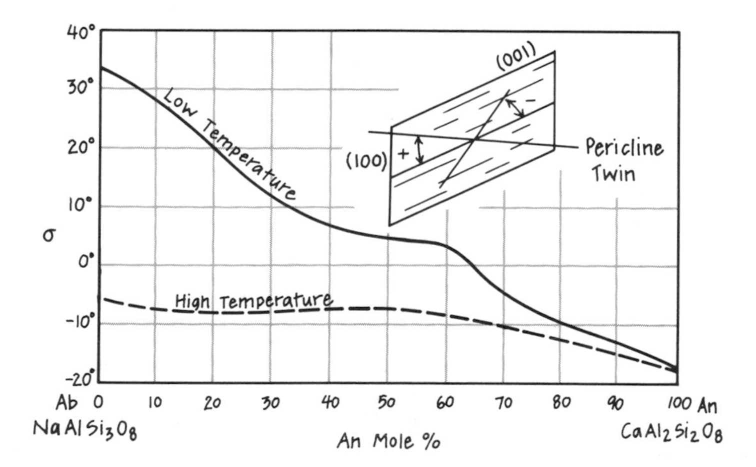

Figure 100. Range of the angle σ between the rhombic section (the composition plane of pericline twins) and {001}, as measured on {010} (Starkey, 1967).

twins are present, a single grain (with an approximate index estimate) provides a simple method of plagioclase identification.

The best method (Morse, 1968) is based upon measurement of the α' and γ' indices on {001} and {010} cleavage fragments. Notice in Figure 101 that the α' index changes in essentially the same way for both the {001} and {010} cleavage fragments. Consequently, it is only necessary to determine the smaller index (α') on any cleavage fragment to determine its composition. It is not necessary to determine the orientation of any particular grain.

Another method (developed by Schuster, 1880, Köhler, 1942, and Burri, Parker, and Wenk, 1967) is based on extinction angle measurements for grains lying on {001} and {010} cleavages (Fig. 102). With this method, the angle between the low (fast) index is measured against the cleavage edge. Note that it is normally necessary to determine whether the grain is lying on {001} or {010}. This can be accomplished by noting the orientation of twins or the type of interference figure. Alternatively, if a number of extinction angles are measured without considering grain orientation, two values should be obtained (except for An_{25-30}). As can be seen in Figure 102, the smaller angle is always obtained on cleavage fragments lying on {001}, and the larger is obtained on grains lying on {010}.

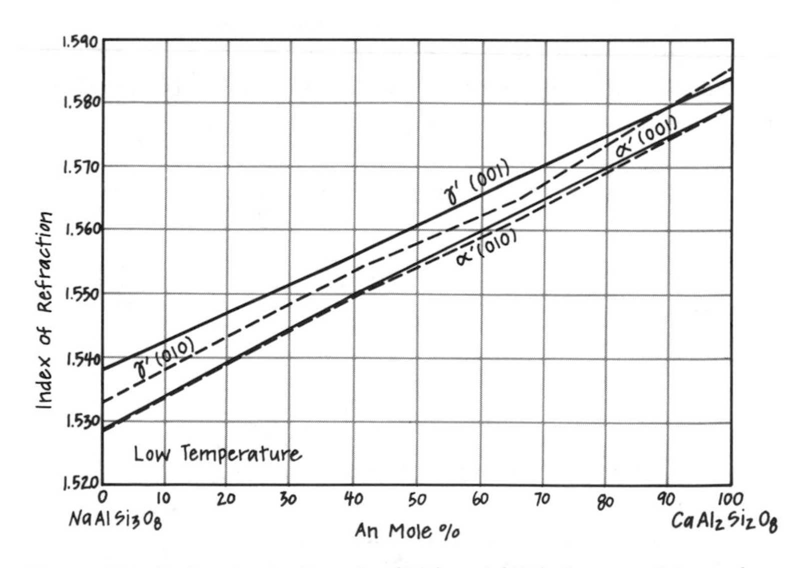

Figure 101. Refractive indices for {001} and {010} cleavage flakes of plagioclase (Morse, 1968). Note the coincidence of α' values for both {001} and {010} cleavage flakes.

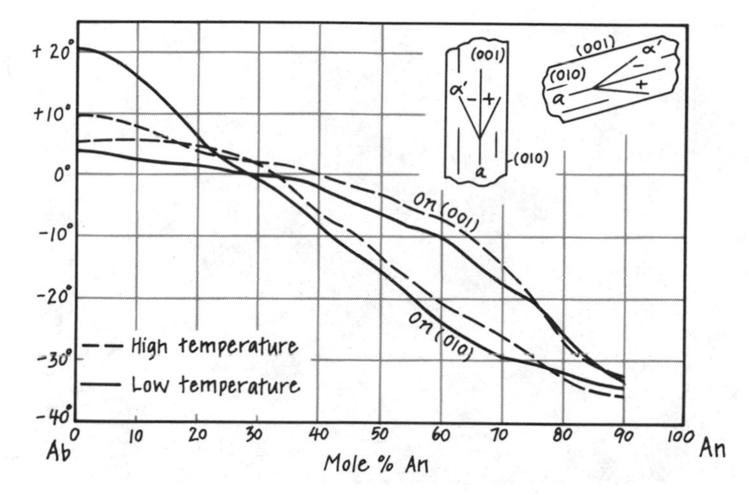

Figure 102. Extinction angles of $\alpha' \wedge a$ for plagioclase cleavage fragments on {001} and {010} (after the data of Burri, Parker, and Wenk, 1967).

FELDSPATHOID GROUP (F2)

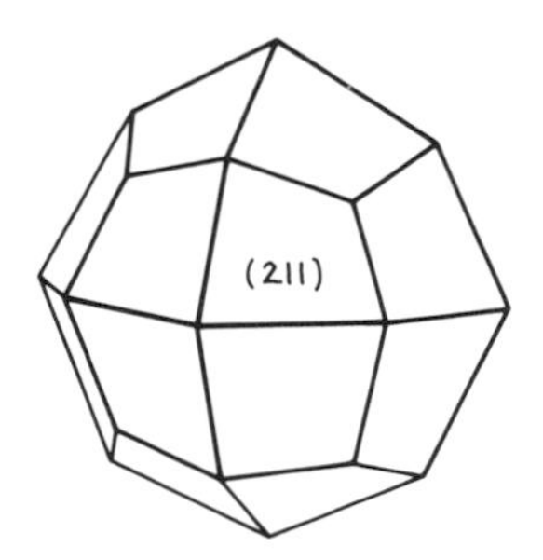

Leucite

$KAlSi_2O_6$
Tetragonal (pseudoisometric)
$\omega = 1.508–1.511$, $\varepsilon = 1.509–1.511$
$\varepsilon - \omega = 0.000–0.001$ (isotropic or 1st-order dark gray)
Uniaxial (+), almost isotropic
Pleochroism and color: colorless.
Habit: euhedral trapezohedral {112} phenocrysts; rarely skeletal microlites, dodecahedrons {110}, or cubes {100}; radially or concentrically arranged inclusions are common.
Cleavage: {110} poor (Fig. 103).
Twinning: several sets of polysynthetic twins that often intersect at 60° (color photo 37); a 1st-order red plate may be necessary for detection.

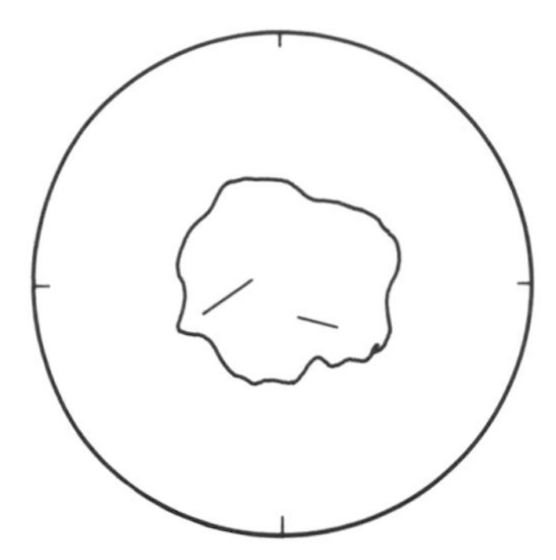

Figure 103. Pseudoisometric leucite grains are in random orientation, permit estimation of ω or ε, and rarely show traces of poor {110} cleavage. The figures are very diffuse.

Figure 104. Leucite phenocryst in fine-grained feldspar matrix. Crossed nicols. Photo length is 2.22 mm.

Hardness: 5½–6.
Specific gravity: 2.47–2.50.
Occurrence: commonly as phenocrysts in silica-poor potassic volcanic or hypabyssal rocks (Fig. 104); unknown in plutonic rocks.
Alteration: to an intergrowth of nepheline and K-feldspar (known as pseudoleucite); also to analcime or clays.
Characteristic features: occurrence; trapezohedral cross-section; very low birefringence; polysynthetic twinning; low indices.
Look-alikes: analcime lacks twinning and has lower indices; sodalite and hauyne are isotropic.

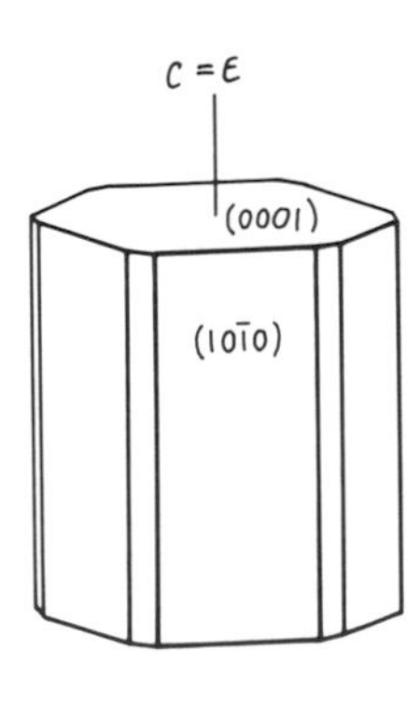

Nepheline

$NaAlSiO_4$, commonly contains up to 27 percent $KAlSiO_4$
Hexagonal
$\omega = 1.529-1.549$, $\varepsilon = 1.526-1.544$
$\omega - \varepsilon = 0.003-0.005$ (1st-order gray to white)
Uniaxial (−)
Pleochroism and color: colorless; often turbid due to inclusions or alteration.
Habit: euhedral to anhedral, with prisms and basal pinacoid dominant.

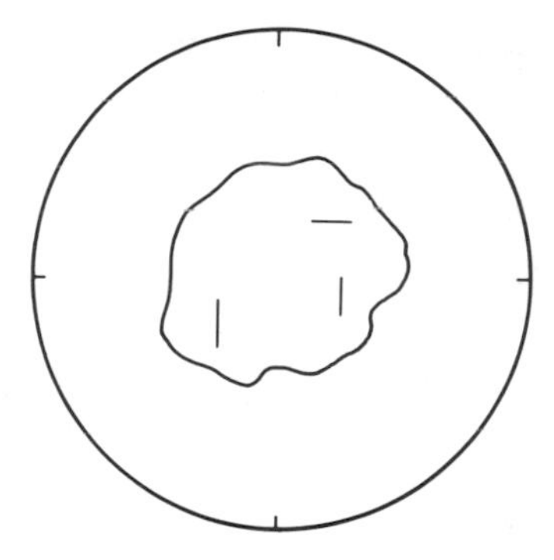

Figure 105. Nepheline grains in random orientation permit estimation of ω or ε, and rarely show traces of poor {10$\bar{1}$0} and {0001} cleavage. The figures vary in orientation from centered O.A. to O.N.

Cleavage: poor {10$\bar{1}$0} and {0001} (Fig. 105).
Extinction in section: parallel to well-developed faces; length high (slow) or low (fast).
Twinning: none.
Hardness: 5½–6.
Specific gravity: 2.56–2.67.
Occurrence: quartz-free alkaline igneous rocks, often with feldspar (Fig. 106), Na-amphiboles and/or Na-pyroxenes, or silica-poor minerals.

Figure 106. Two blocky grains of nepheline in a fine-grained feldspar matrix. The poor nepheline cleavage is visible, as well as typical fine-grained alteration products; the feldspar is essentially unaltered. Crossed nicols. Photo length is 2.22 mm.

Alteration: to a fibrous mica-rich aggregate as a result of deuteric alteration; to cancrinite, sodalite, analcime, or natrolite.

Characteristic features: only in rocks that are quartz-free; low birefringence and relief; uniaxial (−); untwinned; commonly altered. See color photo 38.

Look-alikes: feldspars have good cleavage and are commonly twinned; quartz is unaltered and uniaxial (+); leucite is pseudoisotropic and has polysynthetic twinning; melilite has higher relief; scapolite has higher birefringence.

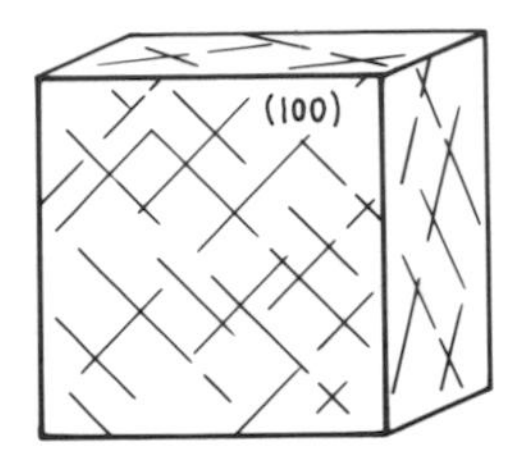

FLUORITE (FLUORSPAR) (F3)

CaF_2

Isometric

$n = 1.433\text{–}1.435$

Pleochroism and color: usually colorless, but may be green to purple; color zoning may be seen in thin section.

Habit: commonly euhedral cubes, but may be octahedrons, dodecahedrons, fine granular, or earthy.

Cleavage: {111} perfect (Fig. 107).

Twinning: {110} penetration common.

Hardness: 4.

Specific gravity: 3.18.

Occurrence: very common in veins and carbonate replacement deposits; also in hot springs, pegmatites, carbonatites, and granites (as a late accessory mineral); in some carbonate sediments with celestite, gypsum, anhydrite, and sulfur (see Figure 108); fairly common as a detrital mineral.

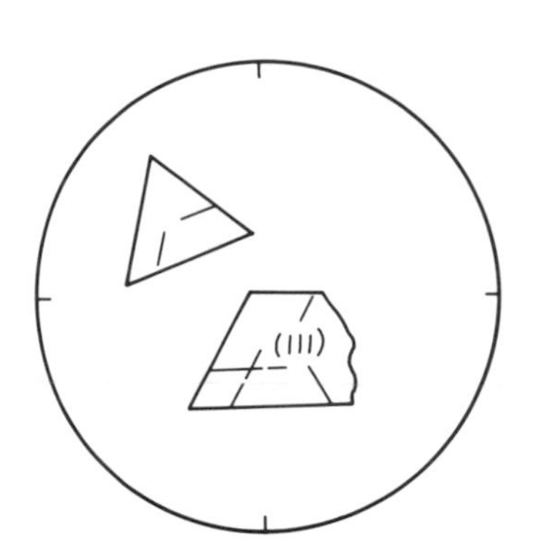

Figure 107. Isotropic fluorite grains lie on the perfect {111} cleavage, permit an estimate of the single index of refraction, and often have a triangular outline.

Figure 108. Anhedral fluorite (lower left) surrounded by larger grains of calcite. The fluorite varies in color from deep purple (black) to pale purple (light gray). Plane-polarized light.

Alteration: relatively soluble in carbonate-rich water; many minerals form pseudomorphs after fluorite.
Characteristic features: isotropic; octahedral cleavage; very low index; occurrence.
Look-alikes: halite has {100} cleavage and a higher index.

Forsterite. See Olivine Group (O1).

THE GARNET GROUP (G1)

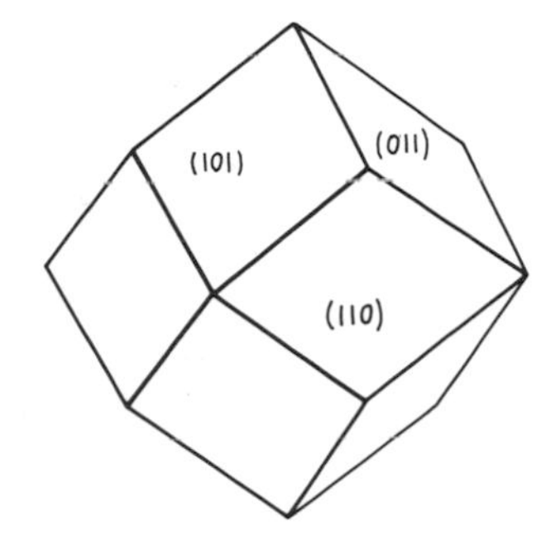

Garnets are isometric and have the general composition $R_3^{+2}R_2^{+3}(SiO_4)_3$, in which R^{+2} is (Mg,Fe,Mn,Ca) and R^{+3} is (Al,Fe,Cr). Ca-rich members compose the *ugrandite* garnets (*u*varovite, $Ca_3Cr_2(SiO_4)_3$; *gr*ossular (grossularite), $Ca_3Al_2(SiO_4)_3$; *and*radite, $Ca_3Fe_2(SiO_4)_3$). Al-rich members (with the exception of grossular) compose the *pyralspite* garnets (*py*rope, $Mg_3Al_2(SiO_4)_3$; *al*mandine, $Fe_3Al_2(SiO_4)_3$; *sp*essartine, $Mn_3Al_2(SiO_4)_3$). Extensive solid solution occurs within each group, but is limited between the two groups. Many naturally occurring garnets contain the three end members, and are named on the basis of the major component.

Most garnets cannot be identified solely on the basis of index of refraction, and often require cell dimension and specific gravity as well (see Winchell, 1958); analysis by electron microprobe has proved most effective. The occurrence and color is sometimes useful in distinguishing garnet species. See color photos 39 and 40.

Pyralspite Garnets

$(Mg,Fe^{+2},Mn)_3Al_2(SiO_4)_3$
Isometric
$n = 1.714$ (pyrope); $n = 1.830$ (almandine); $n = 1.800$ (spessartine)
$\gamma - \alpha = 0.000$ (pyr, alm); 0.000–0.004 (sp)
Pleochroism and color: colorless, pale red or brown.
Habit: euhedral dodecahedral {110} or trapezohedral {112} forms, or combination of both.
Cleavage: none, but almandine may have parting on {110}. See Figure 109.
Extinction in section: pyrope and almandine are isotropic; rarely, spessartine may be weakly birefringent.
Twinning: complex twin patterns may be present in birefringent varieties.
Hardness: 7½ (py); 7–7½ (alm, sp).

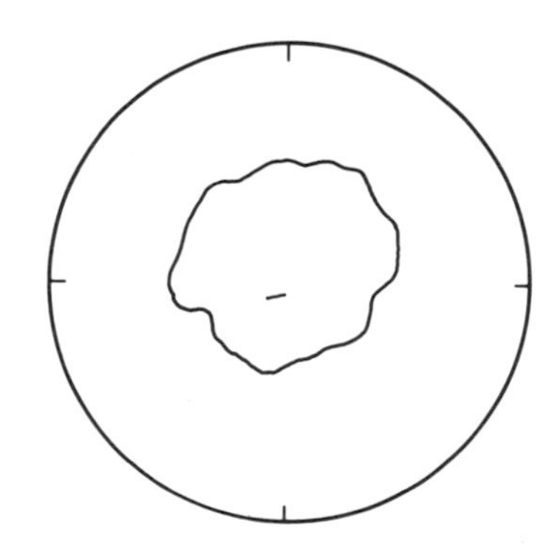

Figure 109. Isotropic garnet grains in random orientation permit estimation of a single index, the value of which is dependent upon composition. Interference figures are not present, except for some birefringent varieties; these may be uniaxial or biaxial.

Figure 110. (A) Garnet porphyroblasts (black) with biotite (gray) in a fine-grained quartz-mica matrix. Crossed nicols. (B) The same in plane-polarized light. The garnet has high relief, and the biotite is deeply colored. Photo length is 3.30 mm. (From J. Hinsch, E. Leitz, Inc.)

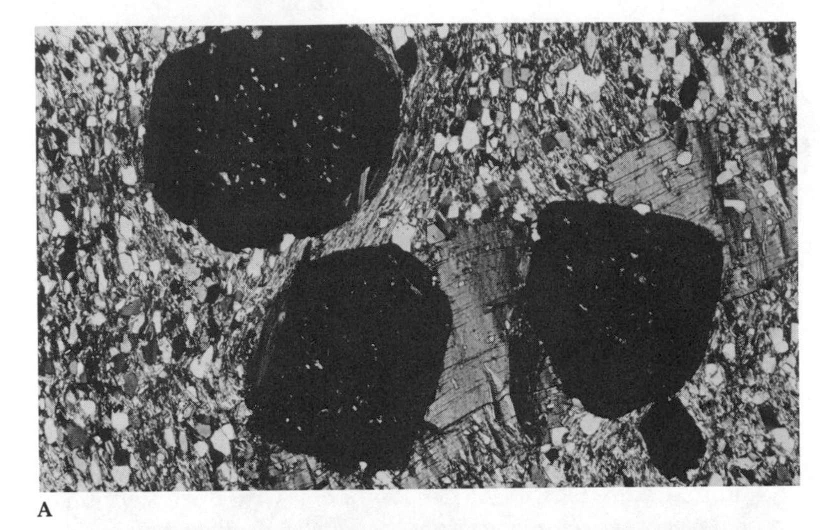

A

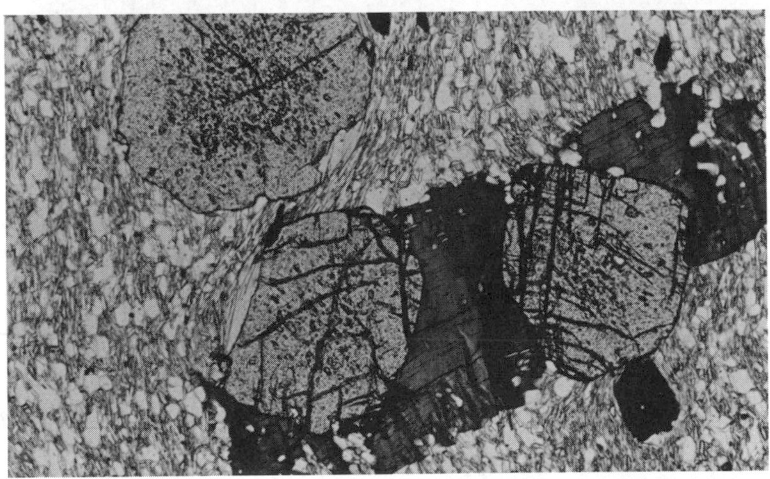

B

Specific gravity: 3.6 (py); 4.3 (alm); 4.2 (sp).

Occurrence: pyrope is found in ultramafic rocks such as peridotite and kimberlite; garnets of pyrope-almandine composition occur in eclogite and amphibolite; almandine is the typical garnet in schist and gneiss of medium to high grade, as well as in contact metamorphic zones (with cordierite and spinel); almandine is uncommon in siliceous plutonic rocks and is common as detrital grains in sediments; spessartine is rare in pegmatites and granite. See Figure 110.

Alteration: commonly to chlorite; hornblende, epidote; Fe- and Mn-oxides may be present in rims.

Characteristic features: euhedral crystals; isotropic or nearly so; high index; occurrence.

Look-alikes: ugrandite garnets occur in limestone contact zones; spinel has an octahedral habit, may be dark green or brown, with {111} parting.

Ugrandite Garnets

$Ca_3(Cr,Al,Fe^{+3})_2(SiO_4)_3$
Isometric
$n = 1.86$ (uvarovite); $n = 1.734$ (grossular); $n = 1.887$ (andradite)
$\gamma - \alpha = 0.000–0.005$ (uv, gr, and)
Pleochroism and color: pale emerald green (uv); colorless (gr); yellow, pale brown (and); Ti-rich varieties are dark brown.
Habit: euhedral dodecahedral {110} or trapezohedral {112} forms, or combination of both; massive in contact zones.
Cleavage: none, but may have {110} parting. See Figure 109.
Extinction in section: isotropic, but often is weakly birefringent. See Figure 110.
Twinning: birefringent varieties commonly show complex wedge-shaped twins.
Hardness: 7½ (uv); 6½–7 (gr, and).
Specific gravity: 3.9 (uv); 3.6 (gr); 3.9 (and).
Occurrence: uvarovite is rare, and is found with chromite in peridotite and serpentinite; grossular and andradite are common in contact metamorphic limestone (with calc-silicate minerals), skarns, and in some schists. See Figure 110.
Alteration: commonly to chlorite; also to calcite, epidote, feldspar, limonite, or nontronite.
Characteristic features: euhedral crystals; isotropic or nearly so; high index; occurrence.
Look-alikes: pyralspite garnets are rare in limestone contact zones and rarely are birefringent; spinel has an octahedral habit, may be dark green or brown, and has {111} parting.

Gedrite. See Amphibole Group (A2).

Gehlenite. See Melilite Group (M1).

Gibbsite. See Clays and Clay Minerals (C6).

GLASS (Volcanic) (G2)

Rhyolitic glass through basaltic glass
Amorphous
$n = 1.48–1.64$ (decreases with hydration and Si content, Fig. 111)
$\gamma - \alpha = 0.000$ (may show weak strain birefringence)

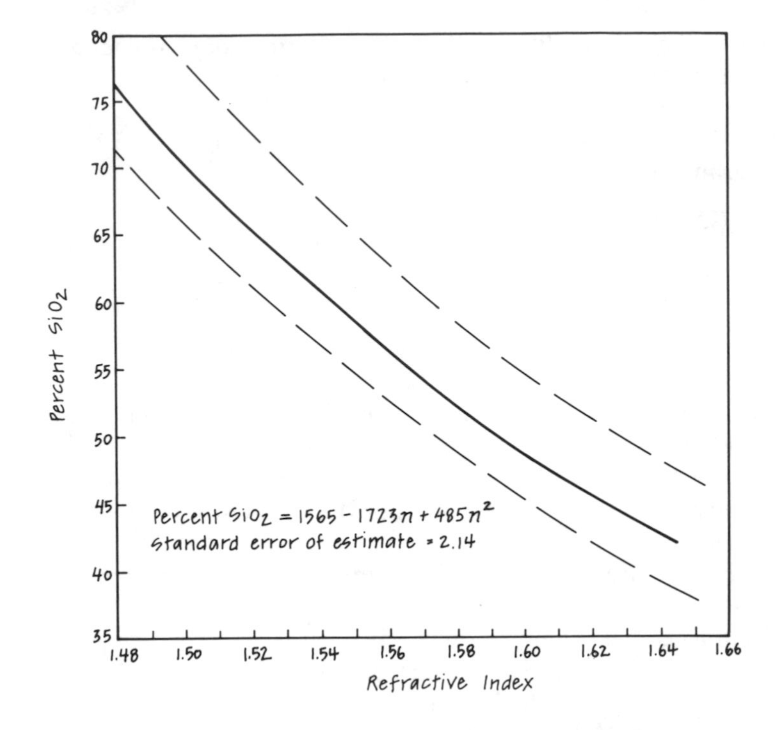

Figure 111. The solid curve represents the average silica refractive index for 178 analyses of volcanic glasses compiled by Huber and Rinehart (1966); the dashed lines (added) show the range of values that were compiled. Differences in values between volcanic suites of different petrologic affinities warrant the construction of individual curves for maximum usefulness of the method.

Pleochroism and color: usually colorless (siliceous glasses), but may be gray, brown, green, or completely opaque (altered or devitrified mafic glasses).

Habit: ranges from massive flows to irregular patches in the matrix of volcanic rocks; often contains crystallites, microlites, spherulites, phenocrysts, flow banding, and vesicles.

Cleavage: none; irregular fractures and spherical hydration cracks (perlite).

Hardness: 5½ (mafic glasses) to 7 (siliceous glass).

Specific gravity: obsidian, ~2.32; perlite, ~2.35; pumice, <1; pitchstone, ~2.34; alkalic glass, ~2.56; tachylite, ~2.76.

Occurrence: most common in siliceous volcanic flows or tuff (often as shards—see color photo 41); less common in mafic volcanic rocks; in meteorites, tektites, meteoritic impact products, and in lunar materials; rare as fused country rock adjacent to mafic igneous intrusions; as all glass is metastable, devitrification has resulted in the absence of terrestrial glass in rocks older than the late Cretaceous. See Figure 112.

Alteration: siliceous glass devitrifies to a uniform microcrystalline aggregate of quartz and feldspar; basaltic glass devitrifies to *palagonite* (a yellow to orange mineraloid whose index is less than balsam); tuff may become silicified or hydrated to form montmorillonite.

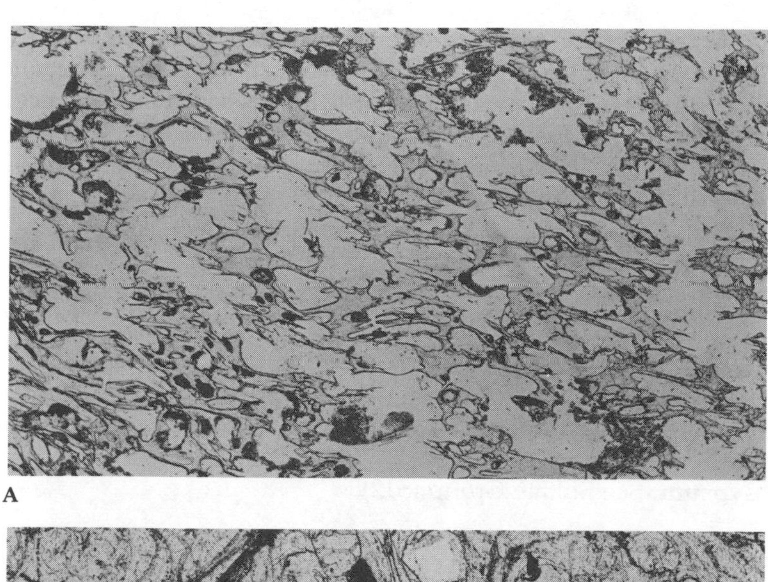
A

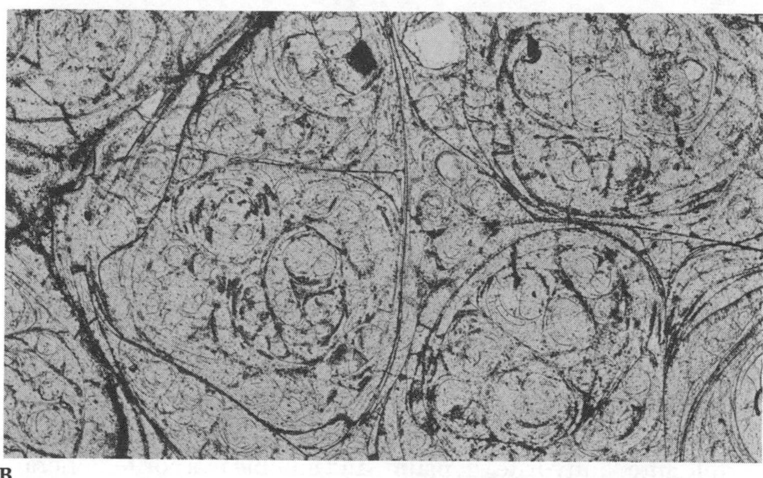
B

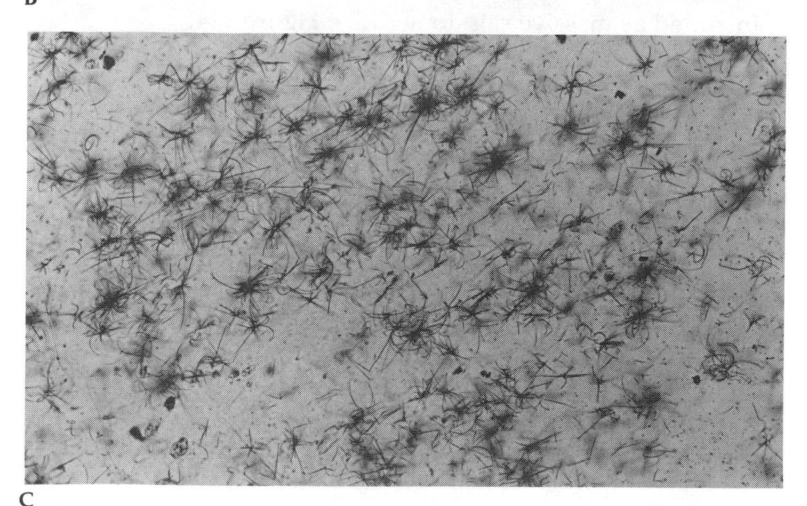
C

Figure 112. (A) Elongated vesicles in pumice. Photo length is 5.16 mm. (B) Concentric perlitic cracks in volcanic glass. Photo length is 5.16 mm. (C) Branched crystallites in volcanic glass. Photo length is 0.82 mm. Plane-polarized light. (From J. Hinsch, E. Leitz, Inc.)

Characteristic features: occurrence and isotropic character.
Look-alikes: opal has a lower index of refraction; analcime occurs in silica-deficient rocks, and may have weak birefringence, polysynthetic twins, and a trapezohedral habit.

Glauconite. See Mica Group (M2).

Glaucophane. See Amphibole Group (A2).

Graphite. See Opaque Minerals (O2).

Grossular. See Garnet Group (G1).

Grunerite. See Amphibole Group (A2).

Gypsum. See Sulfate Group (S12).

HALITE GROUP (H1)

Halite

NaCl
Isometric
$n = 1.544$
Isotropic
Pleochroism and color: colorless.
Habit: cubic, rarely octahedral, coarse to fine granular, stalactitic.
Cleavage: {100} perfect (Fig. 113).
Hardness: 2–2½.
Specific gravity: 2.165.
Occurrence: an evaporite from seawater or saline lakes, with dolomite, anhydrite, gypsum, and soluble evaporite minerals; intruded as massive salt domes. See Figure 114.

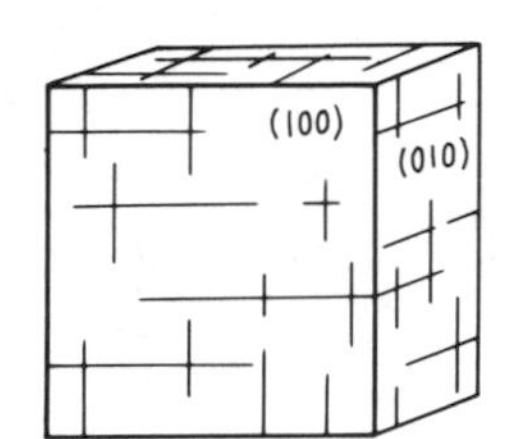

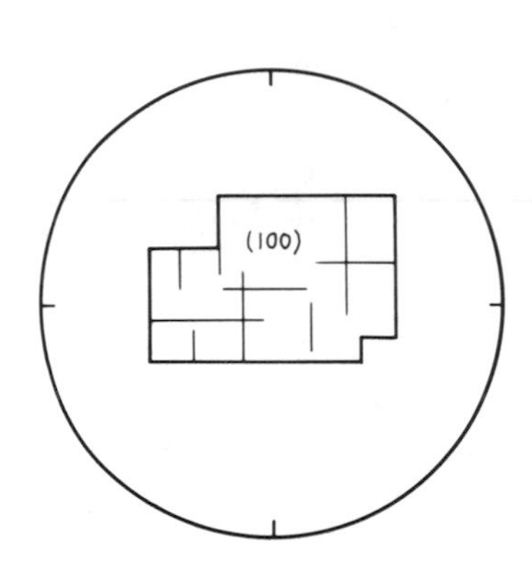

Figure 113. All halite grains lie on {100} cleavage, yield a single index of refraction, and have a square to rectangular outline.

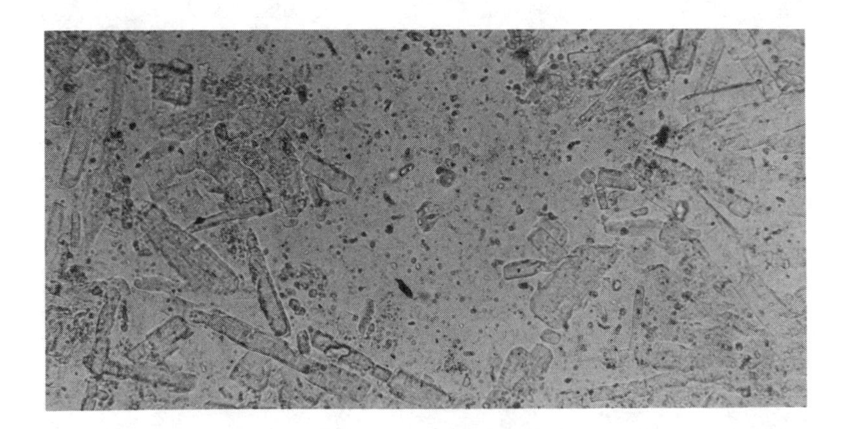

Figure 114. Anhydrite (rectangular) grains in a coarse halite matrix. Halite ($n = 1.544$) has minimum relief and is essentially invisible. Cubic cleavage is equally nonvisible. Plane-polarized light. Photo length is 0.55 mm.

Alteration: very water soluble; other soluble minerals may form pseudomorphs after halite.
Characteristic features: isotropic; perfect cubic cleavage; low relief in balsam; saline taste.
Look-alikes: fluorite has octahedral cleavage and high negative relief; sylvite has moderate negative relief; cryolite (Na_3AlF_6) has weak birefringence.

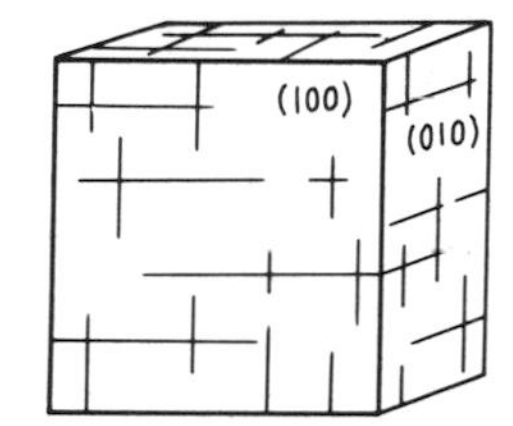

Sylvite

KCl
Isometric
$n = 1.490$
Isotropic
Pleochroism and color: colorless.
Habit: cubic, rarely octahedral, coarse to fine granular.
Cleavage: {100} perfect (Fig. 115).
Hardness: 2.
Specific gravity: 1.99.
Occurrence: uncommon evaporite from seawater or saline lakes, with dolomite, gypsum, clays, anhydrite, and soluble evaporite minerals.
Alteration: very water soluble.
Characteristic features: isotropic; perfect cubic cleavage; moderate negative relief in balsam; bitter saline taste.
Look-alikes: halite has higher indices; fluorite has octahedral cleavage.

Hastingsite. See Amphibole Group (A2).

Hauyne. See Sodalite Group (S5).

Hedenbergite. See Pyroxene Group (P8).

Hematite. See Opaque Minerals (O2).

Hercynite. See Spinel Group (S8).

Heulandite. See Zeolite Group (Z1).

High Sanidine. See Feldspar Group (F1).

Hornblende. See Amphibole Group (A2).

Hortonolite. See Olivine Group (O1).

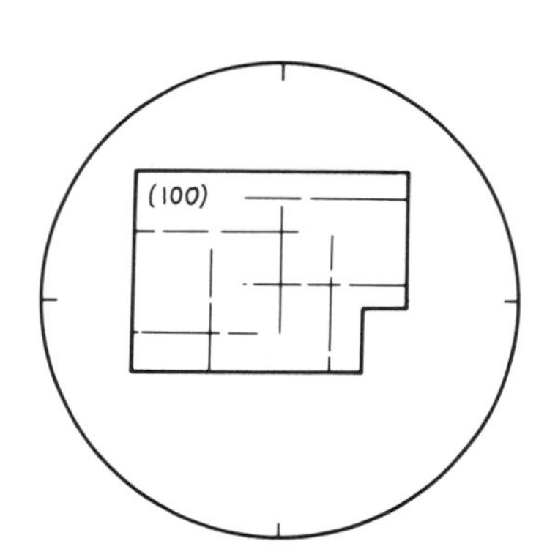

Figure 115. All sylvite grains lie on {100} cleavage, yield a single index of refraction, and have a square to rectangular outline.

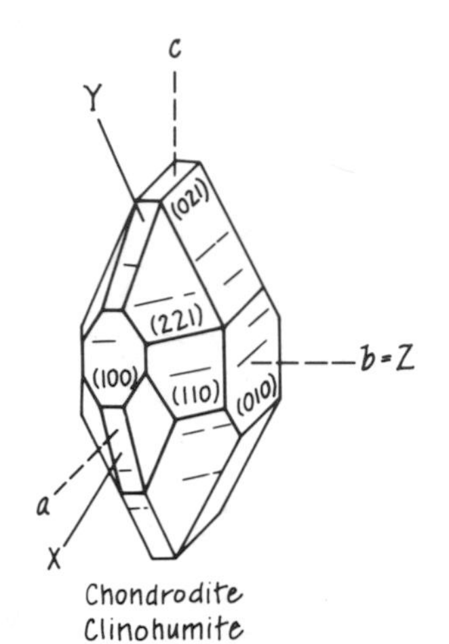

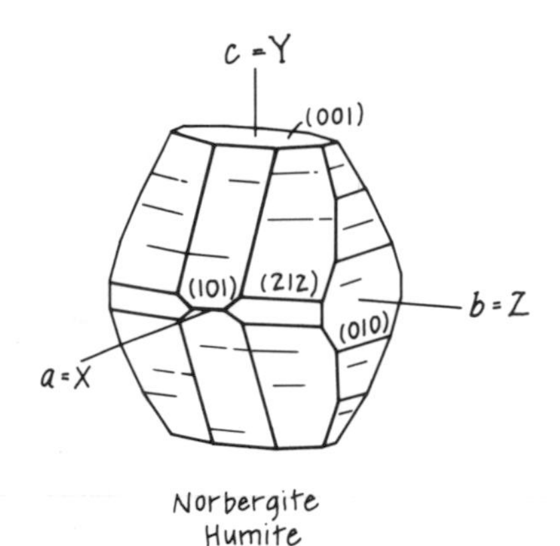

HUMITE GROUP (H2)

Norbergite, Chondrodite, Humite, Clinohumite

Optical, crystallographic, and compositional data for the four members of the humite group are summarized in Table H2.

Pleochroism and color: colorless, yellow, or pale brown; moderate pleochroism with abs. $X > Y = Z$, X = pale to deep yellow, brownish yellow, Y = colorless, yellow, yellow-green, Z = colorless to pale yellow, brown, or green.

Dispersion: $r > v$ weak to strong.

Habit: commonly anhedral (rounded), or plates with {010}, {001}, or {100} dominant.

Cleavage: {001} poor. See Figure 116.

Extinction in section: in orthorhombic varieties, commonly parallel to pinacoidal forms; in monoclinic varieties commonly parallel to pinacoidal forms when viewed normal to *b*, and inclined when viewed parallel to *b*.

Twinning: simple or polysynthetic common on {001} in chondrodite and clinohumite.

Hardness: 6–6½.

Specific gravity: 3.1–3.4.

Occurrence: chondrodite (Fig. 117 and color photos 42 and 66) and clinohumite are most common, humite less common, and norbergite rare; in contact dolomitic marble, with other humite minerals, forsterite, wollastonite, grossular, calcite, fluorite, spinel, and phlogopite; less common in altered peridotite.

Table H2
The Humite Group

Mineral	**Norbergite**	**Chondrodite**	**Humite**	**Clinohumite**
Composition	$Mg_2SiO_4 \cdot Mg(OH,F)_2$	$2Mg_2SiO_4 \cdot Mg(OH,F)_2$	$3Mg_2SiO_4 \cdot Mg(OH,F)_2$	$4Mg_2SiO_4 \cdot Mg(OH,F)_2$
Crystal system	Orthorhombic	Monoclinic	Orthorhombic	Monoclinic
Optical orientation	$a = X, b = Z, c = Y$	$\measuredangle\beta = 109°, b = Z$, $c \wedge Y = 22°–31°$	$a = X, b = Z, c = Y$	$\measuredangle\beta = 101°, b = Z$, $c \wedge Y = 7°–15°$
α	1.561–1.567	1.592–1.643	1.607–1.643	1.623–1.702
β	1.566–1.579	1.602–1.655	1.619–1.653	1.636–1.709
γ	1.587–1.593	1.619–1.675	1.639–1.675	1.651–1.728
$\gamma - \alpha$	0.026–0.027	0.025–0.037	0.028–0.036	0.028–0.045
$2V$ (sign)	44°–50° (+)	64°–90° (+)	65°–84° (+)	52°–90° (+)

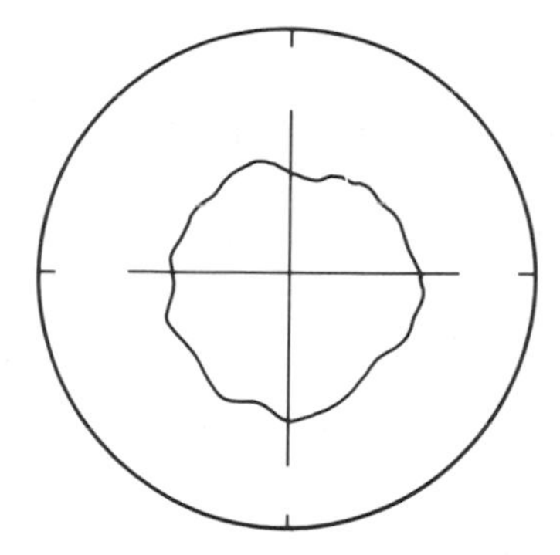

Figure 116. Most humite-group grains lie in random orientation. If, however, the grains lie on the poor {001} cleavage, norbergite and humite permit estimation of α and γ, and yield a bisymmetric O.N. figure. Chondrodite and clinohumite on {001} permit estimation of γ and α' (close to α in value) and yield a figure which appears to be bisymmetric O.N.

Alteration: to serpentine or chlorite.

Characteristic features: occurrence; moderate pleochroism; large $2V$ (+); high relief.

Look-alikes: olivine has higher birefringence and indices, and is often (−); staurolite tends to be euhedral, has lower birefringence and maximum absorption parallel to *Z*; tourmaline is uniaxial (−); biotite has maximum absorption parallel to *Z*.

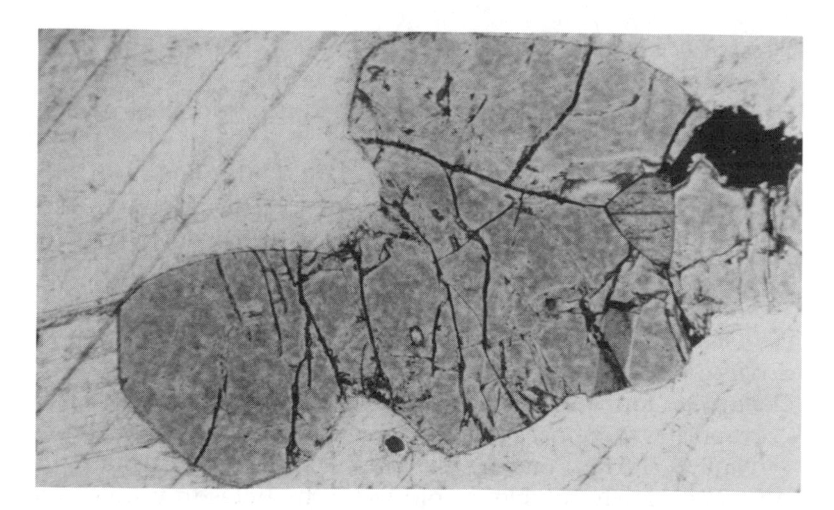

Figure 117. Anhedral chondrodite in marble. Crossed nicols. Photo length is 2.22 mm.

Hyalosiderite. See Olivine Group (O1).

Hypersthene. See Pyroxene Group (P8).

Iddingsite. See Olivine Group (O1).

Idocrase. See Vesuvianite (V1).

Illite. See Clays and Clay Minerals (C6).

Ilmenite. See Opaque Minerals (O2).

Jadeite. See Pyroxene Group (P8).

Kaolinite. See Clays and Clay Minerals (C6).

Katophorite. See Amphibole Group (A2).

KERNITE (K1)

$Na_2B_4O_7 \cdot 4H_2O$
Monoclinic
$\measuredangle\beta \approxeq 109°$
$\alpha = 1.454$, $\beta = 1.472$, $\gamma = 1.488$
$\gamma - \alpha = 0.034$ (2nd-order orange)
$2V = 80° (-)$
$b = Z$, $c \wedge X = 38.5°$
Pleochroism and color: colorless.
Dispersion: $r > v$ moderate.
Habit: anhedral to euhedral; nearly equant or elongate parallel to *c*.
Cleavage: {100} and {001} perfect (Fig. 118), {$\bar{2}01$} good.
Extinction in section: parallel or inclined to cleavage traces as a function of orientation; length high (slow) or low (fast) for grains elongated on *c*.
Twinning: {110}.
Hardness: $\simeq 2\frac{1}{2}$.
Specific gravity: $\simeq 1.93$.
Occurrence: formed from melting and dehydration of playa lake evaporites; in veins, aggregates, and crystals in clay with borax ($Na_2B_4O_7 \cdot 10H_2O$), ulexite ($NaCaB_5O_9 \cdot 8H_2O$), and calcite.
Alteration: slowly soluble in cold water; hydrates to tincalconite ($Na_2B_4O_7 \cdot 5H_2O$) (uniaxial (+), $\omega = 1.461$, $\varepsilon = 1.477$).
Characteristic features: occurrence; two perfect cleavages; solubility; negative relief in section.

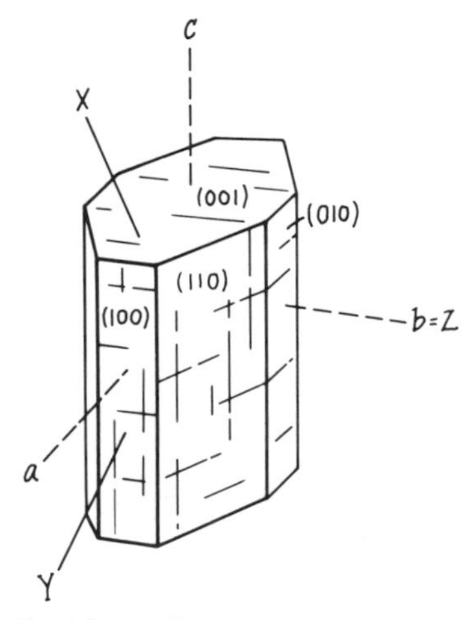

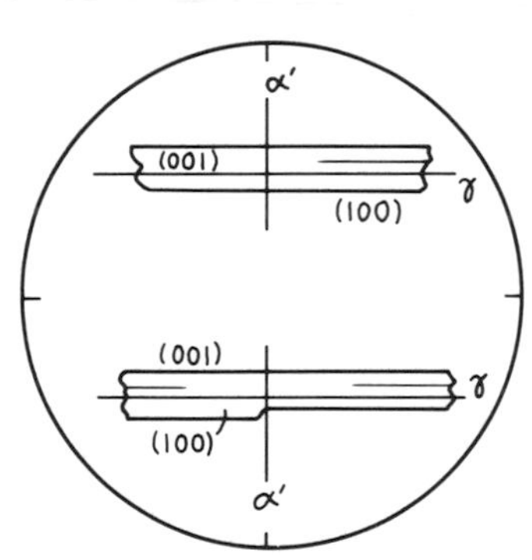

Figure 118. Kernite fragments lying on {001} permit estimation of γ and α' (1.470), have parallel extinction, are length high (slow), and yield a monosymmetric Bxa figure. Fragments lying on {100} permit estimation of γ and α' (1.461), have parallel extinction, are length high (slow), and yield a monosymmetric O.N. figure.

Look-alikes: borax has a moderate $2V$, lower birefringence, and three perfect cleavages; ulexite and colemanite have one perfect cleavage and are biaxial (+).

Kutnahorite. See Carbonate Group (C2).

Kyanite. See Aluminosilicate Group (A1).

Labradorite. See Feldspar Group (F1).

Laumontite. See Zeolite Group (Z1).

LAWSONITE (L1)

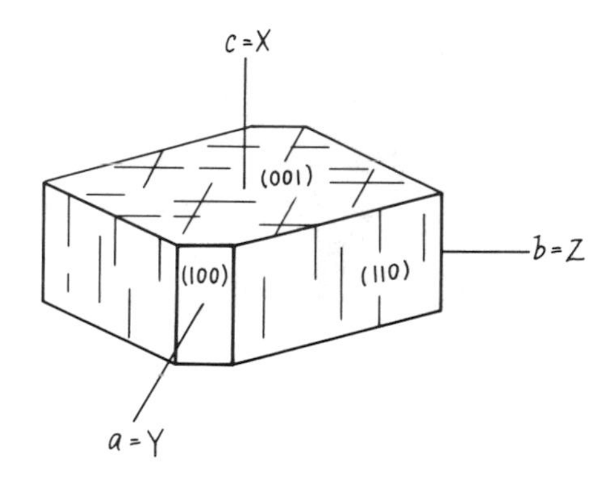

$CaAl_2Si_2O_7(OH)_2 \cdot H_2O$
Orthorhombic
$\alpha = 1.665$, $\beta = 1.672–1.676$, $\gamma = 1.684–1.686$
$\gamma - \alpha = 0.019–0.021$ (up to 2nd-order blue)
$2V = 76°–87°$ (+)
$a = Y, b = Z, c = X$
Pleochroism and color: usually colorless, but thick sections may be pleochroic with X = blue, Y = yellow, and Z = colorless.
Dispersion: $r > v$ strong.
Habit: commonly euhedral and tabular on {001}.
Cleavage: {100} and {010} perfect, {101} poor (Fig. 119).
Extinction in section: commonly parallel to pinacoidal faces and cleavage traces; length high (slow) for tabular {001} grains.

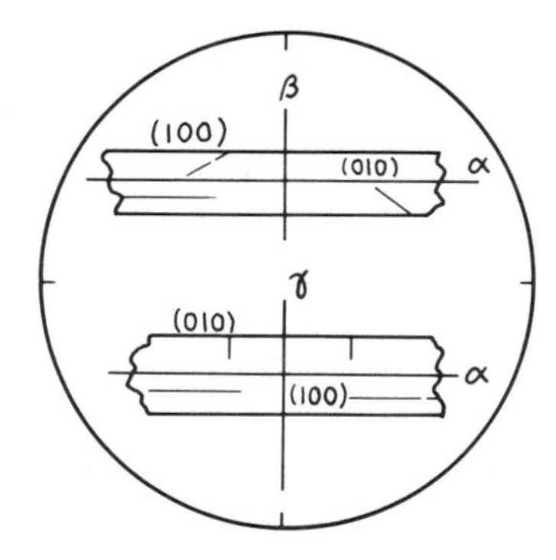

Figure 119. Lawsonite fragments lying on {010} permit estimation of α and β. Extinction is parallel and grains are length low (fast). Traces of poor {101} cleavage, when present, intersect at 67°. The figure is bisymmetric Bxa. Fragments lying on {100} permit estimation of α and γ. The extinction is parallel and grains are length low (fast). The figure is bisymmetric O.N.

Figure 120. Several rectangular grains of lawsonite (center) in a muscovite matrix. Glaucophane is nearby, but not in the photo. Crossed nicols. Photo length is 2.22 mm.

Twinning: simple or polysynthetic {110} common.
Hardness: 7–8.
Specific gravity: ~3.1.
Occurrence: rare; limited to low-temperature high-pressure metamorphic rocks; associated with glaucophane, aegirine, pumpellyite, and jadeite (Fig. 120).
Alteration: relatively stable, but may be converted to pumpellyite or plagioclase.
Characteristic features: occurrence; moderately high relief; generally parallel extinction on cleavage traces; biaxial (+).
Look-alikes: members of the epidote group may have (−) sign, anomalous interference colors, or may be green; prehnite has higher birefringence and lacks two perfect cleavages; andalusite is biaxial (−); scapolite is uniaxial; orthopyroxenes have lower birefringence; clinopyroxenes have higher birefringence, and usually inclined extinction.

Lazurite. See Sodalite Group (S5).

Lepidolite. See Mica Group (M2).

Leucite. See Feldspathoid Group (F2).

Limonite. See Opaque Minerals (O2).

Lizardite. See Serpentine Group (S3).

Lussatite. See Silica Group (S4).

Magnesite. See Carbonate Group (C2).

Magnetite. See Opaque Minerals (O2).

Margarite. See Mica Group (M2).

Marialite. See Scapolite Group (S2).

Meionite. See Scapolite Group (S2).

MELILITE GROUP (M1)

Gehlenite-Akermanite Series

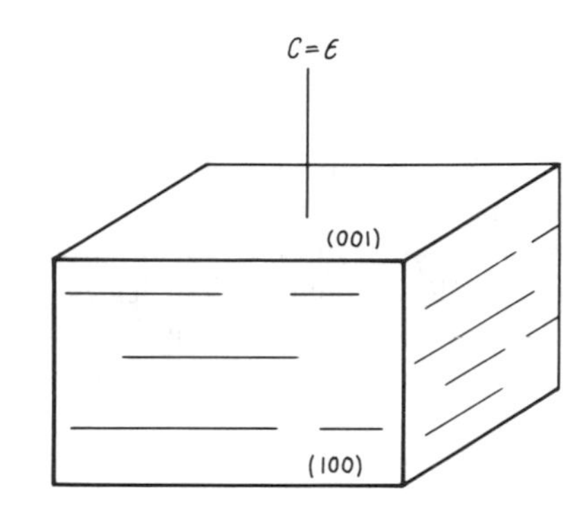

$Ca_2Al_2SiO_7$ to $Ca_2MgSi_2O_7$
Tetragonal
ω = 1.670–1.651, ε = 1.658–1.651 (gehlenite)
ω = 1.651–1.632, ε = 1.651–1.640 (akermanite). As shown in Figure 121, indices of refraction decrease with akermanite content. The sign and mineral name change at $Ge_{48}Ak_{52}$, at which point melilite is isotropic.
$\omega - \varepsilon$ = 0.000–0.012 (up to 1st-order yellow) (gehlenite); $\varepsilon - \omega$ = 0.000–0.008 (up to 1st-order white) (akermanite)
Uniaxial (−) (gehlenite); uniaxial (+) (akermanite)
Pleochroism and color: colorless to pale yellow, with abs. $O > E$ weak; commonly shows anomalous interference colors with Berlin blue replacing first-order gray.
Habit: subhedral to euhedral; tabular with {001} dominant and {100} less so (color photo 43); may contain isotropic rodlike inclusions parallel to *c*.

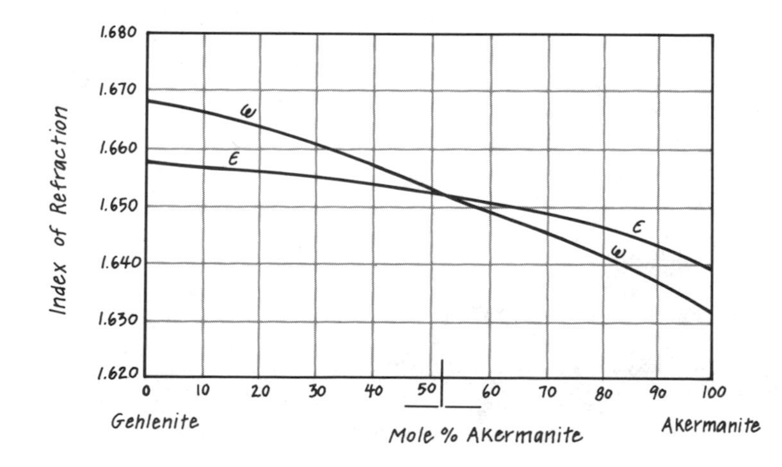

Figure 121. Range of indices of refraction in the gehlenite-akermanite series (Nurse and Midgley, 1953).

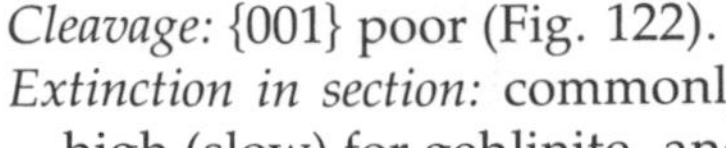

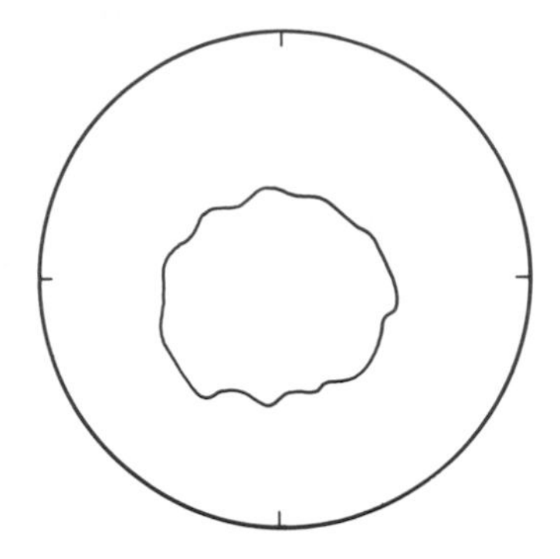

Figure 122. Most melilite-group fragments are in random orientation and have an irregular outline. Uncommon fragments lie on the poor {001} cleavage, permit estimation of only ω, and have a centered O.A. figure.

Cleavage: {001} poor (Fig. 122).
Extinction in section: commonly parallel to largest faces; length high (slow) for gehlinite, and length low (fast) for akermanite.
Twinning: none.
Hardness: 5–6.
Specific gravity: 2.94–3.04.
Occurrence: rare; in mafic Si-poor, Al- and Ca-rich extrusive rocks (Figs. 123 and 124), with nepheline, leucite, olivine, and perovskite; also in the rare rocks alnoite, jacupirangite, and uncompahgrite, and rarely in contact metamorphic limestone and the sanidinite metamorphic facies; common in furnace slags and portland cement.
Alteration: to zeolites and carbonates.

Figure 123. Elongate grain of melilite (center) adjacent to two large grains of clinopyroxene. Crossed nicols. Photo length is 0.55 mm.

Figure 124. Rectangular grains of melilite within clinopyroxene. Crossed nicols. Photo length is 0.55 mm.

Characteristic features: occurrence; low birefringence; anomalous interference colors.

Look-alikes: vesuvianite has higher indices and greater relief; zoisite with anomalous interference colors has a perfect cleavage and is biaxial (+); apatite has a hexagonal outline.

MICA GROUP (M2)

Biotite

$K_2(Mg,Fe^{+2})_{6-4}(Fe^{+3},Al,Ti)_{0-2}[Si_{6-5}Al_{2-3}O_{20}](OH,F)_4$

Biotite forms a complete solid solution with phlogopite, but is here arbitrarily differentiated (as by Deer, Howie, and Zussman, 1966) in having Mg:Fe < 2:1.

Monoclinic

$\measuredangle\beta = 90°-100°$

$\alpha = 1.565-1.625$, $\beta = 1.605-1.696$, $\gamma = 1.605-1.696$

$\gamma - \alpha = 0.04-0.07$ (3rd-order colors most common)

$2V = 0°-25° (-)$, usually <10°

$b = Y, c \wedge X = 0°-9°$

Pleochroism and color: strongly pleochroic in shades of brown, red-brown, or green; abs. $Z \simeq Y > X$, with X = buff, light brown, pale green, $Z \simeq Y$ = reddish brown, golden yellow-brown, dark brown, green, opaque. See color photo 44. Note that in immersion mounts grains lie on {001} and pleochroism is not apparent.

Dispersion: $v > r$ weak (Fe-rich); $r > v$ weak (Mg-rich), inclined.

Habit: tabular {001} and pseudohexagonal ({110} and {010}) outline; zircon with pleochroic haloes may occur as inclusions.

Cleavage: {001} perfect (Fig. 125).

Extinction in section: parallel or slightly inclined (0°–9°) to cleavage traces and principal faces; length high (slow); grains near extinction commonly show bird's-eye maple structure; may show undulatory extinction due to deformation.

Twinning: may show complex twins parallel to {001}.

Hardness: 2½–3.

Specific gravity: 2.7–3.3.

Occurrence: the most common ferromagnesian mineral in most igneous and metamorphic rocks; may be found in almost any plutonic or volcanic rock, but is more common in plutonic varieties; common in both contact and regionally metamorphosed sediments; common in detrital sediments.

Alteration: to vermiculite; extreme weathering yields kaolin or montmorillonite.

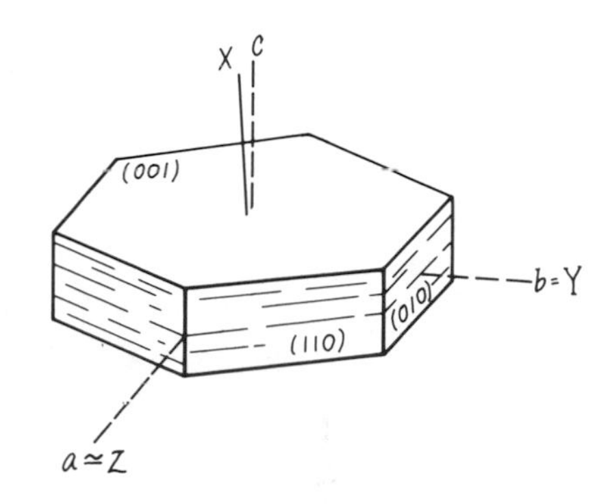

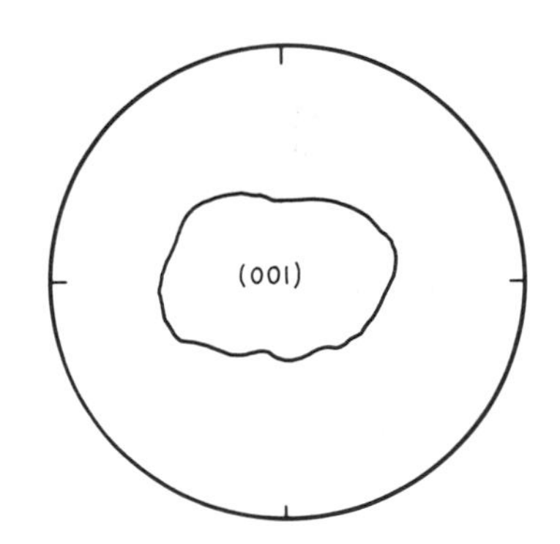

Figure 125. All biotite fragments lie on the perfect {001} cleavage and permit estimation of β and a value of γ' that is identical or almost identical with γ. Pleochroism appears weak to nonevident in this orientation. The figure closely approaches either a bisymmetric Bxa or uniaxial O.A. type (when the $2V$ is small).

Figure 126. Porphyroblast of biotite in quartz-mica schist. The somewhat mottled appearance of the biotite is due to bird's-eye maple structure. Crossed nicols. Photo length is 3.3 mm. See also Figure 110. (From J. Hinsch, E. Leitz, Inc.)

Characteristic features: strong pleochroism (color photos 44 and 45); excellent cleavage with parallel or near-parallel extinction; bird's-eye maple structure (Fig. 126).

Look-alikes: phlogopite has paler colors, weaker pleochroism, and lower indices; amphiboles have a different habit and two perfect cleavages; tourmaline has its direction of maximum absorption normal to its length; chlorites have lower birefringence; stilpnomelane may be indistinguishable optically from biotite, but it can sometimes be identified by a poor cleavage normal to {001}, a strong yellow color in the X direction, and a lack of bird's-eye maple structure.

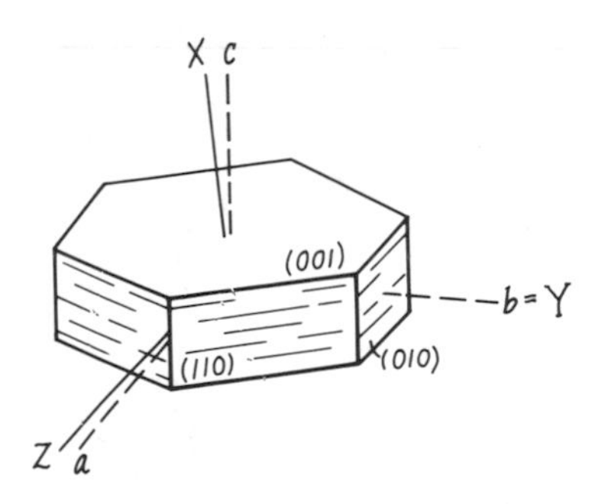

Glauconite

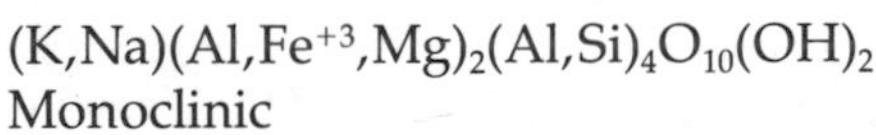

$(K,Na)(Al,Fe^{+3},Mg)_2(Al,Si)_4O_{10}(OH)_2$

Monoclinic

$\measuredangle\beta \simeq 100°$

$\alpha = 1.56–1.61$, $\beta \simeq \gamma \simeq 1.61–1.65$ (increases with Fe^{+3} content and decreases with interlayered water)

$\gamma - \alpha = 0.014–0.032$ (up to 2nd-order orange)

$2V = 0°–20° (-)$

$b = Y$, $c \wedge X \simeq 10°$

Pleochroism and color: yellow-green to olive-green, but may be colorless; abs. $Z = Y > X$; X = pale yellow to green, $Y = Z$ = pale to dark green, olive-green, blue-green; brown when altered.

Dispersion: $r > v$.

Habit: often with clay or chlorite in granules, pellets, or pseudomorphs after foraminifera; as aggregates of minute overlapping "hexagonal" plates (which are rarely observed).

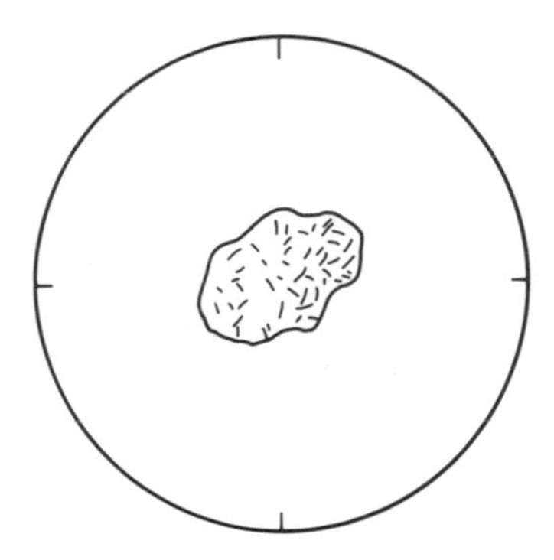

Figure 127. Glauconite grains are commonly a fine-grained aggregate with nonuniform extinction. An average index (~1.61) can be estimated, and interference figures are generally not obtainable.

Cleavage: {001} perfect (Fig. 127).
Extinction in section: commonly appears incomplete in fine-grained aggregates (due to random orientation); larger grains have parallel extinction to basal faces and cleavage traces, and are length high (slow).
Twinning: not described.
Hardness: 2.
Specific gravity: 2.4–3.0.
Occurrence: rounded pellets (Fig. 128), granules, and pseudomorphs after foraminifera in marine detrital sands, with quartz, feldspar, clay minerals, and calcite.
Alteration: to montmorillonite, limonite, or goethite.
Characteristic features: habit; occurrence; color; microcrystalline character.
Look-alikes: most chlorites have lower birefringence.

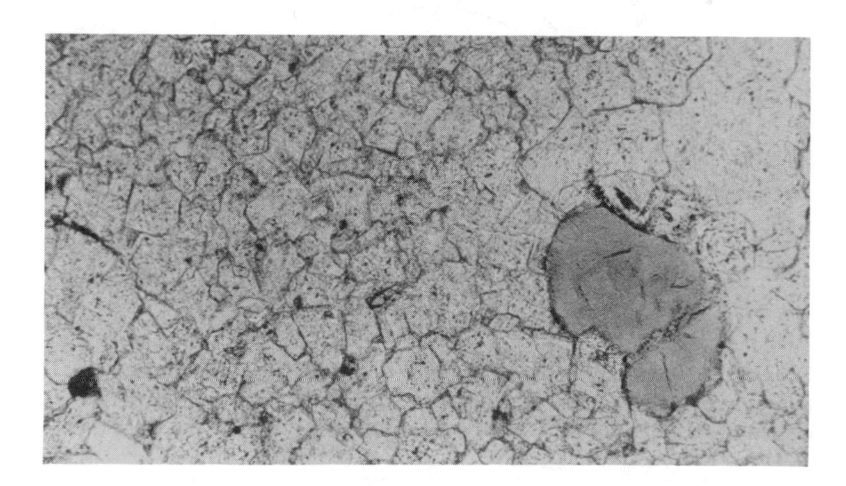

Figure 128. Glauconite pellet (gray) in a carbonate matrix. Crossed nicols. Photo length is 2.2 mm.

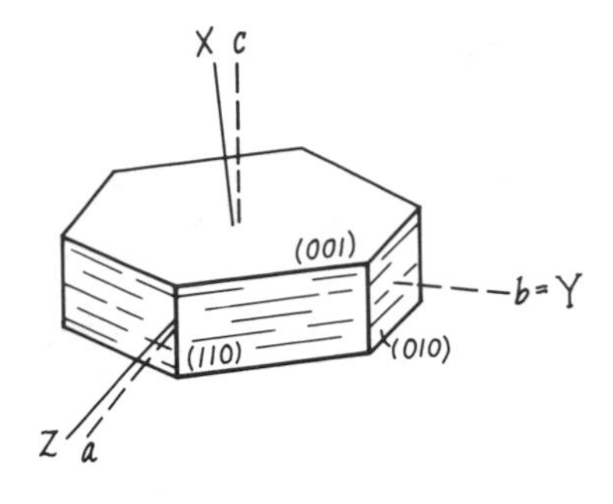

Lepidolite

$K_2(Li,Al)_{5-6}[Si_{6-7}Al_{2-1}O_{20}](OH,F)_4$
Monoclinic
$\measuredangle\beta = 100°$
$\alpha = 1.525-1.548$, $\beta = 1.551-1.585$, $\gamma = 1.554-1.587$ (indices increase with Mn^{+2} and Fe^{+3} and decrease with F^-)
$\gamma - \alpha = 0.018-0.038$ (up to 2nd-order yellowish green)
$2V = 0°-58°$ (usually $30°-50°$) (−); $2V$ decreases with increase in Mn^{+2} and Fe^{+3}
$b = Y, c \wedge X = 3°-10°$
Pleochroism and color: colorless; thick sections may be pale violet or green with $Y = Z > X$.
Dispersion: $r > v$ weak, inclined.
Habit: pseudohexagonal plates ({110} and {010}) with {001} dominant; zircon with pleochroic haloes may occur as inclusions.
Cleavage: {001} perfect (Fig. 129).
Extinction in section: parallel or slightly inclined (0°–7°) to cleavage traces and principal faces; length high (slow).
Twinning: may show complex twins on {001}.
Hardness: 2½–4.
Specific gravity: 2.80–2.90.
Occurrence: mainly confined to granitic pegmatites, with spodumene, amblygonite, tourmaline, beryl, and topaz; rare in granite and high-temperature veins with cassiterite.
Alteration: not known.
Characteristic features: pink or purple color in hand specimens; occurrence; similar to muscovite.
Look-alikes: easily confused with muscovite and phlogopite, both of which have higher indices.

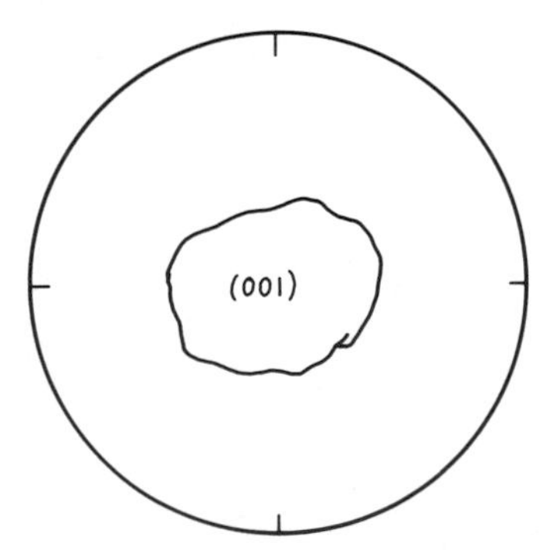

Figure 129. All lepidolite fragments lie on the perfect {001} cleavage and permit estimation of β and a value of γ' that is essentially identical with γ. The figure is, or closely approaches being, bisymmetric Bxa.

Margarite

$Ca_2Al_4[Si_4Al_4O_{22}](OH)_4$
Monoclinic
$\measuredangle\beta = 95°–100°$
$\alpha = 1.630–1.638$, $\beta = 1.642–1.648$, $\gamma = 1.644–1.650$
$\gamma - \alpha = 0.012–0.014$ (up to 1st-order orange)
$2V = 40°–67°$ (−)
$b = Z$, $c \wedge X = 11°–13°$
Pleochroism and color: colorless.
Dispersion: $v > r$ moderate, parallel.
Habit: pseudohexagonal plates ({110} and {010}) with {001} dominant.
Cleavage: {001} near perfect (Fig. 130).
Extinction in section: slightly inclined (6°–8°) to cleavage traces and principal faces; length high (slow).
Twinning: polysynthetic {001}.
Hardness: 3½–4½.
Specific gravity: 3.0–3.1.
Occurrence: associated with corundum, diaspore, staurolite, and tourmaline in metamorphic rocks.
Alteration: to vermiculite.
Characteristic features: occurrence; one perfect cleavage with near-parallel extinction; colorless; moderate $2V$ (−).
Look-alikes: muscovite has lower indices, higher birefringence, and smaller $2V$; chlorite and chloritoid are green and may be optically (+).

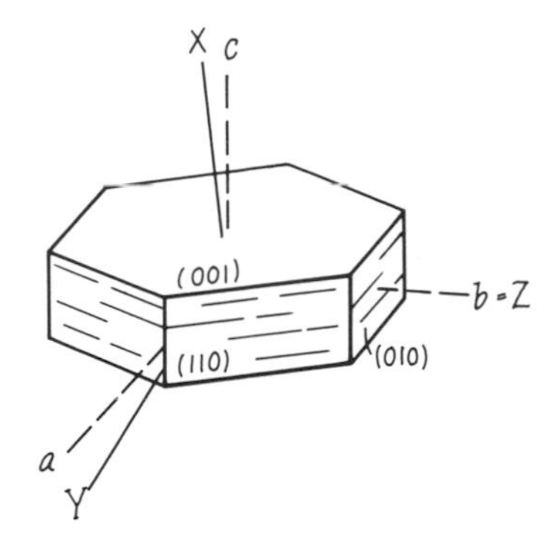

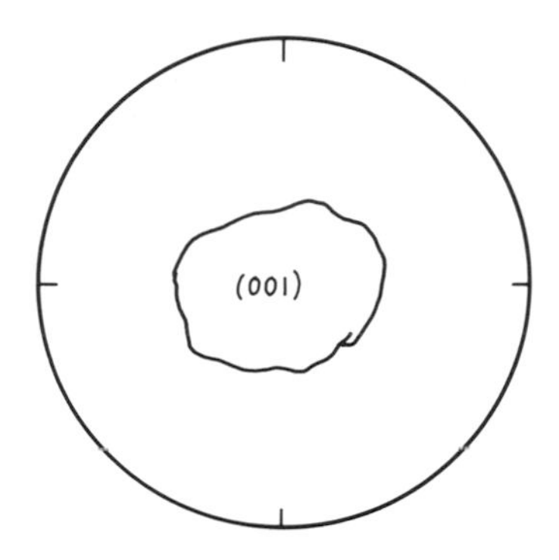

Figure 130. All margarite fragments lie on the near-perfect {001} cleavage and permit estimation of γ and a value of β' that is slightly less than β. The figure is effectively bisymmetric Bxa.

Muscovite

$K_2Al_4[Si_6Al_2O_{20}](OH,F)_4$. When formed under low-temperature conditions, up to about 50 percent substitution of the celadonite component, $K_2(Mg,Fe^{+2})_2Si_2[Si_6(Al,Fe^{+3})_2O_{20}](OH)_4$, can occur: this variety is called *phengite.*
Monoclinic
$\measuredangle\beta = 95°30'$
$\alpha = 1.552–1.574$, $\beta = 1.582–1.610$, $\gamma = 1.587–1.616$ (indices increase with deviation from ideal formula)
$\gamma - \alpha = 0.036–0.049$ (up to 3rd-order yellow)
$2V = 30°–47°$ (−); $2V$ decreases with deviation from ideal formula
$b = Z$, $c \wedge X = 0°–5°$
Pleochroism and color: colorless.
Dispersion: $r > v$ moderate, parallel.
Habit: irregular or pseudohexagonal plates ({110} and {010}) with {001} dominant; secondary muscovite appears as fine scales or

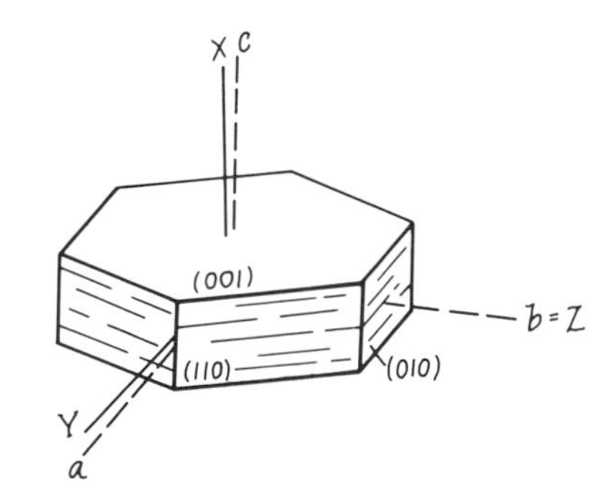

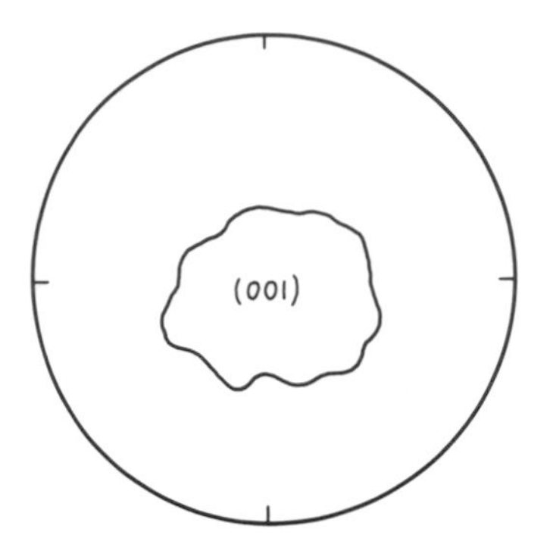

Figure 131. All muscovite fragments lie on the perfect {001} cleavage and permit estimates of γ and a value of β' that is slightly less than β. The figure is effectively bisymmetric Bxa.

fibrous aggregates that are called *sericite* (a general term for any fine-grained, colorless, and unidentified mica; sericite is usually muscovite, but may be paragonite or illite); zircon with pleochroic haloes may occur.

Cleavage: {001} perfect (Fig. 131).

Extinction in section: near-parallel (1°–3°) to cleavage traces and principal faces; length high (slow); bird's-eye maple structure common; deformed grains show undulatory extinction.

Twinning: polysynthetic {001} or simple {110}.

Hardness: 2½–3.

Specific gravity: 2.77–2.88.

Occurrence: very common in prograde metamorphic rocks, from low grade into the amphibolite facies (where dehydration may produce K-feldspar and sillimanite); retrograde metamorphism commonly yields sericite; in igneous rocks, primary muscovite is found in intrusive rocks, veins, dikes, and pegmatites; deuteric or hydrothermal alteration commonly yields sericite (particularly within feldspars); fine-grained muscovite is common in argillaceous and sandy sediments. See Figure 132.

Alteration: quite stable, but may be converted to illite or hydromuscovite.

Characteristic features: moderately high birefringence and indices; one perfect cleavage with essentially parallel extinction; bird's-eye maple structure.

Look-alikes: nearly colorless phlogopite has a 2V of 0°–15°; virtually identical in appearance to talc, which can sometimes be distinguished on the basis of occurrence, smaller 2V (when

Figure 132. Muscovite (center and top) and quartz. The perfect {001} cleavage is visible in most orientations. Crossed nicols. Photo length is 0.55 mm.

obtainable), or a somewhat smeared aspect when viewed between crossed nicols; paragonite is uncommon, and cannot be distinguished optically from muscovite.

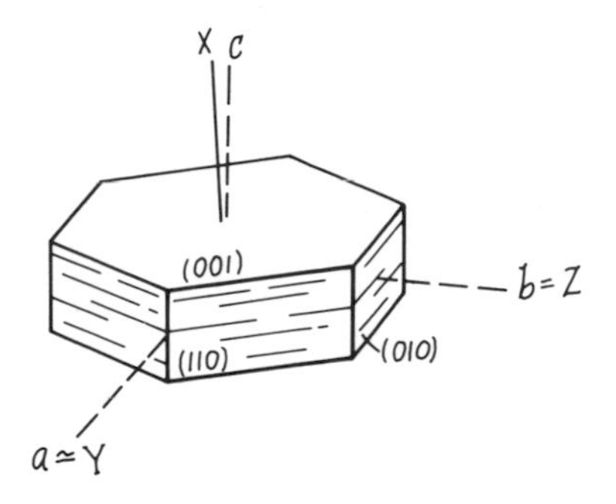

Paragonite

$Na_2Al_4[Si_6Al_2O_{20}](OH)_4$
Monoclinic (rarely trigonal)
$\measuredangle\beta = 95°$
$\alpha = 1.564–1.580$, $\beta = 1.594–1.609$, $\gamma = 1.600–1.609$
$\gamma - \alpha = 0.028–0.038$ (up to 3rd-order blue)
$2V = 0°–40° (-)$
$b = Z, c \wedge X \simeq 5°$
Pleochroism and color: colorless.
Dispersion: $r > v$ moderate.
Habit: massive scaly aggregates with {001} dominant.
Cleavage: {001} perfect (Fig. 133).
Extinction in section: parallel to cleavage traces and principal faces; length high (slow).
Twinning: unknown, but probably similar to muscovite.
Hardness: 2½.
Specific gravity: 2.85.
Occurrence: in phyllites, schists, and gneisses, as well as fine-grained sediments; the frequency of occurrence of paragonite is not well established, as it is optically similar to muscovite, and is commonly identified as such.
Alteration: stable.
Characteristic features: moderately high birefringence and indices; extinction parallel to one perfect cleavage; bird's-eye maple structure.
Look-alikes: muscovite is optically similar, and can only be distinguished by nonoptical methods (such as X-ray diffraction or chemical analysis).

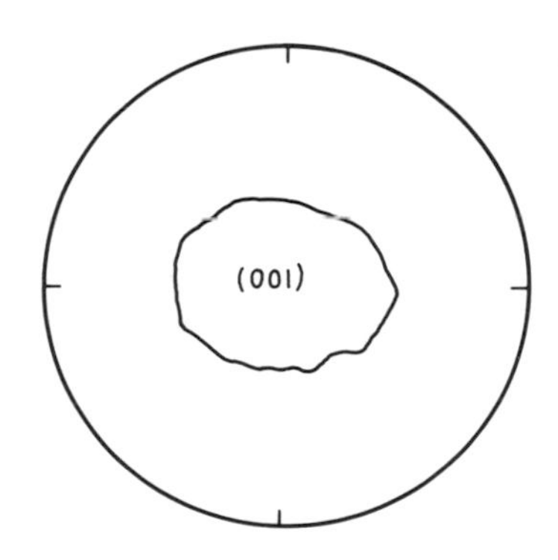

Figure 133. All paragonite fragments lie on the perfect {001} cleavage and permit estimation of γ and β (as Y is parallel to *a*). The figure is bisymmetric Bxa.

Phlogopite

$K_2(Mg,Fe^{+2})_6[Si_6Al_2O_{20}](OH,F)_4$. Phlogopite forms a complete solid solution with biotite.
Monoclinic
$\measuredangle\beta = 100°$
$\alpha = 1.530–1.590$, $\beta = 1.557–1.637$, $\gamma = 1.558–1.637$
$\gamma - \alpha = 0.028–0.049$ (up to 2nd-order yellow)
$2V = 0°–15° (-)$
$b = Y, c \wedge X = 5°–10°$
Pleochroism and color: colorless, pale yellow, pale green; pleochroism weak to moderate, with $Z = Y > X$; X = yellow, colorless, $Y = Z$ = brownish red, green, or yellow.

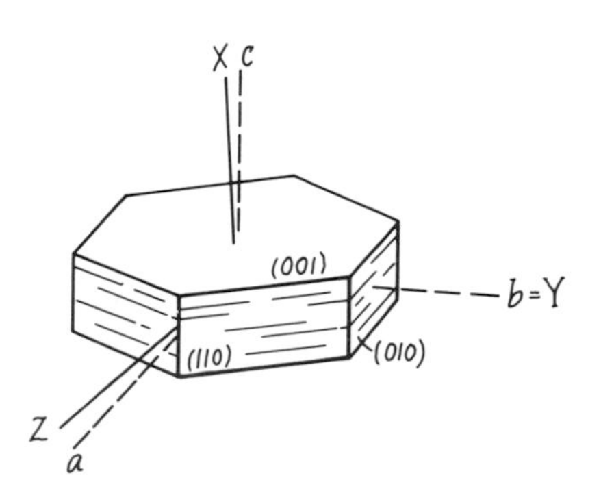

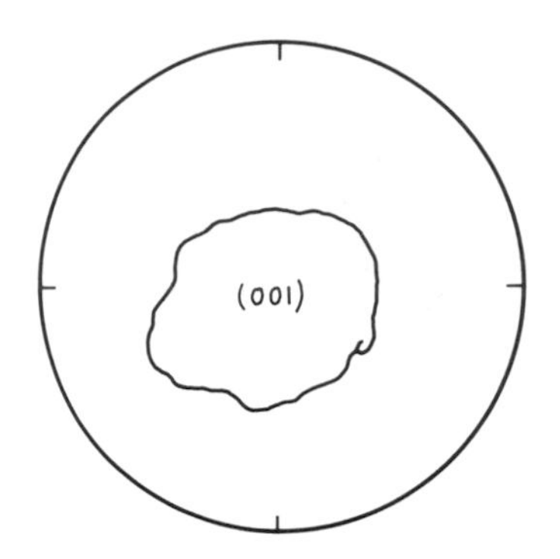

Figure 134. All phlogopite fragments lie on the perfect {001} cleavage and permit estimates of β and a value of γ' that is slightly less than γ. The figure is effectively bisymmetric Bxa.

Dispersion: $v > r$ weak to moderate, inclined.
Habit: commonly pseudohexagonal euhedral to subhedral tablets with {001} dominant; inclusions of rutile needles may produce asterism; inclusions of zircon may produce pleochroic haloes.
Cleavage: {001} perfect (Fig. 134).
Extinction in section: parallel or near-parallel (0°–5°) to cleavage traces and principal faces; length high (slow); bird's-eye maple structure.
Twinning: {001}.
Hardness: 2–2½.
Specific gravity: 2.76–2.90.
Occurrence: fairly common in ultramafic igneous rocks and their serpentinized equivalents; in impure marble with tremolite, diopside, forsterite, talc, chondrodite, and spinel; in pyroxenite with apatite and calcite.
Alteration: to vermiculite.
Characteristic features: small $2V$ (−); weak to moderate pleochroism; extinction near-parallel to the perfect {001} cleavage; bird's-eye maple structure.
Look-alikes: muscovite is not pleochroic and has a moderate $2V$; biotite has higher indices and strong pleochroism; lepidolite commonly has a moderate $2V$, and is essentially restricted in occurrence to granitic pegmatites.

Microcline. See Feldspar Group (F1).

Mizzonite. See Scapolite Group (S2).

Molybdenite. See Opaque Minerals (O2).

MONAZITE (M3)

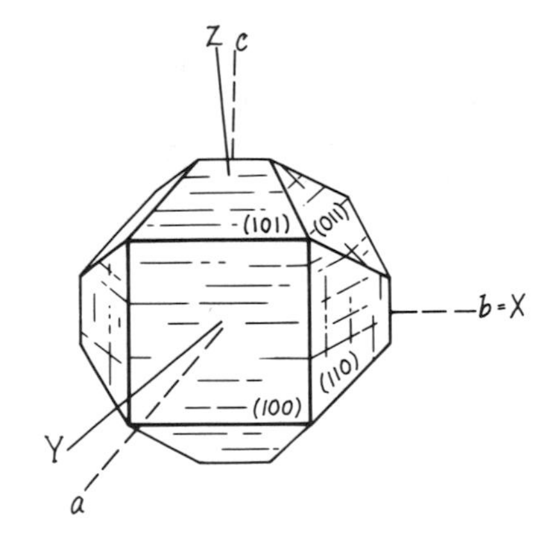

$(Ca,La,Th)PO_4$
Monoclinic
$\measuredangle\beta \simeq 104°$
$\alpha = 1.770–1.800$, $\beta = 1.777–1.801$, $\gamma = 1.828–1.851$
$\gamma - \alpha = 0.045–0.075$ (upper 3rd- and 4th-order colors)
$2V = 6°–19°$ (+)
$b = X$, $c \wedge Z = 2°–7°$
Pleochroism and color: yellow-brown, gray, or colorless; detrital grains may be yellow or various shades of brown; weak pleochroism with $Y > X = Z$; X = light yellow, Y = dark yellow, Z = greenish yellow.
Dispersion: $v > r$ or $r > v$ weak, parallel.

Habit: small euhedral crystals, either elongated on *b*, or tabular with {100} dominant.

Cleavage: {100} good, {001} parting variable (Fig. 135).

Extinction in section: grains elongate on *b* commonly have parallel extinction against major faces and cleavage traces, and are length low (fast); tabular {100} grains have parallel or inclined extinction against major faces and cleavage traces and may be length high (slow) or length low (fast) as a function of orientation.

Twinning: simple {100} common.

Hardness: 5.

Specific gravity: 5.0–5.3.

Occurrence: a rare accessory mineral in a great variety of siliceous igneous rocks and veins; may be present in large crystals in some pegmatites; an accessory mineral in many metamorphic rocks; may be concentrated as a detrital mineral in the heavy fraction of sands. See Figure 136.

Alteration: relatively stable, but may acquire an opaque coating of questionable composition.

Characteristic features: high positive relief; yellow-brown color; weak pleochroism; strong birefringence.

Look-alikes: sphene has higher birefringence, extreme dispersion, and a distinctive crystal outline; zircon is uniaxial and has higher indices.

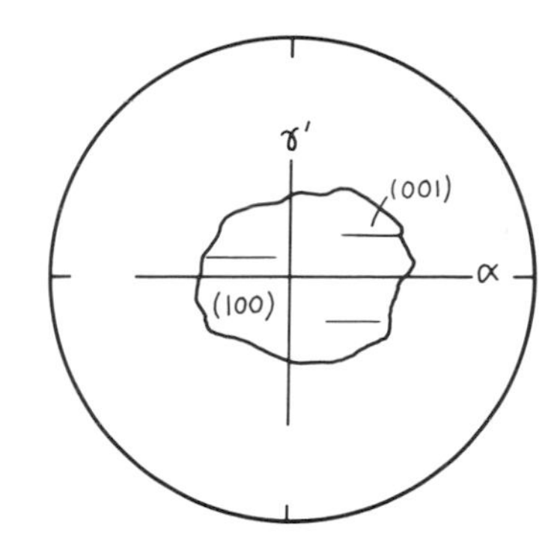

Figure 135. Many monazite fragments lie on the good {100} cleavage and permit estimation of α (parallel to traces of {001} parting) and γ' (quite close in value to γ). Extinction is parallel to traces of the {001} parting. If the fragment is elongated, it is length low (fast). The figure is effectively bisymmetric O.N. Uncommon fragments (not shown) that lie on the {001} parting permit estimation of α and β' (slightly greater than β), have parallel extinction against {100} cleavage traces, are length low (fast), and have an almost bisymmetric Bxa figure.

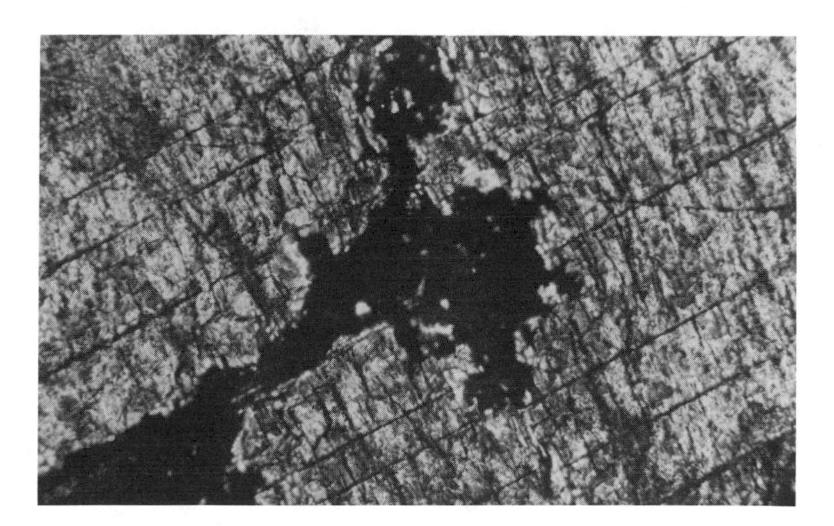

Figure 136. Single grain of monazite showing good {100} cleavage. The black area is epoxy cement. Crossed nicols. Photo length is 0.55 mm.

Monticellite. See Olivine Group (O1).

Montmorillonite. See Clays and Clay Minerals (C6).

Mullite. See Aluminosilicate Group (A1).

Muscovite. See Mica Group (M2).

Natrolite. See Zeolite Group (Z1).

Nepheline. See Feldspathoid Group (F2).

Norbergite. See Humite Group (H2).

Noselite. See Sodalite Group (S5).

Obsidian. See Glass (G2).

Octahedrite. See Rutile Group (R2).

Oligoclase. See Feldspar Group (F1).

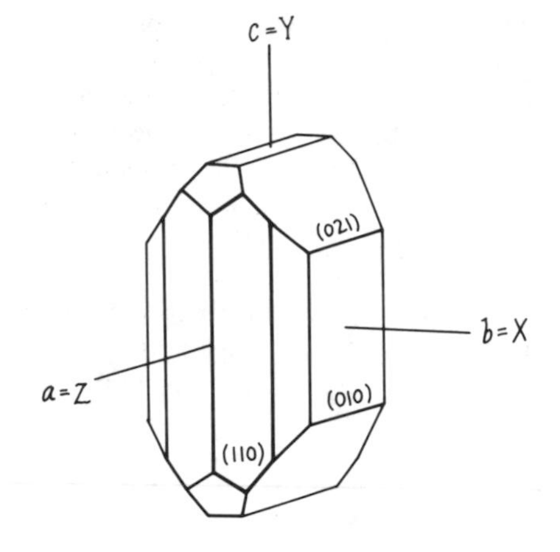

OLIVINE GROUP (O1)

Forsterite-Fayalite Series

Mg_2SiO_4 to Fe_2SiO_4. Compositions commonly stated in terms of end-members (for example, $Fo_{90}Fa_{10}$, or simply Fo_{90})
Orthorhombic
$\alpha = 1.635-1.827$, $\beta = 1.651-1.869$, $\gamma = 1.670-1.879$ (See Figure 137)
$\gamma - \alpha = 0.035-0.052$ (See Figure 137)
$2V = 82°-90° (+)$; $46°-90° (-)$ (See Figure 137)
$a = Z$, $b = X$, $c = Y$
Pleochroism and color: Mg-rich members are colorless; Fe-rich members are yellow-green or pale yellow; pleochroism weak, with $Y > X = Z$.
Dispersion: $v > r$ weak ($Fo_{100}-Fo_{80}$); $r > v$ weak ($Fo_{80}-Fo_{0}$).
Habit: anhedral to subhedral in plutonic and metamorphic rocks (color photo 46); some phenocrysts in volcanic rocks are euhedral, somewhat elongated on *c*, and have a four- or six-sided outline as a result of dominant {010}, {110}, and {011} or {021} forms (color photo 47); skeletal crystals in volcanic rocks are not common; compositional zoning (with Mg-rich cores and Fe-rich rims) is fairly common.
Cleavage: {010} and {100} poor in Mg-rich members, and good in Fe-rich members (Fig. 138).
Extinction in section: commonly parallel or symmetrical to crystal faces, and parallel to cleavage traces.
Twinning: simple {100}, {011}, or {012} uncommon.

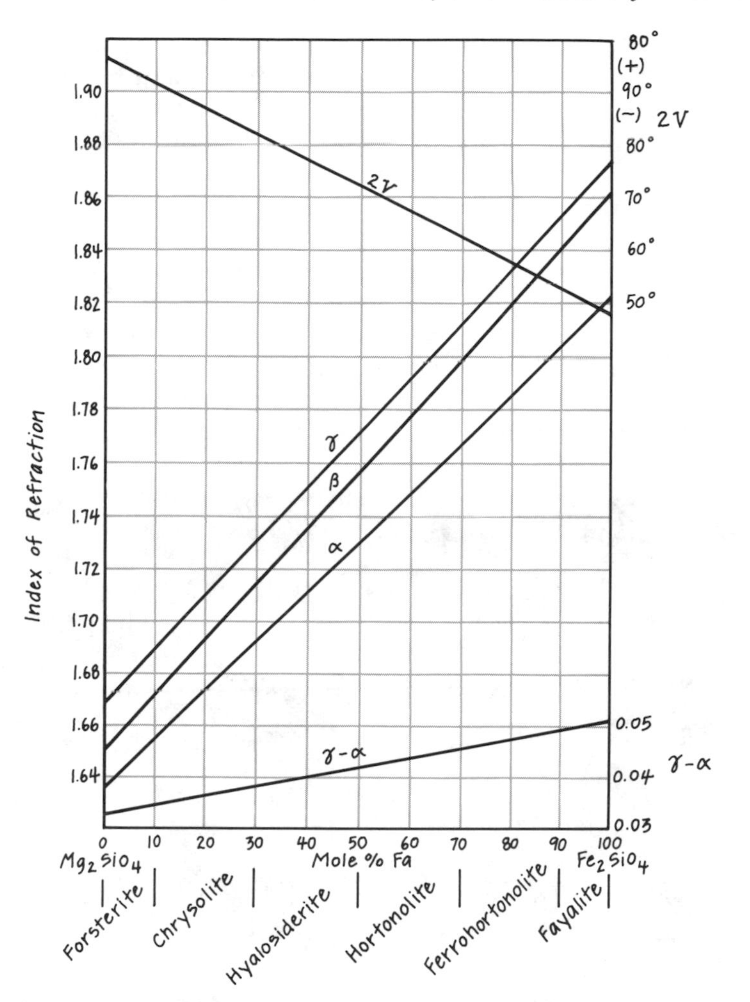

Figure 137. Variation in $2V$, α, β, γ, and $\gamma - \alpha$ with composition in the olivine series (Poldervaart, 1950). Species names between forsterite and fayalite currently receive little usage.

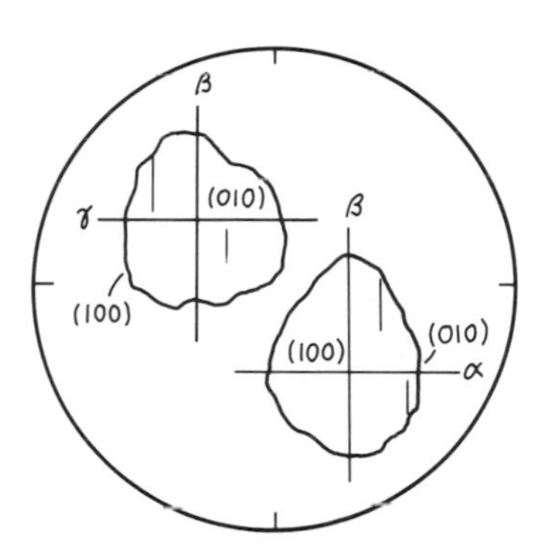

Figure 138. The large majority of olivine fragments lie in random orientation. Fragments lying on the generally poor {010} cleavage permit estimation of β and γ, have parallel extinction against poor {100} cleavage traces, are length low (fast), and yield a bisymmetric Bxo or Bxa figure (as a function of sign). Fragments lying on the generally poor {100} cleavage permit estimation of α and β, have parallel extinction against poor {010} cleavage traces, are length high (slow), and yield a bisymmetric Bxa or Bxo figure (as a function of sign).

Hardness: 6½–7.

Specific gravity: 3.22–4.39.

Occurrence: olivine of composition Fo_{92-88} is common in ultramafic rocks; mafic igneous rocks often contain olivine in the range Fo_{85-40} (with Mg-rich members most common); magmatic differentiation results in an increase of Fe : Mg ratio; for most of its compositional range, olivine (being a Si-deficient mineral) is incompatible with SiO_2 polymorphs; however, fayalite is relatively common in quartz-bearing syenite; olivine is found in impure dolomitic marble (color photo 48) associated with phlogopite, garnet, and calcite, as well as in stony meteorites; it is easily altered, and hence is rare in detrital sediments.

Alteration: commonly to serpentine in both igneous and metamorphic rocks; also to talc and magnesite, and to *iddingsite* and *bowlingite,* both of which are ill-defined submicroscopic scaly aggregates of iron-oxides and chlorite; iddingsite is blood-red, has $n = 1.76-1.89$, and a high birefringence; bowlingite is similar, but yellowish green.

Characteristic features: high indices and birefringence; large $2V$ (+) or (−); poor cleavage; partial alteration to serpentine or iddingsite is common. See Figure 139.

Look-alikes: pyroxenes have good cleavage, inclined extinction, generally slightly lower relief, and lower birefringence; epidote has good cleavage and perhaps anomalous interference colors; monticellite has lower birefringence.

A

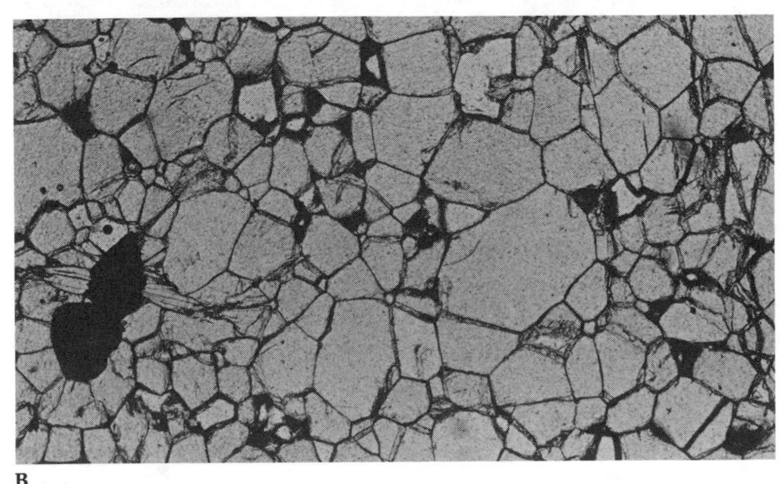

B

Figure 139. (A) Olivine in dunite. Crossed nicols. (B) The same in plane-polarized light. Note the lack of cleavage and high relief. Photo length is 5.16 mm. (From J. Hinsch, E. Leitz, Inc.)

Monticellite

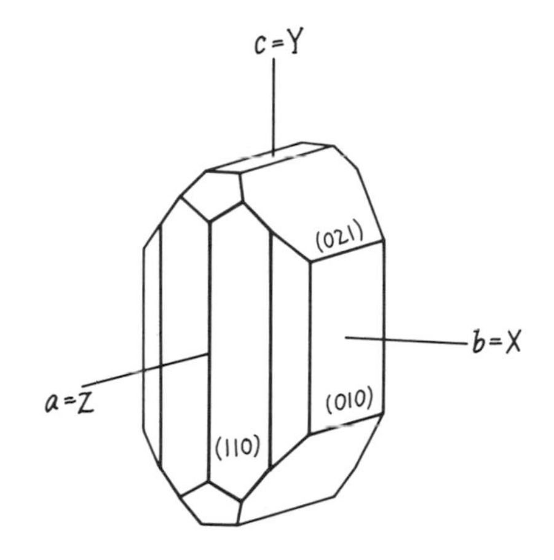

$CaMgSiO_4$. Fe^{+2} and Mn replace Mg, but most naturally occurring varieties are 90 percent Mg.
Orthorhombic
$\alpha = 1.639-1.663$, $\beta = 1.646-1.674$, $\gamma = 1.653-1.680$
$\gamma - \alpha = 0.014-0.020$ (up to 2nd-order violet)
$2V = 70°-90°$ (−)
$a = Z$, $b = X$, $c = Y$
Pleochroism and color: colorless.
Dispersion: $r > v$ moderate.
Habit: usually anhedral-granular, but may be somewhat prismatic with elongation on *c*.
Cleavage: {010} poor (Fig. 140).
Extinction in section: commonly parallel to cleavage traces and major faces (note that these are usually not present); length high (slow) or low (fast), depending upon orientation.
Twinning: cyclic {031}, may yield star-shaped crystals composed of six individuals.
Hardness: 5½.
Specific gravity: 3.1–3.3.
Occurrence: in dolomitic marble contact zones with vesuvianite, gehlenite, diopside, wollastonite, and rare high-temperature calc-silicate minerals; may be found in alnoite, peridotite, nepheline basalt, and melilite-diopside rocks; common in furnace slags and magnesite refractories.
Alteration: to serpentine and augite.
Characteristic features: occurrence; large 2*V* (−); low birefringence and lack of cleavage.
Look-alikes: other olivines have higher birefringence; most pyroxenes have a (+) sign, two perfect cleavages, and inclined extinction.

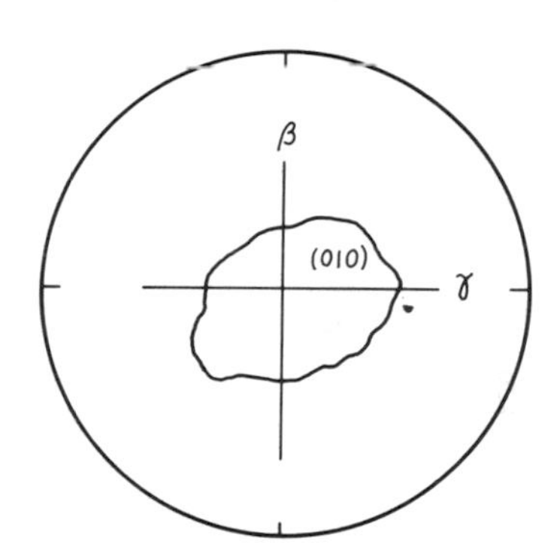

Figure 140. Most monticellite fragments lie in random orientation. A few may lie on the poor {010} cleavage, permit estimation of β and γ, and yield a bisymmetric Bxa figure.

Omphacite. See Pyroxene Group (P8).

Opal. See Silica Group (S4).

THE OPAQUE MINERALS (O2)

The nine minerals described next are commonly listed as opaque; under certain circumstances, however, some are, in fact, semi-opaque or translucent when very thin fragments are examined (e.g., hematite flakes). The translucency is best observed with maximum light intensity and insertion of the upper condenser element.

The opaque minerals can be identified with the use of an ore microscope, which uses reflected light from a polished uncovered specimen (see Ramdohr, 1969); under the standard polarizing microscope, which relies on transmitted light, all opaque minerals appear black.

Opaque minerals in thin section can be examined by means of a strong light reflection from the upper surface of the section, or with a hand lens or binocular microscope. The color of the opaque mineral can be determined; with information on grain shape, alteration characteristics, associated transparent minerals, and occurrence, an approximate identification can be made. It is often better (and more realistic) to avoid the identification problem, and simply list the opaque phases as "iron ores" or "opaques."

Chalcopyrite

$CaFeS_2$
Tetragonal
Pleochroism and color: golden yellow and metallic luster in reflected light.
Habit: anhedral grains, aggregates; rare euhedral crystals are tetrahedral.
Cleavage: none.
Hardness: 3½–4.
Specific gravity: 4.1–4.3.
Occurrence: a widely disseminated ore of copper; in veins with other copper minerals, and disseminated in minor amounts in metamorphic rocks; not common in igneous rocks.
Alteration: to sulfates, carbonates (malachite, azurite), and hydrated silicate (chrysocolla) and oxide (limonite).
Characteristic features: color; occurrence.
Look-alikes: pyrite commonly has a square or rectangular outline.

Chromite (a member of the spinel mineral group)

$FeCr_2O_4$
Isometric
Pleochroism and color: gray to black with submetallic luster in reflected light; thin edges may be semiopaque and brown.
Habit: octahedral, with square or rectangular outlines; also anhedral.
Cleavage: none.
Hardness: 6.
Specific gravity: 4.5–4.8.

Figure 141. Chromite crystals with interstitial chlorite. Crossed nicols. Photo length is 2.22 mm.

Occurrence: common in ultramafic and serpentinized rocks (where it may be concentrated in cumulate layers), and uncommon as a late magmatic mineral. See Figure 141.
Alteration: stable.
Characteristic features: occurrence; habit.
Look-alikes: chromite is weakly magnetic, and magnetite is strongly magnetic.

Graphite

C
Hexagonal
Opaque in thin section; $\omega = 1.93-2.07$ in extremely thin flakes
Pleochroism and color: black in reflected light; gray in extremely thin flakes, abs. $O > E$ strong.
Habit: rounded, anhedral, or thin flakes with {0001} dominant.
Cleavage: {0001} perfect.
Extinction in section: opaque; in extremely thin flakes, parallel to major faces and cleavage traces.
Hardness: 1–2.
Specific gravity: $\simeq 2.23$.
Occurrence: in slates, schists, gneisses, and marble, as a result of regional or contact metamorphism.
Alteration: stable.
Characteristic features: opaque; black in reflected light; occurrence.
Look-alikes: the habit of magnetite is octahedral; tabular ilmenite is commonly altered to white leucoxene; *molybdenite* (MoS_2) forms gray flakes but is rare, except in mineralized rocks, and

has violet internal reflections; the minute opaque particles in sedimentary materials, referred to as "carbonaceous material," are organic matter that has not reached a sufficiently high temperature to form graphite.

Hematite

Fe_2O_3
Hexagonal
$\omega = 3.15-3.22$, $\varepsilon = 2.87-2.94$
$\omega - \varepsilon \simeq 0.28$ (white of the higher order)
Uniaxial (−)

Pleochroism and color: usually opaque (black in reflected light with red internal reflections) except at thin edges, which are deep blood red to red-brown (insertion of the condenser may be necessary to see this); pleochroism very strong with abs. $O > E$; ω = deep red-brown, ε = yellow-brown.
Dispersion: extreme.
Habit: anhedral, flakes, or scales, reniform or botryoidal; may be oolitic or fossiliferous as a result of carbonate replacement; thin plates or microlites as inclusions in other minerals.
Cleavage: none; parting common on {0001} or $\{10\bar{1}1\}$.
Extinction in section: interference colors masked by mineral color; extinction in thin edges may not be seen in fine-grained aggregates.
Hardness: 5–6.
Specific gravity: 5.26.
Occurrence: very common in metamorphosed Fe-rich rocks (as in Precambrian banded iron ores), alternating with quartz layers, and associated with grunerite and magnetite; rare as a primary igneous mineral, but common as an alteration product, and in fumaroles; common as a cement and alteration product in soils and sedimentary materials.
Alteration: stable.
Characteristic features: almost opaque; platy habit; red color of minute flakes; red internal reflections.
Look-alikes: goethite is biaxial, yellow-brown, and more translucent; limonite is yellow-brown and isotropic.

Ilmenite

$FeTiO_3$ (with $MgTiO_3$, $MnTiO_3$, or Fe_2O_3 in solid solution)
Hexagonal
$\omega \simeq 2.7$

Pleochroism and color: opaque in thin section; gray with metallic luster in reflected light; very thin fragment edges may be deep red to orange-red with a violet tinge.
Habit: hexagonal plates or tablets, often skeletal, or anhedral.

Cleavage: none, but may have {0001} or {10$\bar{1}$1} parting.
Hardness: 5–6.
Specific gravity: 4.72.
Occurrence: in veins and disseminated deposits associated with gabbro, diorite, and anorthosite; a common accessory mineral in many types of rocks; in ore veins, pegmatites; a detrital mineral, especially in sands.
Alteration: to *leucoxene,* a fine-grained aggregate (white, gray, yellowish, or red-brown in reflected light) composed of one or more titanium oxides (usually rutile).
Characteristic features: opaque; gray in reflected light, often with bands or patches of leucoxene; rectangular or skeletal outline.
Look-alikes: magnetite has an octahedral habit, lacks leucoxene alteration, and is magnetic; graphite and molybdenite (MoS_2) are of limited occurrence.

Limonite

$FeO(OH) \cdot nH_2O$
Amorphous
n = 2.0–2.1 (may increase due to absorption of immersion liquids)
Pleochroism and color: opaque to translucent, yellow, yellow-brown, or red-brown.
Habit: colloform textures, such as botryoidal, mammillary, or stalactitic.
Hardness: 4–5½.
Specific gravity: 2.7–4.3.
Occurrence: always secondary from weathering of Fe-bearing minerals; always with goethite, and commonly with hematite and Mn-hydroxides; in swamp or spring colloidal deposits, or in weathered sulfide deposits; may form pseudomorphs after pyrite and other minerals.
Alteration: stable.
Characteristic features: occurrence; texture; amorphous character.
Look-alikes: cannot be distinguished optically from other fine-grained iron oxides.

Magnetite[1]

Fe_3O_4
Isometric
Pleochroism and color: steel blue with metallic luster in reflected light; thin edges may be semiopaque and dark reddish brown.

[1] A member of the spinel mineral group.

Figure 142. Euhedral magnetite porphyroblast in chlorite matrix. Crossed nicols. Photo length is 2.22 mm.

Habit: commonly octahedral, with square or rectangular outlines; also dodecahedral, with cubes rare, massive, or granular.
Cleavage: none; octahedral parting may be well developed.
Twinning: {111} common.
Hardness: 6.
Specific gravity: 5.2.
Occurrence: a common accessory mineral in igneous and metamorphic rocks; hydration of mafic minerals may produce secondary magnetite; a common detrital mineral in sands. See Figure 142.
Alteration: to hematite or limonite.
Characteristic features: color; octahedral habit; strongly magnetic.
Look-alikes: graphite and ilmenite tend to form hexagonal plates; graphite is very soft, and ilmenite may be partially altered to leucoxene.

Pyrite

FeS_2
Isometric
Pleochroism and color: brass yellow with metallic luster in reflected light; opaque.
Habit: euhedral cubes and pyritohedra common; outlines are commonly square, rectangular, triangular, or hexagonal; also massive or granular.
Cleavage: none.
Twinning: interpenetration with twin axis [110].
Hardness: 6½.
Specific gravity: 5.0.

Occurrence: the commonest sulfide mineral; primary and secondary in many igneous and metamorphic rocks; common in veins, fumaroles, and skarns; may be diagenetic in sedimentary rocks.
Alteration: to limonite.
Characteristic features: color; habit; common occurrence.
Look-alikes: marcasite (FeS_2) greatly resembles pyrite, but tends to have a radiating or fibrous texture, and is restricted to sediments and veins; pyrrhotite is magnetic and has a bronze color in reflected light; chalcopyrite has a golden color in reflected light.

Pyrrhotite

$Fe_{1-0.8}S$
Hexagonal
Pleochroism and color: bronze with metallic luster in reflected light; opaque.
Habit: usually massive or granular; euhedral crystals are basal plates.
Cleavage: none, but may have {0001} parting.
Twinning: $\{10\bar{1}2\}$ rare.
Hardness: 4.
Specific gravity: 4.60–4.65.
Occurrence: principally in mafic igneous rocks as an accessory, but may be concentrated into large masses; may be in pegmatites, hydrothermal veins, contact metamorphic rocks, and meteorites.
Alteration: to limonite.
Characteristic features: anhedral habit; color; magnetic.
Look-alikes: pyrite tends to be euhedral and is brassy yellow in reflected light; chalcopyrite has a golden yellow color in reflected light.

Orthite. See Epidote Group (E1).

Orthoclase. See Feldspar Group (F1).

Orthoferrosilite. See Pyroxene Group (P8).

Oxyhornblende. See Amphibole Group (A2).

Palagonite. See Glass (G2).

Paragonite. See Mica Group (M2).

Pargasite. See Amphibole Group (A2).

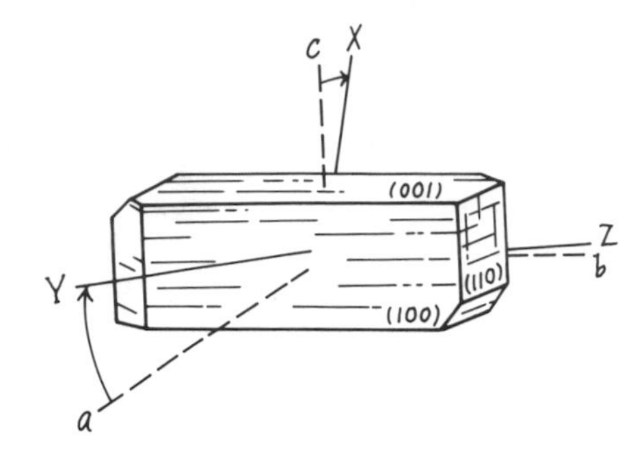

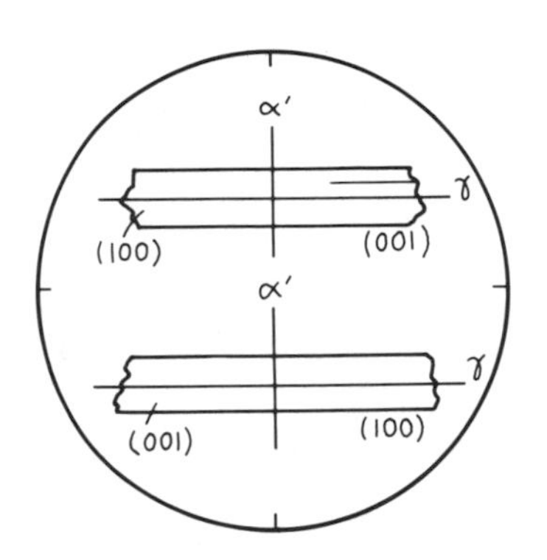

Figure 143. About half of pectolite fragments lie on the perfect {100} cleavage and permit estimation of γ and α' ($\sim$1.60). Extinction is near-parallel and grains are length high (slow). The figure is close to being bisymmetric O.N. The other fragments lie on the perfect {001} cleavage and permit estimation of γ and α' ($\simeq$1.61). Extinction is near-parallel and grains are length high (slow). The figure is close to being monosymmetric Bxo.

PECTOLITE (P1)

$NaCa_2Si_3O_8(OH)$
Triclinic
$\measuredangle\alpha = 90°31'$, $\measuredangle\beta = 95°11'$, $\measuredangle\gamma = 102°28'$
$\alpha = 1.595–1.610$, $\beta = 1.605–1.615$, $\gamma = 1.632–1.645$
$\gamma - \alpha = 0.030–0.038$ (2nd-order orange)
$2V = 50°–63°$ (+)
$a \wedge Y = -10°$ to $-16°$, $b \wedge Z \simeq 2°$, $c \wedge X \simeq -5°$ to $-11°$
Pleochroism and color: colorless.
Dispersion: $r > v$ weak.
Habit: radiating groups of acicular crystals (elongated on *b*); rarely fine-grained and massive.
Cleavage: {100} and {001} perfect at 85° and 95° (Fig. 143).
Extinction in section: essentially parallel to cleavage traces and dominant faces (as *b* is almost parallel to *Z*); grains are length high (slow).
Twinning: {100} simple fairly common.
Hardness: 4½–5.
Specific gravity: 2.75–2.90.
Occurrence: most common in cavities and veins in mafic volcanic rocks, with prehnite, datolite, zeolites, and calcite; less common in phonolite, nepheline syenite, and contact metamorphic marble.
Alteration: not reported, except to stevensite ($Mg_3Si_4O_{10}(OH)_2$).
Characteristic features: occurrence; moderate birefringence and relief; two perfect cleavages with parallel extinction.
Look-alikes: wollastonite has lower birefringence and is length low; pyroxenes with moderate birefringence have extinction inclined to cleavage traces.

Penninite. See Chlorite Group (C4).

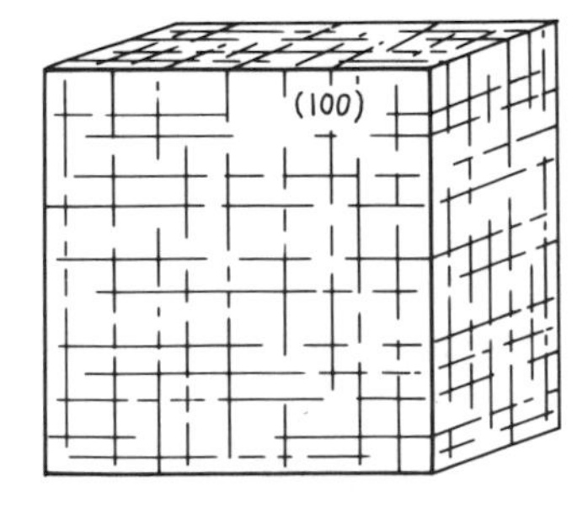

PERICLASE (P2)

MgO
Isometric
$n = 1.736–1.745$
Isotropic
Pleochroism and color: colorless; iron-rich varieties may be pale yellow or brown.
Habit: euhedral cubes and octahedra; also anhedral granular masses.

Cleavage: {100} perfect (Fig. 144).
Twinning: {111}.
Hardness: 5½–6.
Specific gravity: 3.56–3.65.
Occurrence: in thermally metamorphosed carbonate rocks due to decarbonation of dolomite or magnesite; commonly with forsterite, serpentine, calcite, and CaMg silicates.
Alteration: easily altered to brucite, which commonly forms rims about unaltered periclase relics.
Characteristic features: high relief; isotropic; cubic cleavage; occurrence; brucite alteration.
Look-alikes: spinel and garnet do not have perfect {100} cleavage; spinel may have prominent {111} partings that intersect at angles other than 90°.

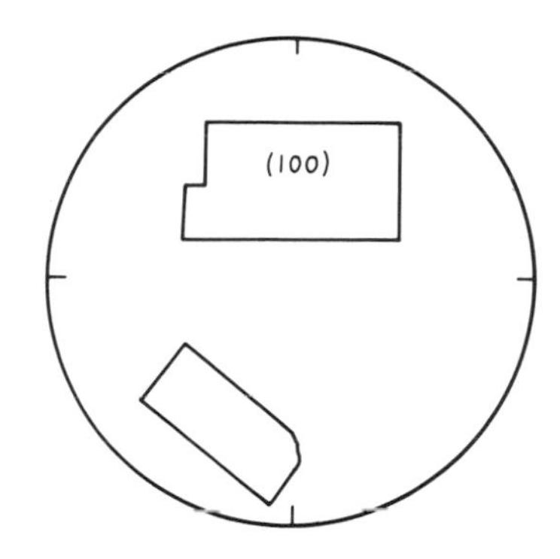

Figure 144. All periclase fragments lie on {100}, have a square to rectangular outline, have a single index (1.736–1.745), and are isotropic.

Pericline. See Feldspar Group (F1).

Perlite. See Glass (G2).

PEROVSKITE (P3)

$CaTiO_3$
Orthorhombic (essentially isometric)
$\alpha, \beta, \gamma \simeq 2.30–2.38$
$\gamma - \alpha = 0.000–0.002$ (1st-order gray)
$2V$ is essentially indeterminate, but reported as 0°–90° (−)
$a = X, b = Z, c = Y$
Pleochroism and color: yellow, orange-brown, or light brownish red; abs. $Z > X$ weak; darker-colored varieties are semiopaque and transmit light only in very thin fragments.
Habit: tiny euhedral cubic or octahedral crystals.
Cleavage: cubic {100} poor (Fig. 145).
Extinction in section: commonly appears isotropic; birefringent varieties have inclined extinction due to complex twinning.
Twinning: polysynthetic {111} very common.
Hardness: 5½.
Specific gravity: 3.98–4.26.
Occurrence: a common accessory mineral in silica-poor mafic, ultramafic, and highly alkaline igneous rocks; also in metamorphic limestones, chlorite schists, and talc schists.
Alteration: to leucoxene.
Characteristic features: small euhedral cubic or octahedral crystals; extreme relief; yellowish or brownish to semiopaque; polysynthetic twinning.

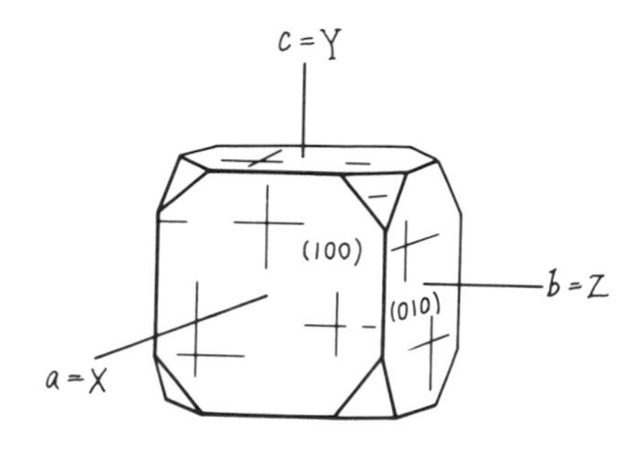

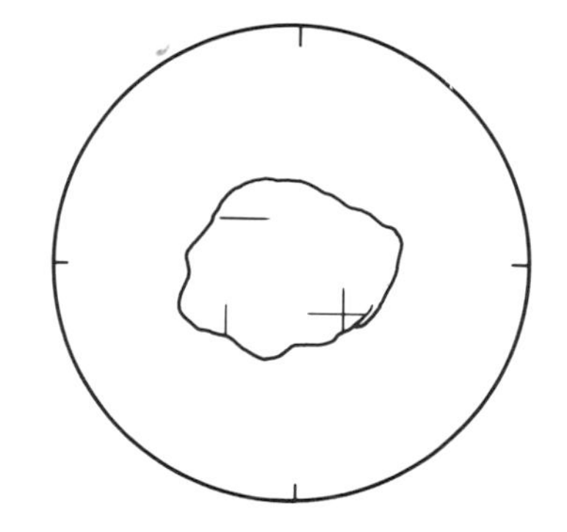

Figure 145. Perovskite fragments lie in random orientation, are transparent to opaque, permit estimation of a single index (2.30–2.38), and commonly appear isotropic.

Look-alikes: spinels and garnets have a different habit and lower indices; ilmenite has a different habit.

Phengite. See Mica Group (M2).

Phlogopite. See Mica Group (M2).

Picotite. See Spinel Group (S8).

Piemontite. See Epidote Group (E1).

Pigeonite. See Pyroxene Group (P8).

Pitchstone. See Glass (G2).

Plagioclase. See Feldspar Group (F1).

Pleonaste. See Spinel Group (S8).

POLYHALITE (P4)

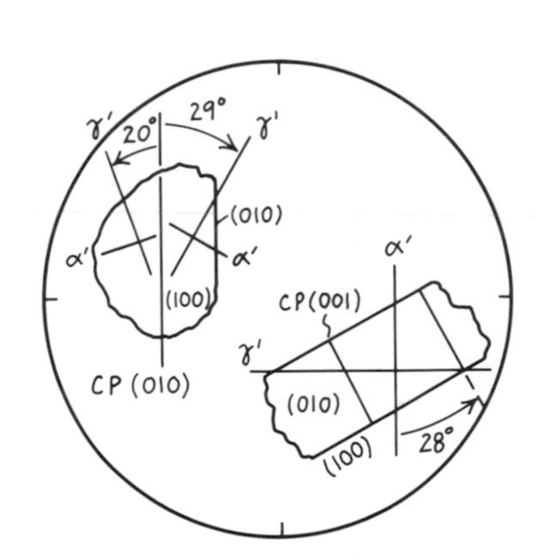

Figure 146. Polyhalite fragments lying on the good {100} cleavage (NW quadrant) commonly show cross-hatch twins ({010} and {001}). Extinction angles of γ' against the {010} composition plane (C.P.) are 20° and 29°. The interference figure shows one optic axis just outside of the field of view (objective, N.A. = 0.85). Fragments lying on {010} cleavage (SE quadrant) show polysynthetic twins with composition plane {001}. Extinction of $\gamma' \wedge$C.P. = 28°.

$K_2Ca_2Mg(SO_4)_4 \cdot 2H_2O$
Triclinic
$\measuredangle\alpha \simeq 91°39'$, $\measuredangle\beta = 90°6.5'$, $\measuredangle\gamma = 91°53'$
$\alpha = 1.547–1.548$, $\beta = 1.560–1.562$, $\gamma = 1.567$
$\gamma - \alpha = 0.019–0.020$ (2nd-order violet)
$2V = 62°–70°\ (-)$
Optical orientation: uncertain; one optic axis is nearly parallel to *b*.
Pleochroism and color: colorless to pale red (due to hematite inclusions).
Dispersion: $v > r$.
Habit: usually massive; fibrous (elongated on *b*) to foliated; spherulitic; {010} tablets; crystals rare.
Cleavage: {100} good, {010} parting good (Fig. 146).
Extinction in section: variable according to habit and orientation.
Twinning: polysynthetic {010} very common; polysynthetic {001} common.
Hardness: 3½.
Specific gravity: 2.78.
Occurrence: in evaporite deposits with anhydrite, halite, magnesite, and kieserite ($MgSO_4 \cdot H_2O$).
Alteration: easily decomposed by water to form gypsum.
Characteristic features: occurrence; moderate birefringence; cross-hatch twinning.

Look-alikes: gypsum has lower indices and birefringence and is optically (+); *glauberite* ($Na_2Ca(SO_4)_2$) has very strong dispersion, a small 2V, and yields a bisymmetric Bxa figure on cleavage surfaces.

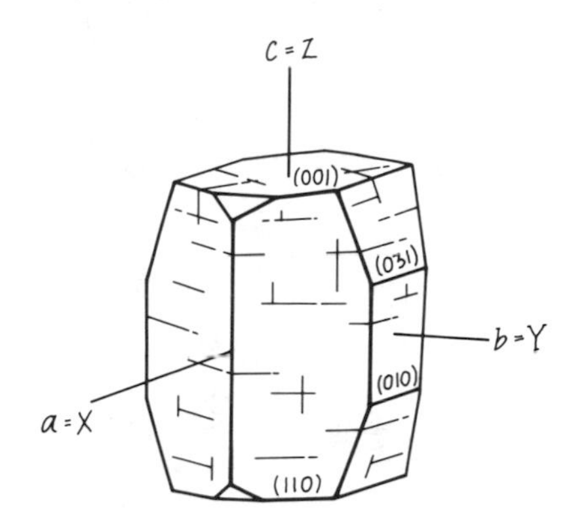

PREHNITE (P5)

$Ca_2Al(AlSi_3)O_{10}(OH)_2$
Orthorhombic
$\alpha = 1.611$–1.630, $\beta = 1.617$–1.641, $\gamma = 1.632$–1.669 (indices increase with substitution of Fe^{+3} for Al)
$\gamma - \alpha = 0.021$–0.039 (2nd-order red, but may show anomalous interference colors)
$2V = 65°$–$69°$, but may be anomalously low, (+)
$a = X, b = Y, c = Z$
Pleochroism and color: colorless.
Dispersion: usually $r > v$ weak, but may be $v > r$ weak to strong.
Habit: tabular parallel to {001} or arranged in spherulitic, sheaflike, or fan-shaped aggregates; some individuals show hourglass structure, in which opposing sectors of a crystal have a radial texture.
Cleavage: {001} good, {110} poor (Fig. 147).
Extinction in section: generally parallel to cleavage traces and dominant crystal faces; length low (fast).
Twinning: one or two sets of fine polysynthetic twins may be present.
Hardness: 6–6½.
Specific gravity: 2.90–2.95.

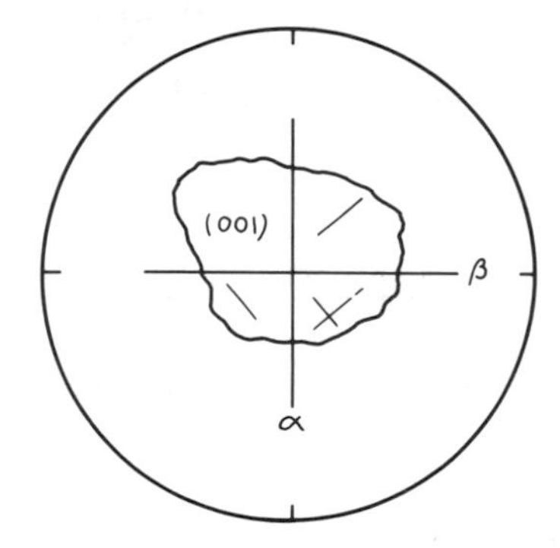

Figure 147. Most prehnite fragments lie on the good {001} cleavage and permit estimation of α and β. Extinction is symmetric to traces of the poor {110} cleavage. The extinction angle of $X \wedge \{110\} = 50°$. The figure is bisymmetric Bxa.

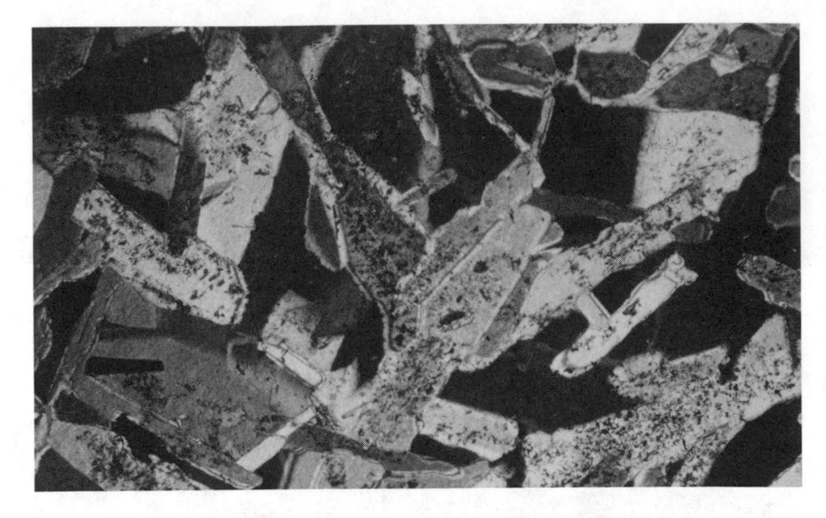

Figure 148. Aggregate of prehnite crystals. Crossed nicols. Photo length is 0.55 mm.

Occurrence: commonly in veins and cavities in ultramafic, mafic, and intermediate igneous rocks, with zoisite, epidote, actinolite, chlorite, zeolites, pectolite, and chalcedony; often in low-grade facies of regional metamorphic rocks, as well as in contact metamorphic carbonate rocks; may be a product of saussuritic alteration of plagioclase. See Figure 148.
Alteration: to zeolites or chlorite.
Characteristic features: anomalous interference colors and hourglass structure; occurrence; moderate birefringence.
Look-alikes: topaz, andalusite, wollastonite, lawsonite, datolite, and the zeolite minerals have lower birefringence.

Prochlorite. See Chlorite Group (C4).

Pseudothuringite. See Chlorite Group (C4).

Pseudowollastonite. See Wollastonite (W1).

Pumice. See Glass (G2).

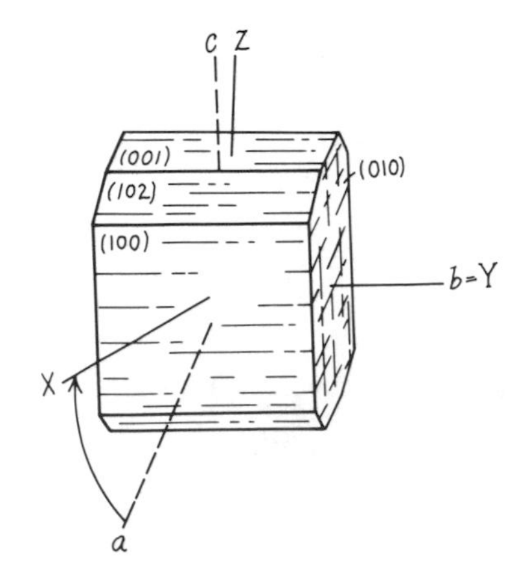

PUMPELLYITE (P6)

$Ca_4(Mg,Fe^{+2},Mn)(Al,Fe^{+3},Ti)_5O(OH)_3[Si_2O_7]_2[SiO_4]_2 \cdot 2H_2O$
Monoclinic
$\measuredangle\beta = 97°36'$
$\alpha = 1.674–1.702$, $\beta = 1.675–1.715$, $\gamma = 1.688–1.722$[1]
$\gamma - \alpha = 0.012–0.022$ (up to 2nd-order blue)
$2V = 26°–85°$ (+); rarely (−)
$b = Y$, $c \wedge Z = 4°$ to $-24°$

[1] A member of the spinel mineral group.

Pleochroism and color: various shades of green, yellow, or (rarely) brown: strongly pleochroic with $Y > Z > X$; X = colorless, pale greenish yellow, pale blue-green; Y = blue-green, deep green; Z = colorless, yellow, yellow-brown, red-brown.
Dispersion: $v > r$ strong.
Habit: fibrous, bladed, or acicular crystals that may be oriented in parallel, radial, or random orientation; direction of elongation is parallel to b.
Cleavage: {001} and {100} good (Fig. 149).
Extinction in section: commonly parallel to major faces and cleavage traces on elongated grains; length low (fast) or high (slow) as a function of orientation (Fig. 149).
Twinning: {100} and {001} relatively common.
Hardness: 6.
Specific gravity: 3.18–3.23.
Occurrence: fairly common in greenschist- and blueschist-facies metamorphic rocks, with glaucophane, lawsonite, epidote, zoisite, prehnite, and chlorite; also forms by hydrothermal and deuteric activity in altered mafic rocks (Fig. 150); common in the heavy mineral fraction of detrital sediments.
Alteration: stable.
Characteristic features: fairly high indices; moderate $2V$ (+); pleochroism; strong dispersion.
Look-alikes: epidote is (−) and usually lacks the deeper color of pumpellyite; zoisite and clinozoisite are colorless and dispersion is $r > v$.

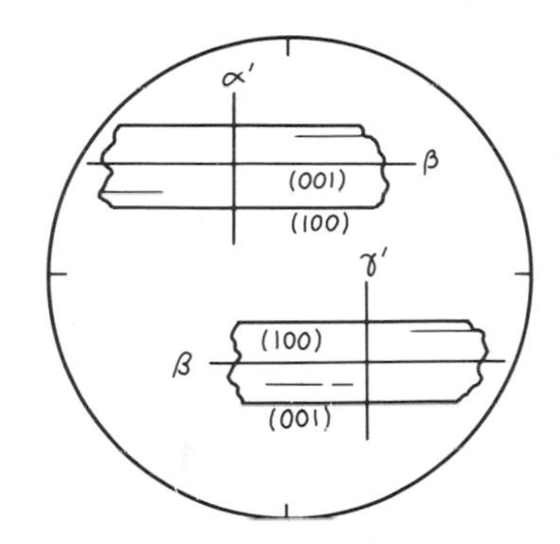

Figure 149. Pumpellyite fragments lying on {001} permit estimation of α' (1.67–1.71) and β. Grains have parallel extinction and are length high (slow). The figure is monosymmetric Bxa. Fragments lying on {100} permit estimation of β and γ' (1.69–1.72). Extinction is parallel and the grain is length low (fast). The figure is monosymmetric Bxo.

Figure 150. Euhedral green high-relief pumpellyite within an amygdule. The gray areas are quartz, and the rock is a highly altered basalt. Plane-polarized light. Photo length is 0.82 mm.

Pycnochlorite. See Chlorite Group (C4).

Pyrite. See Opaque Minerals (O2).

Pyrope. See Garnet Group (G1).

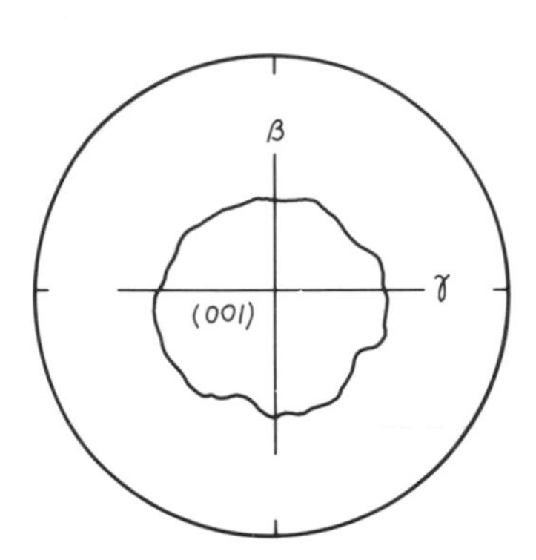

Figure 151. All pyrophyllite grains lie on the perfect {001} cleavage and permit estimation of β and γ. As $\measuredangle\beta$ = 100° and $c \wedge X = 10°$, a bisymmetric Bxa figure is obtained.

PYROPHYLLITE (P7)

$Al_2Si_4O_{10}(OH)_2$
Monoclinic
$\measuredangle\beta = 100°$
$\alpha = 1.534-1.556$, $\beta = 1.586-1.589$, $\gamma = 1.596-1.601$
$\gamma - \alpha \simeq 0.045-0.062$ (up to 4th-order diffuse green)
$2V = 53°-62°$ (−)
$b = Y$, $c \wedge X = 10°$
Pleochroism and color: colorless.
Dispersion: $r > v$ weak.

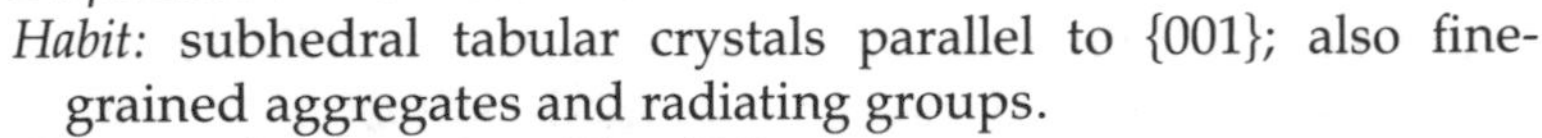

Habit: subhedral tabular crystals parallel to {001}; also fine-grained aggregates and radiating groups.
Cleavage: {001} perfect (Fig. 151).
Extinction in section: parallel to major faces and cleavage traces; length high (slow); bird's-eye maple structure at near-extinction positions; undulatory extinction common.
Twinning: uncommon.
Hardness: 1–1½.
Specific gravity: 2.84.
Occurrence: uncommon; in low-grade metamorphic rocks; also forms as a hydrothermal alteration product of feldspar and aluminum silicates; may be associated with sericite, zoisite, chlorite, chloritoid, rutile, or alunite. See Figure 152.

Figure 152. An aggregate of pyrophyllite. Pyrophyllite can be distinguished from talc and muscovite by its higher optic angle. Crossed nicols. Photo length is 2.22 mm.

Characteristic features: perfect {001} cleavage; bird's-eye maple structure; high birefringence; occurrence.

Look-alikes: talc and muscovite have smaller $2V$s, but may be impossible to distinguish optically when fine grained.

THE PYROXENE GROUP (P8)

Composition

The general formula for the pyroxene group is $(W)_{1-p}(X,Y)_{1+p}$ Z_2O_6, in which W is (Ca,Na), X is (Mg,Fe^{+2},Mn,Ni,Li), Y is (Al,Fe^{+3},Cr,Ti), and Z is (Si,Al). For orthorhombic pyroxenes, $p = 1$; for monoclinic, p ranges from 0 to 1. Most of the rock-forming pyroxenes have compositions that can be located within the *pyroxene quadrilateral* (Fig. 154). The quadrilateral is a portion of the $MgSiO_3$–$FeSiO_3$–$CaSiO_3$ compositional triangle, and includes those compositions that contain 50 percent or less of the $CaSiO_3$ component.

Studies of synthetic and naturally occurring pyroxenes have established the presence of a solvus (miscibility gap) within the pyroxene quadrilateral (shown as the clear area in Figure 153). The solvus, which enlarges with decreasing temperature, represents a compositional area where two pyroxenes, rather than one, exist under equilibrium conditions (with the exception of the $FeSiO_3$ corner, where fayalite and quartz are stable). A consequence of the presence of the solvus is that pyroxenes with a $CaSiO_3$ content between about 15 and 25 percent are virtually unknown. Using a simplified nomenclature, this permits the pyroxenes to be subdivided into a calcium-rich (25–50 percent) mono-

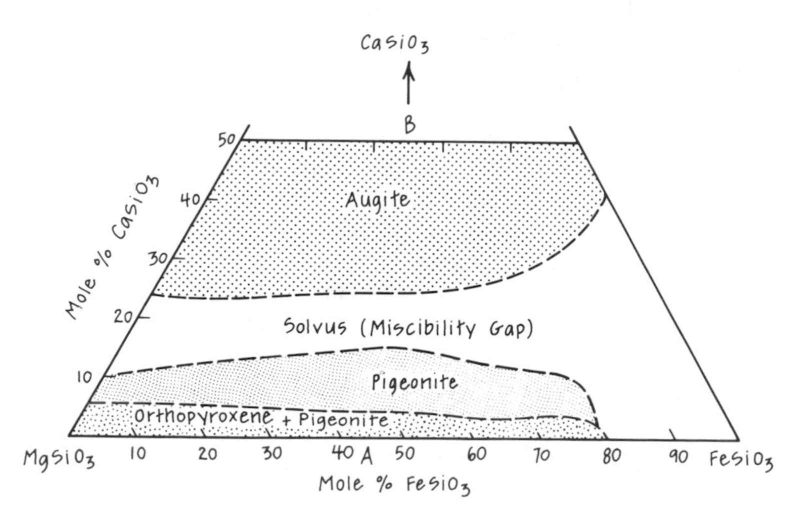

Figure 153. Compositional range of the three principal pyroxene members within the pyroxene quadrilateral (after Huebner, 1980). The clear area is the result of a solvus, within which a single pyroxene cannot stably exist. Temperature decreases from Mg-rich compositions to Fe-rich compositions. The points A and B refer to the cross-section given in Figure 159.

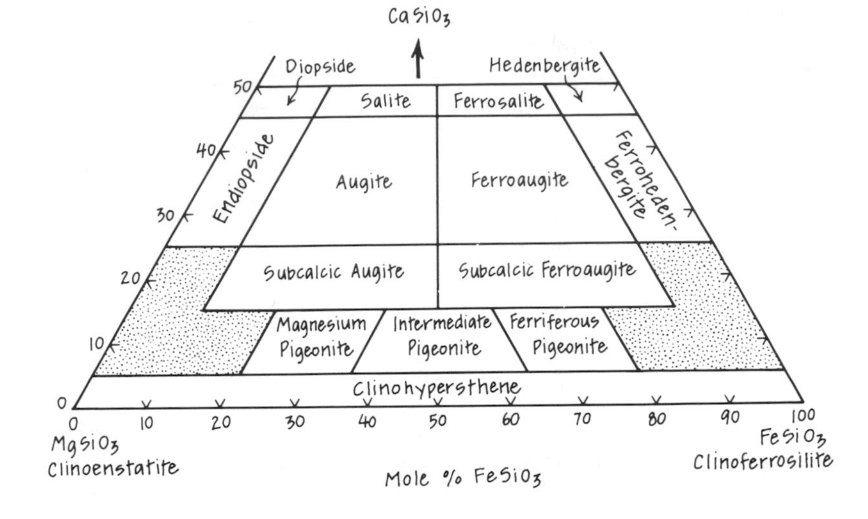

Figure 154. Nomenclature of the clinopyroxenes (after Poldervaart and Hess, 1951).

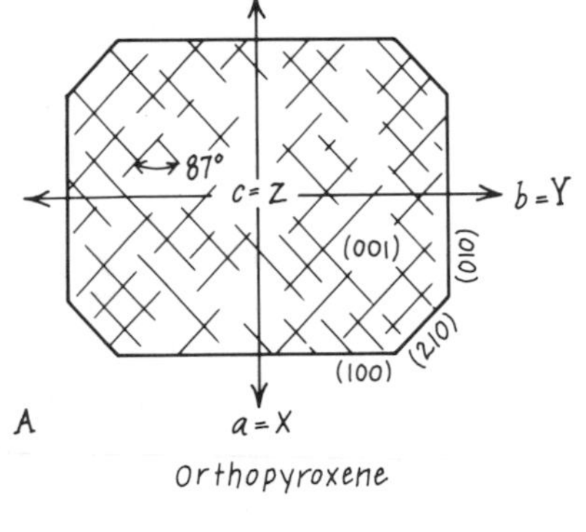

Figure 155. Cross-sections of orthopyroxenes (A) and clinopyroxenes (B) as viewed along the *c* axis. Both present an eight-sided crystal outline. The cleavage traces ({210} in orthopyroxenes and {110} in clinopyroxenes) intersect at 87° and 93°.

clinic variety, called augite, and two calcium-poor varieties—one, monoclinic, called pigeonite (with 0–15 percent $CaSiO_3$), and the other, orthorhombic, called (simply) orthopyroxene (with 0–5 percent $CaSiO_3$). A more detailed nomenclature of the (monoclinic) clinopyroxenes is given in Figure 154.

Crystallography

Pyroxenes may have either monoclinic or orthorhombic symmetry. Orthorhombic varieties are limited in composition to the $MgSiO_3$–$FeSiO_3$ side of the pyroxene quadrilateral, whereas monoclinic varieties cover a wide range of compositions.

Clinopyroxene crystals are usually stubby prismatic (elongated on *c*). The presence of {100}, {010}, and {110} faces typically yields an eight-sided cross-section when viewed along the *c* axis (Fig. 155). The good {110} cleavage surfaces intersect at 87° and 93°, generally permitting the pyroxenes to be easily distinguished from the amphiboles, whose {110} cleavages intersect at about 56° and 124° (Figs. 156–158). However, the presence of {100} partings in pyroxene with unequal development of {110} cleavage may give the illusion of amphibole cleavage both microscopically and macroscopically.

Orthopyroxenes have the same general characteristics, with the exception that the {210} faces and cleavages are well developed; the interfacial angles between orthorhombic {210} faces is essentially the same as the interfacial angle between the monoclinic {110} faces (due to the doubling of the monoclinic *a* unit cell dimension in orthorhombic pyroxenes).

Figure 156. Hypersthene (untwinned) and plagioclase (twinned) in norite. Crossed nicols. Photo length is 2.06 mm. (From J. Hinsch, E. Leitz, Inc.)

Figure 157. Clinopyroxene phenocrysts in a mafic lava. The number of visible cleavages ranges from two to zero as a function of grain orientation. Crossed nicols. Photo length is 0.55 mm.

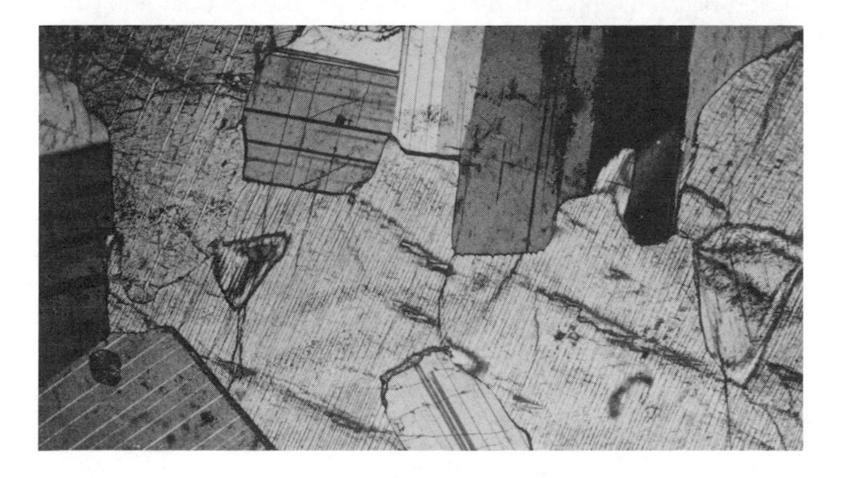

Figure 158. A single grain of augite (center) showing one direction of cleavage. The surrounding grains are Ca-rich plagioclase. Crossed nicols. Photo length is 2.22 mm.

Phase Relations

Melts, whose compositions place them within the pyroxene quadrilateral, show a decrease in crystallization temperatures from Mg-rich compositions toward Fe-rich compositions; in addition, a thermal valley extends downward to the right across the center of the diagram (Fig. 153). Although the thermal crys-

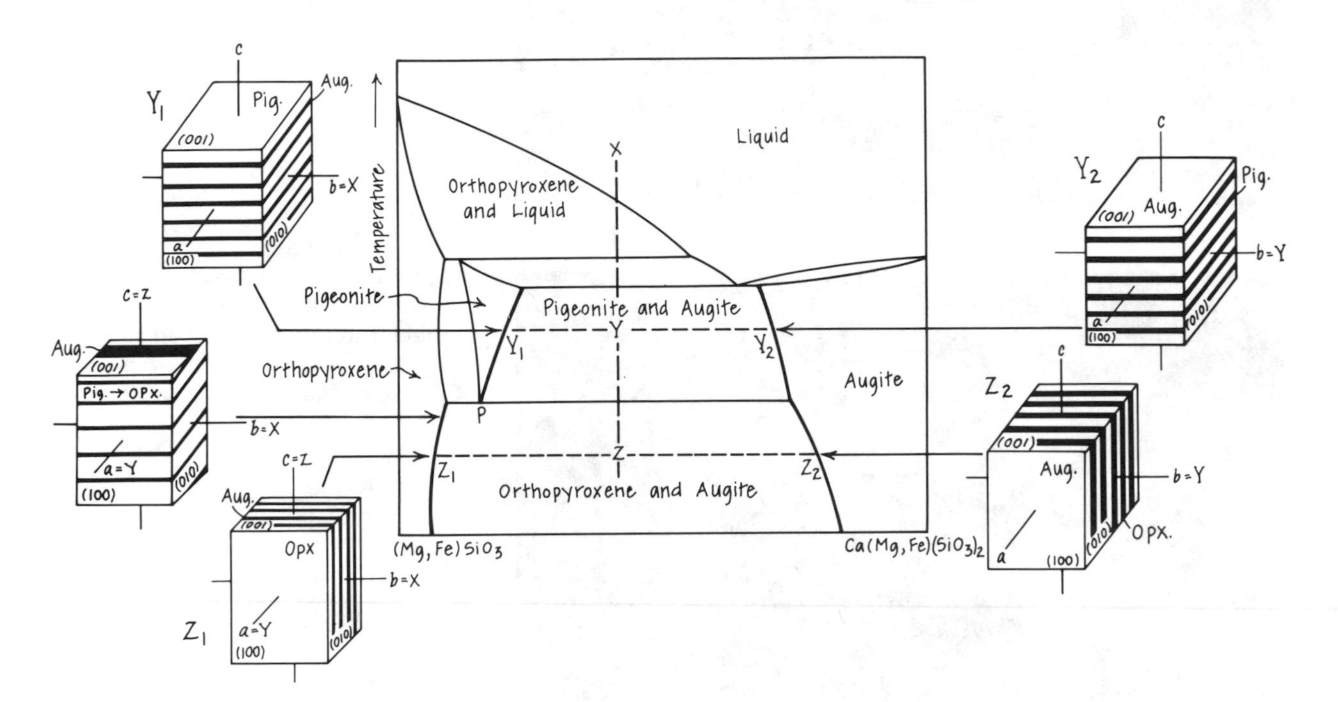

Figure 159. A schematic phase diagram showing the relations between the Ca-rich and Ca-poor pyroxenes (a section such as from A to B in Figure 153). A melt *X*, cooled to *Y*, consists of pigeonite Y_1 and augite Y_2. With slow cooling, Y_1 and Y_2 change in composition along the solvus (heavy lines); this change is accomplished by exsolution of pigeonite from augite, and exsolution of augite from pigeonite (see inserts). Although the inserts show exsolution only on (001) surfaces, such exsolution may occur on surfaces parallel to the *b* axis that are within 20° of either (001) or (100) (Robinson and others, 1971; Jaffe and others, 1975). With additional cooling, pigeonite converts to orthopyroxene (at temperature P); earlier exsolution lamellae of augite may be retained (as shown in insert just above Z_1). Slow cooling of composition *Y* to temperature *Z* yields orthopyroxene Z_1 and augite Z_2. Continued cooling results in compositional changes of both phases along the solvus, with additional exsolution (see inserts). Note that two or even three sets of lamellae may be present within the same grain; these lamellae reveal the cooling history of the enclosing grain.

tallization surface (liquidus) may be lowered (or modified) by the addition of other chemical components (as in a magma), or raised by an increase in the confining (lithostatic) pressure, the general decrease in crystallization temperatures toward Fe-rich compositions is maintained. With fractionation, basaltic magmas change in composition along the liquidus surface to become more Fe-rich.

The presence of a solvus often results in the primary crystallization of both Ca-rich and Ca-poor pyroxenes (somewhat similar to the crystallization of alkali feldspars). With slow cooling (and widening of the solvus with decreasing temperature) unmixing often occurs; augite may exsolve lamellae of orthopyroxene and/or pigeonite, and Ca-poor pyroxene (either orthopyroxene or pigeonite) may exsolve lamellae of augite (Fig. 159 and color photo 49).

The situation is complicated by the existence of two Ca-poor pyroxenes. The relationship between monoclinic pigeonite and orthopyroxene is shown in the incomplete phase diagram in Figure 160 (as well as Fig. 159). Note that for any chosen Mg:Fe

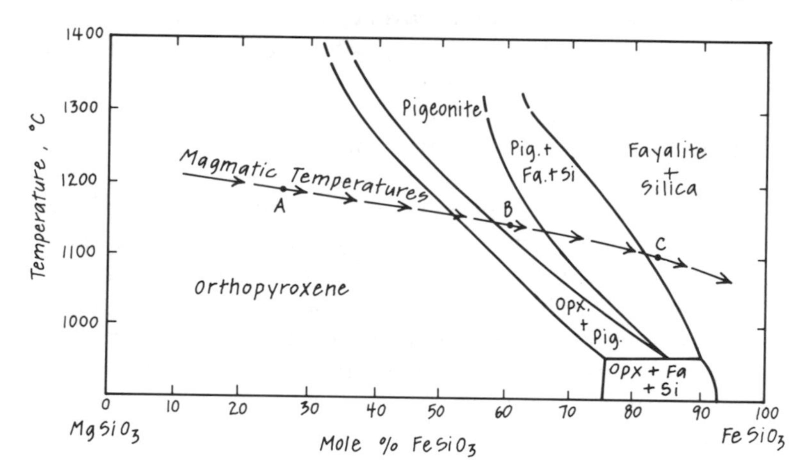

Figure 160. A portion of the binary system $MgSiO_3$–$FeSiO_3$ (after Huebner and Turnock, 1980). At any composition, pigeonite is stable at higher temperature than orthopyroxene. Magmatic crystallization temperatures (added schematically), depressed by the presence of many components, fall within the solid phase fields of this two-component system. Mg-rich magmas (point A) crystallize orthopyroxene without prior crystallization of pigeonite. Intermediate compositions (point B) crystallize pigeonite as a primary phase; pigeonite may convert to orthopyroxene with slow cooling. In extremely Fe-rich melts (point C) the compositional limits of both Ca-poor pyroxenes are exceeded, and fayalitic (Fe-rich) olivine crystallizes with silica.

ratio, pigeonite is stable at a higher temperature than the corresponding orthopyroxene. When cooled slowly, pigeonite converts to orthopyroxene; as this conversion is not usually isochemical (because most pigeonites contain more $CaSiO_3$ than orthopyroxene), it is accompanied by exsolution of irregular blebs of augite (as at point P in Figure 159).

Addition of components other than $MgSiO_3$, $FeSiO_3$, and $CaSiO_3$ to the pyroxene quadrilateral causes a lowering of pyroxene crystallization temperatures. In a typical magmatic situation (see Fig. 160), Mg-rich melts crystallize *within* the (binary) orthopyroxene stability field (point A); orthopyroxene forms without prior crystallization of pigeonite. With decrease in the Mg:Fe ratio (point B), pigeonite crystallizes from the magma. In the rare case of an extremely low Mg:Fe ratio (point C) neither of the Ca-poor pyroxenes is stable, and the melt crystallizes a mixture of fayalitic olivine and silica.

Occurrence

The pyroxenes that crystallize at low pressure from mafic magma commonly consist of orthopyroxene, pigeonite, and/or augite. Shown in Figure 161 are the compositions of a variety of pyroxene pairs, compiled from natural occurrences. The $CaSiO_3$ content of coexistent Ca-poor and Ca-rich pyroxenes reveals the width of the solvus at magmatic temperatures.

Considering the relationships shown between the Ca-poor pyroxenes (Fig. 160), it is reasonable to suppose that the three principal types of pyroxenes should occasionally exist in equilibrium with a mafic melt. Such naturally occurring three-pyroxene assemblages are shown in Figure 162. As the pigeonite-to-orthopyroxene conversion temperature is strongly dependent on composition (Fig. 160), it follows that the compositions of the three-pyroxene assemblages represent a large range in temperature of crystallization.

The compositions of pyroxenes that have formed (or equilibrated) under low- to medium-grade metamorphic conditions are shown in Figure 163. The large difference in $CaSiO_3$ content of the coexisting pyroxene pairs reveals that the solvus is very broad under these relatively low temperature conditions. Note that pigeonite is not found in such environments; instead, orthopyroxene ranges in composition to a maximum of about 80 percent Fe/Fe+Mg (unless extended by either high pressure or the $MnSiO_3$ component).

The pyroxenes that are characteristic of high-pressure environments are omphacite, jadeite, diopside, and enstatite. Substitution of some Al in pyroxene is common at high pressures.

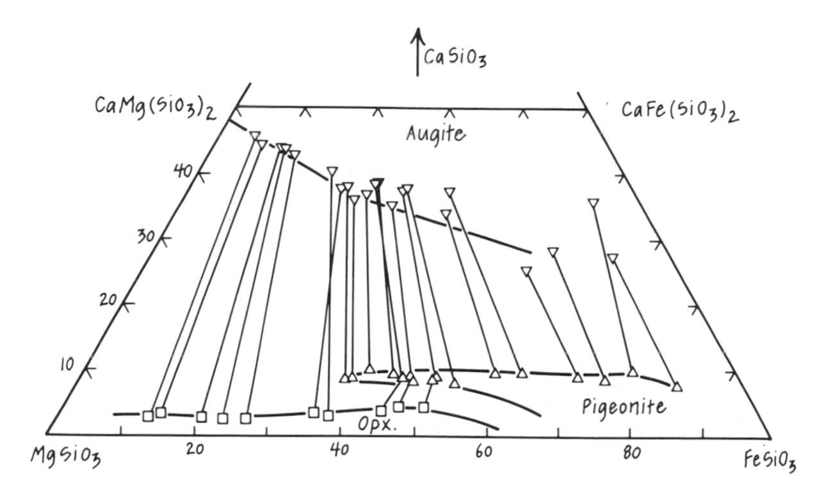

Figure 161. The compositions of pyroxenes that are inferred to coexist with mafic magma at low pressure (after Huebner, 1980, from earlier sources).

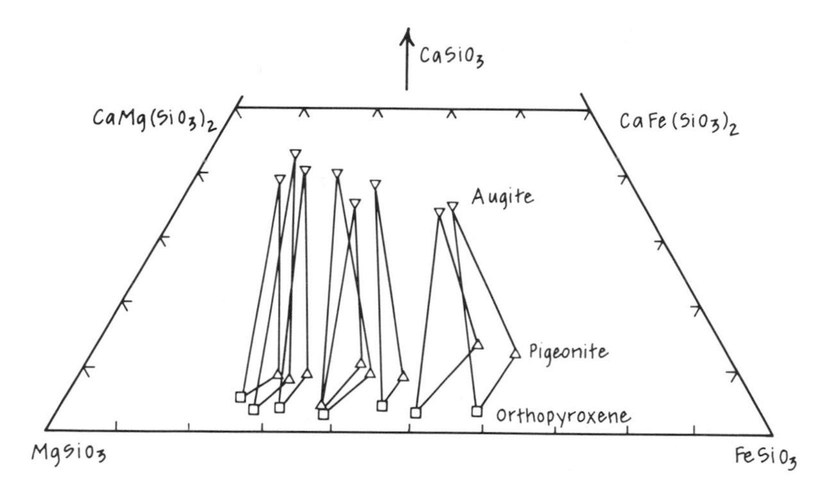

Figure 162. The compositions of three-pyroxene assemblages that coexist with mafic magma at low pressures (after Huebner, 1980, from earlier sources).

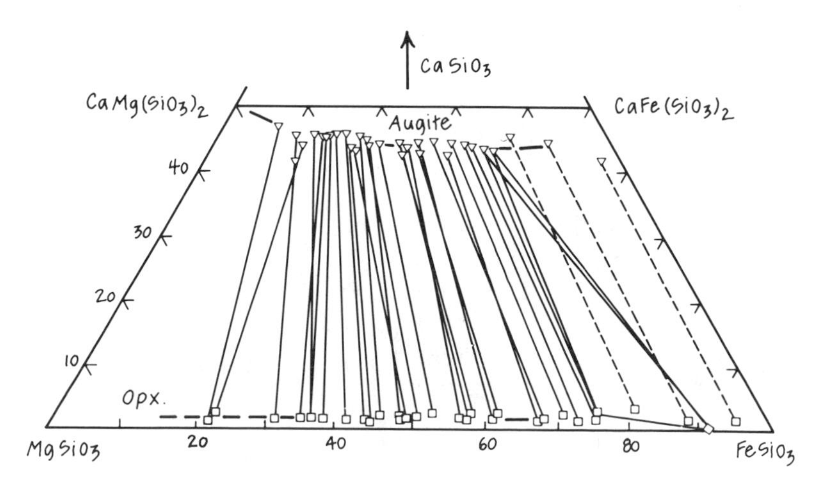

Figure 163. The compositions of coexisting pyroxenes that exist at several hundred degrees below melting temperatures. Note the absence of pigeonite (after Huebner, 1980, from earlier sources).

Optical Characteristics of Clinopyroxenes within the Pyroxene Quadrilateral

The optical orientation of the clinopyroxenes changes as a function of composition. As shown in Figure 164A, the optic axial plane of the more Ca-rich varieties is parallel to {010}, whereas

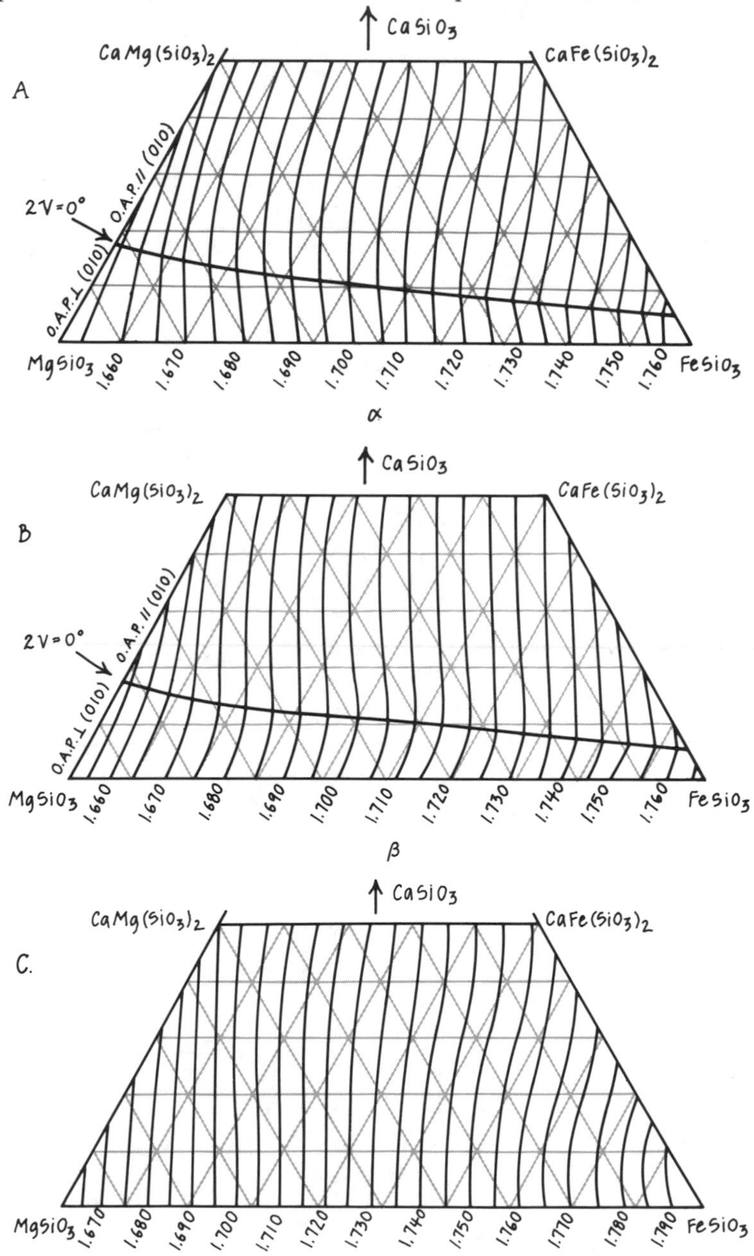

Figure 164. (A) Variation in optic orientation, and α in Ca-Mg-Fe clinopyroxenes as a function of composition (Hess, 1949). (B) Variation in β in Ca-Mg-Fe clinopyroxenes as a function of composition (Hess, 1949). (C) Variation in γ in Ca-Mg-Fe clinopyroxenes as a function of composition (Hess, 1949).

the optic axial plane is perpendicular to {010} in the Ca-poor varieties. This is related to the fact that the α and β refractive indices become identical for certain pigeonite compositions (along the diagonal line); such pigeonite compositions have a $2V$ of 0°. Note that a fair estimate of the Mg : Fe ratio of a clinopyroxene can be obtained by determination of one of the three refractive indices (see also Fig. 164B and C).

A single index determination combined with an estimate of $2V$ is often sufficient to determine the pyroxene variety. Figure 165A shows the range in $2V$ and $\gamma \wedge c$ with composition, and Figure 165B shows the range in birefringence with composition.

When working with immersion mounts it is often convenient to determine refractive indices and extinction angles on {110} cleavage surfaces. As all pyroxenes have a good {*hk*0} cleavage, reasonable approximations of composition can be obtained by measurement of α',γ', and $\gamma' \wedge c$ (Fig. 166). Note that $\gamma \wedge c$ and $\gamma' \wedge c$ vary considerably from the listed values as a result of minor compositional substitution. Thus the clinopyroxenes are best identified by a combination of index determination and $2V$.

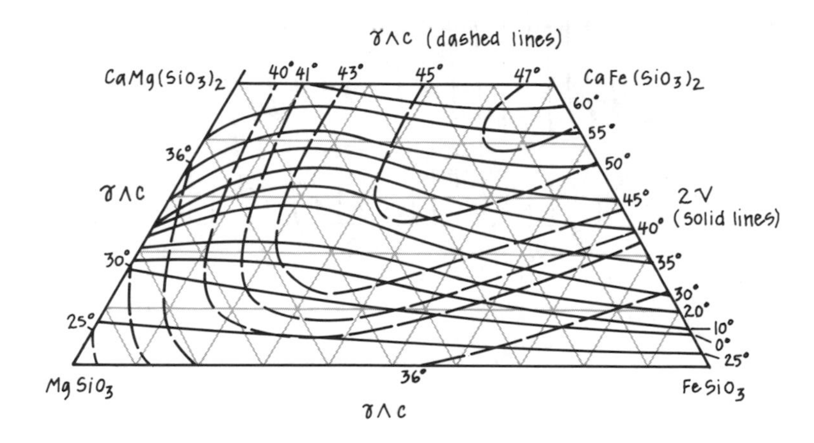

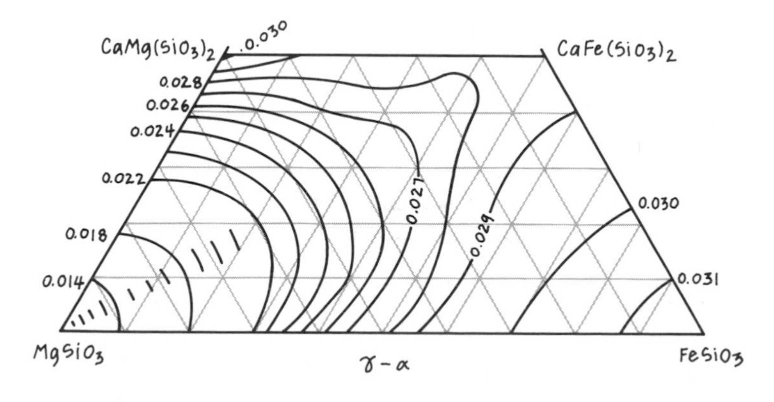

Figure 165. (A) Variation in $2V$ and $\gamma \wedge c$ in Ca-Mg-Fe clinopyroxenes with composition (modified from Tomita, 1934, Deer and Wager, 1938, and Muir, 1951). (B) Variation of birefringence in the Ca-Mg-Fe clinopyroxenes with composition (Hess, 1949).

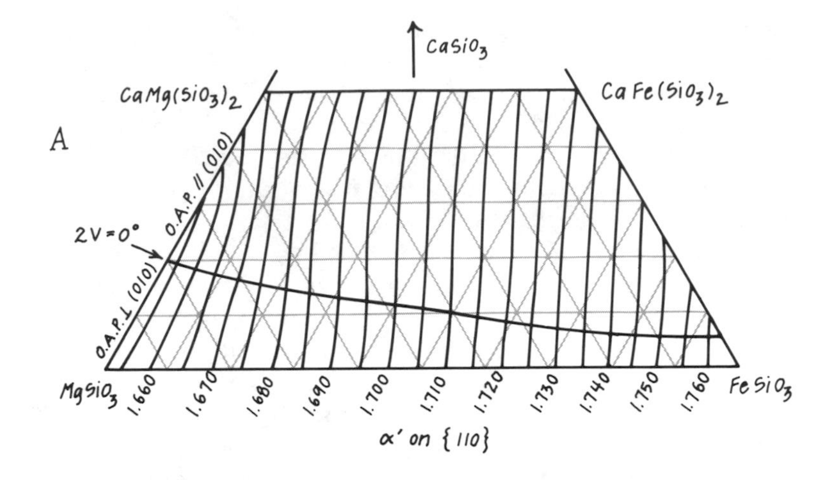

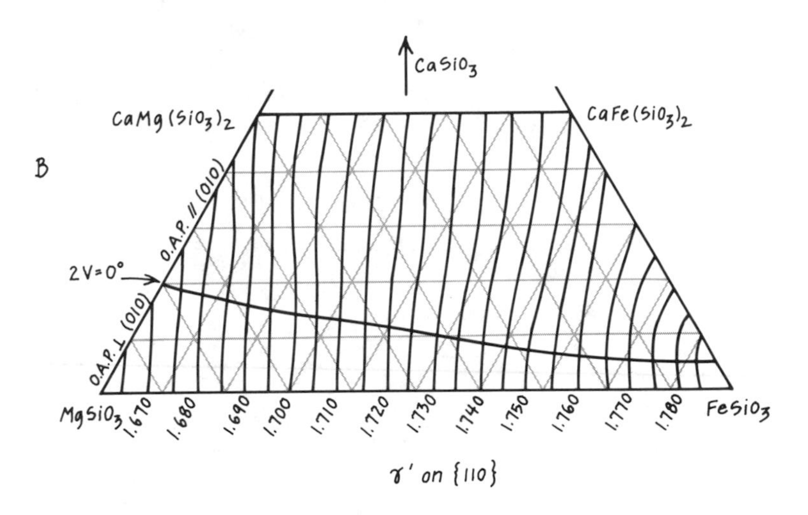

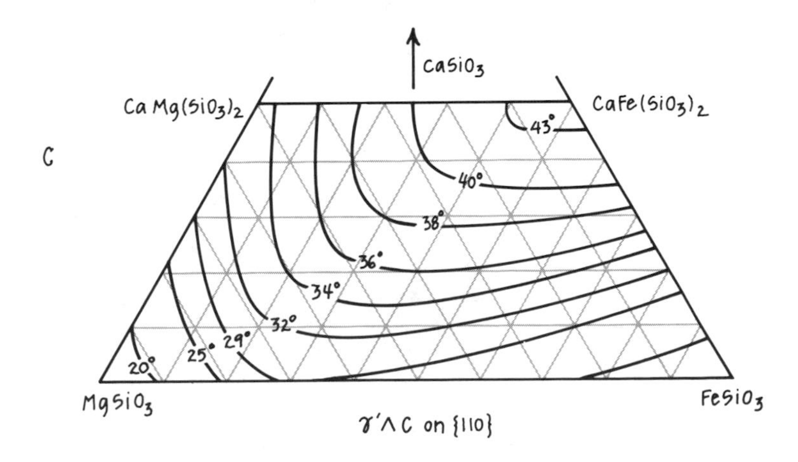

Figure 166. (A) Variation of α' on {110} cleavage surfaces in the Ca-Mg-Fe clinopyroxenes. (B) Variation in γ' on {110} cleavage surfaces in the Ca-Mg-Fe clinopyroxenes. (C) Variation in $\gamma' \wedge c$ on {110} cleavage surfaces in the Ca-Mg-Fe clinopyroxenes. These diagrams have been calculated from the data of Tomita (1934), Deer and Wager (1938), Hess (1949), and Muir (1951).

Enstatite-Orthoferrosilite Series

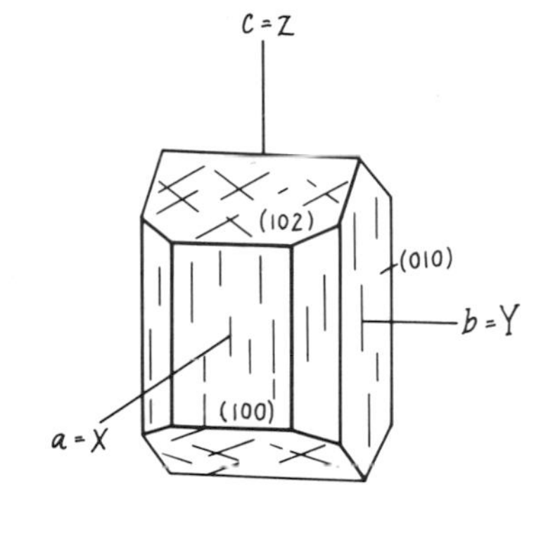

$MgSiO_3$–$FeSiO_3$. Compositions commonly stated in terms of end-members (for example, $En_{60}Fs_{40}$, or simply En_{40}).

Orthorhombic

α = 1.651–1.768, β = 1.652–1.771, γ = 1.660–1.790 (increasing with Fe content; see Figure 167)

$\gamma - \alpha$ = 0.008–0.022 (up to 2nd-order blue, but usually no higher than 1st-order orange—see color photo 50)

$2V$ = medium to large, sign (+) or (−); see Figure 167.

$a = X, b = Y, c = Z$

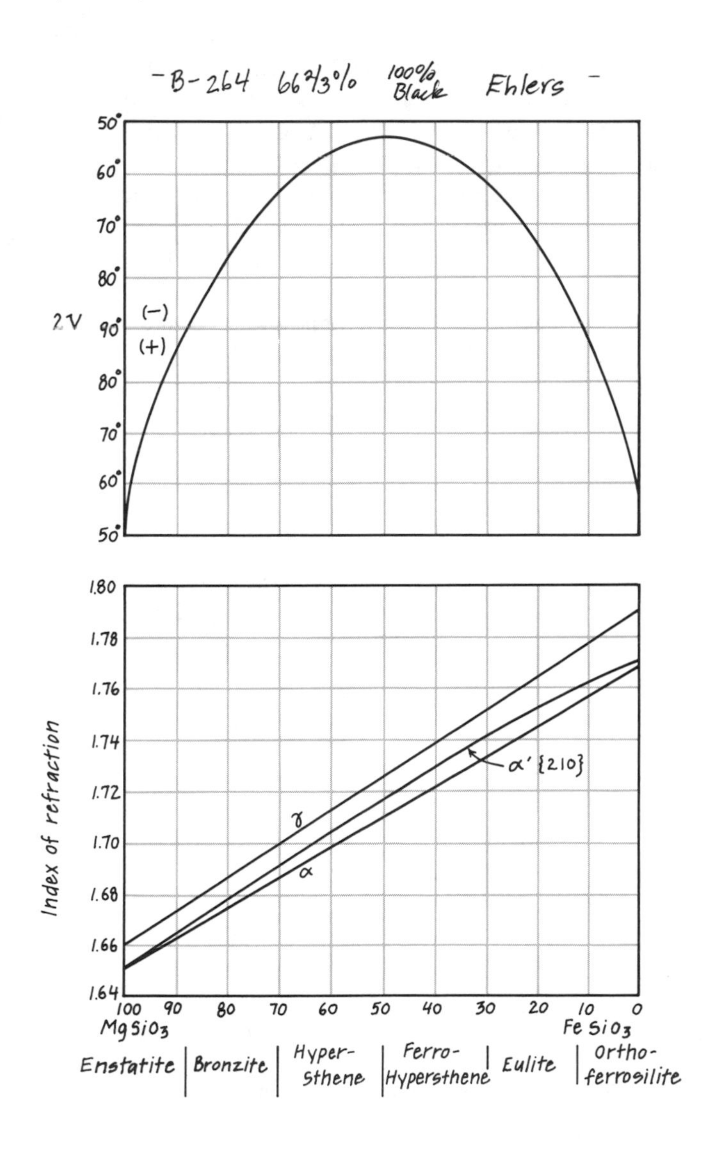

Figure 167. Variation in $2V$ and indices of refraction in the orthopyroxene series (after Leake, 1968). Note that the compositional limits of enstatite and orthoferrosilite have been chosen so as to make both optically positive. In addition, the curve labeled α'\{210\} gives the α' index of refraction for fragments lying on a \{210\} cleavage surface.

Pleochroism and color: enstatite is colorless in thin section. With increasing iron content the color may become some shade of pink or pale green. Pleochroism is independent of composition, and may be weak to strong. Hypersthene may be X = pink, brownish pink, pale yellow or red, or brownish red, Y = pinkish or greenish yellow, yellow, pale green, or greenish gray, and Z = light green, bluish green, grayish green, or blue (color photo 51).

Dispersion: En_{100}–En_{50} $r > v$ weak to moderate; En_{50}–En_{12} $v > r$ weak to moderate; En_{12}–En_{0} $r > v$ strong.

Habit: euhedral to subhedral stubby prismatic (elongated on *c*) in volcanic rocks; eight-sided ({100}, {010}, {210}) cross-section when viewed along *c;* subhedral to anhedral in plutonic rocks; one or more sets of exsolution lamellae of augite may be present in slowly cooled primary orthopyroxenes (Fig. 159), or equally common orthopyroxenes that have formed from inversion of pigeonite; closely spaced exsolved opaque plates or blades (schiller structure) may be present.

Cleavage: {210} good (color photo 52), {100}, {010} parting (Fig. 168).

Extinction in section: parallel to principal faces and cleavage when viewed normal to *c;* length high (slow). When viewed parallel to *c*, extinction is parallel to {100} and {010} faces and symmetric to {210} faces and cleavages.

Twinning: simple and polysynthetic {100} common.

Hardness: 5–6.

Specific gravity: 3.21–3.96.

Occurrence: Mg-rich varieties are common in ultramafic rocks; orthopyroxenes of En_{100} to En_{50} are common in mafic to intermediate igneous rocks, as well as charnockites, granulites,

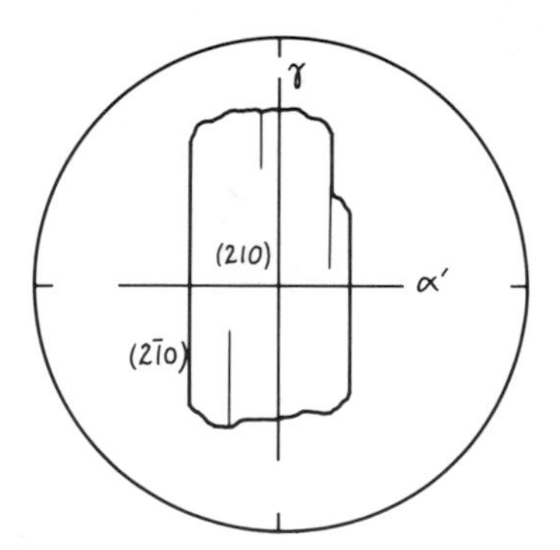

Figure 168. Orthopyroxene fragments on the {210} cleavage surface permit estimation of γ and α' (see Fig. 167). Extinction is parallel and grains are length high (slow). The figure is monosymmetric O.N.

and pelitic metamorphic rocks; Fe-rich orthopyroxene, olivine, and amphibole may be present in metamorphosed iron-rich sediments.

Alteration: fairly easily to serpentine, chlorite, or amphibole (uralite).

Characteristic features: eight-sided outline with two almost perpendicular cleavages; parallel extinction in prismatic section; moderate to large $2V$; low birefringence; occurrence. See color photo 53.

Look-alikes: clinopyroxenes generally have moderate to small $2V$s and higher birefringence; andalusite is length low (fast); anthophyllite has a higher birefringence; sillimanite has a single cleavage and a small $2V$.

Diopside-Hedenbergite Series

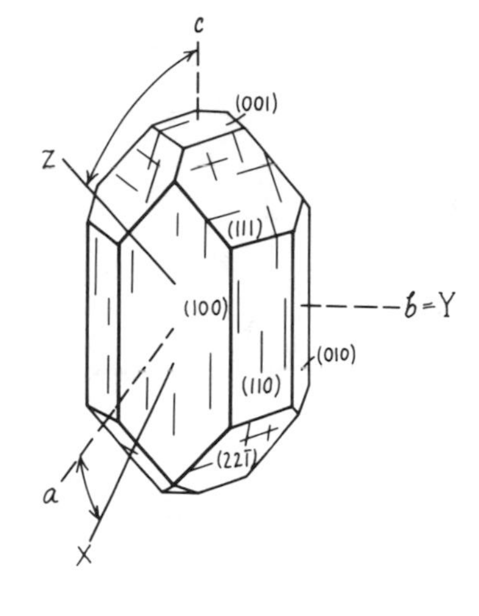

$CaMg(SiO_3)_2$–$CaFe(SiO_3)_2$
Monoclinic
$\measuredangle\beta \simeq 105°$
$\alpha = 1.664–1.732$, $\beta = 1.672–1.739$, $\gamma = 1.694–1.757$ (Fig. 169)
$\gamma - \alpha = 0.025–0.030$ (up to 2nd-order yellow)
$2V = 58°–63°$ (+)
$b = Y$, $c \wedge Z = 38°–48°$

Pleochroism and color: diopside is colorless; with increasing iron content the series becomes pale green to green and strongly pleochroic, with X = pale green, green, or blue-green, Y = brownish green to blue-green, and Z = yellow-green or yellowish blue-green.

Dispersion: $r > v$ weak to strong with increasing iron content.

Habit: euhedral to anhedral; euhedral crystals are stubby prismatic, elongated on c, with an eight- or four-sided outline.

Cleavage: {110} good, at 87° and 93° (Fig. 170); {001} or {100} parting may be good; when fine-grained magnetite is abundant on {001} partings, the variety has been called *diallage.*

Extinction in section: symmetric to cleavage traces, and symmetric or parallel to faces when viewed along the c axis; parallel to major faces and cleavages when viewed along a; inclined ($c \wedge Z = 38°–48°$) to major faces and cleavage traces when viewed along b; with the exception of pure hedenbergite, Z is either parallel or acutely inclined to major faces and cleavage traces when viewed in prismatic section, making the grains length high (slow).

Twinning: {100} or {001} polysynthetic or simple, fairly common.

Hardness: 5½–6½.

Specific gravity: 3.2–3.5.

Figure 169. The range of $2V$, $c \wedge Z$, and indices of refraction in the diopside-hedenbergite series (from the data of Deer, Howie, and Zussman, 1962).

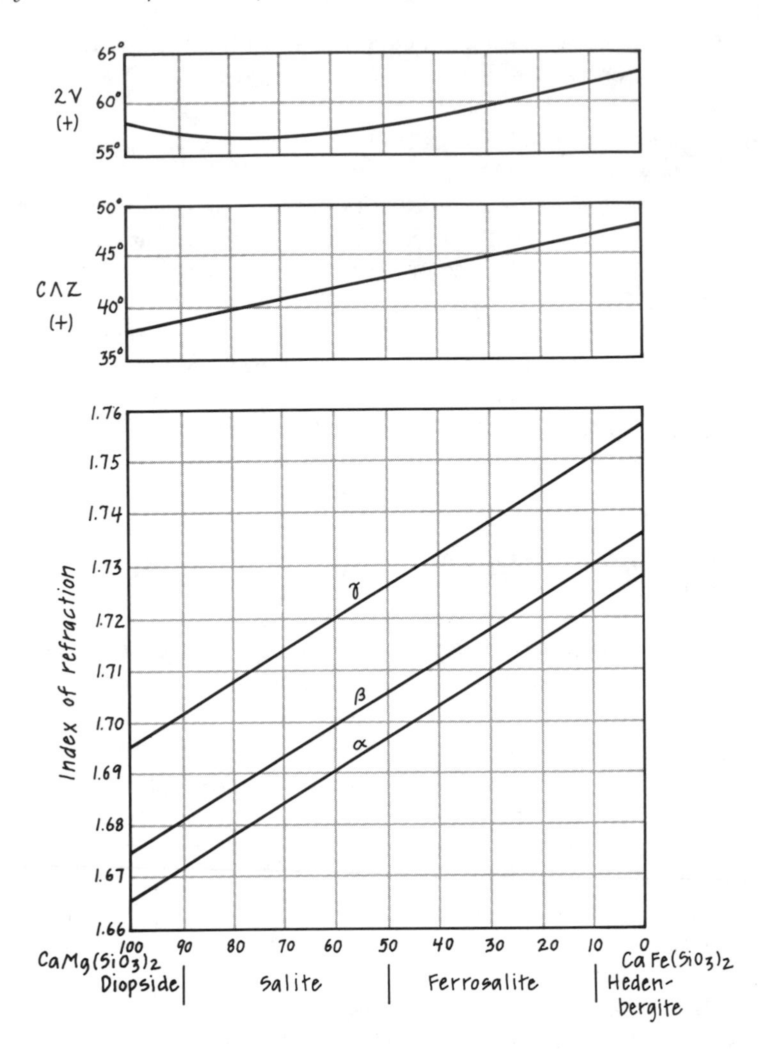

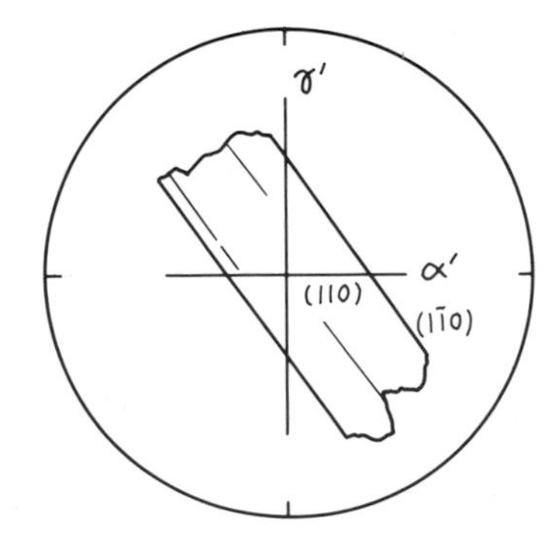

Figure 170. Diopside-hedenbergite-series fragments on the {110} cleavage permit estimation of α' and γ' (see Fig. 166A, B). Extinction is inclined (see Fig. 166C) for all compositions; $\gamma' \wedge \{110\}$ is between 33° and 45°. Grains are length high (slow) and the figure is nonsymmetric.

Occurrence: diopside is often in Ca- and Mg-rich contact and regionally metamorphosed rocks, with forsterite, calcite, wollastonite, grossular, tremolite, or idocrase, as well as in hornfels, gneisses, and granulites, with plagioclase, hypersthene, almandine, or epidote; hedenbergite is found in Fe-rich metasomatic calc-silicate rocks, as well as various ore deposits, and Fe-rich metamorphic rocks; in igneous rocks this series may be found as phenocrysts in alkali-rich mafic extrusives (color photo 54) or in anorthosite, peridotite, lamprophyre dikes, and rarely in pegmatites (through limestone assimilation).

Alteration: commonly to fibrous tremolite or actinolite (uralite).

Characteristic features: commonly in metamorphic environments; moderate $2V$; typical pyroxene cleavages and grain outline.

Look-alikes: essentially indistinguishable from augite and ferroaugite by optical techniques; pigeonite has a small $2V$, and orthopyroxene generally has a parallel extinction in prismatic section; wollastonite has lower birefringence and refractive indices.

Augite

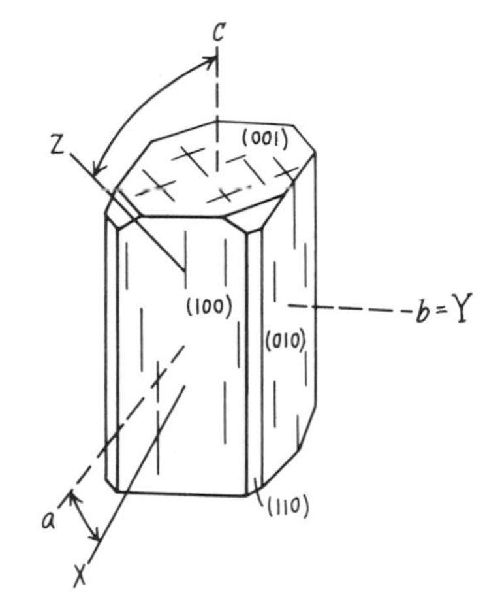

$Ca(Mg,Fe^{+2},Al)[(Si,Al)O_3]_2$ (see Fig. 155)

Monoclinic

$\measuredangle\beta = 105°$

$\alpha = 1.671–1.735$, $\beta = 1.672–1.741$, $\gamma = 1.703–1.761$

$\gamma - \alpha = 0.018–0.030$ (up to 2nd-order yellow)

$2V = 25°–60°$ (+)

$b = Y$, $c \wedge Z = 35°–50°$

Pleochroism and color: colorless to pale green, with weak pale blue-green or yellow-green pleochroism in iron-rich varieties, with Z or $X > Y$; *titanoaugite* has moderate pleochroism, with X = pale violet-brown, Y = violet-pink, and Z = pale brown-violet; preferential absorption of titanium by certain faces may result in wedge-shaped growth sectors of different color and extinction angle (called hourglass structure).

Dispersion: $r > v$ weak to moderate, inclined.

Habit: euhedral to subhedral phenocrysts form stubby prismatic crystals elongated on c; viewed along the c axis direction the crystal outline is eight- or four-sided due to the common presence of {100}, {010}, and {110} faces (color photo 55). Matrix augite is anhedral and interstitial. In rapidly cooled lavas, augite may be present as microlites or as fernlike branching groups. Slowly cooled augite may contain exsolution lamellae of pigeonite or orthopyroxene (see Fig. 159). Overgrowths of amphibole are common.

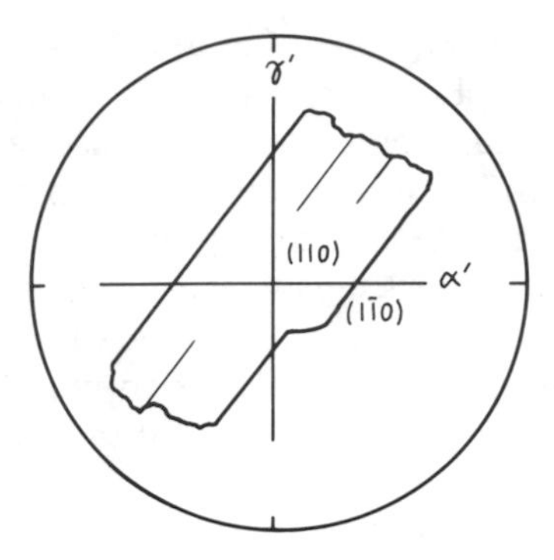

Figure 171. Augite fragments on the {110} cleavage permit estimation of α' and γ' (see Fig. 166A, B). Extinction is inclined (see Fig. 166C); for all compositions $\gamma' \wedge \{110\}$ is between 32° and 40°. Grains are length high (slow) and the figure is nonsymmetric.

Cleavage: {110} good at 87° and 93° (Fig. 171 and color photo 56); parting may be well developed on {100} or {001}.

Extinction in section: when viewed along *c*, extinction is symmetrical to {110} cleavage traces, and parallel and symmetric to major faces; when viewed along *a*, extinction is parallel to cleavage traces and major faces, and grains are usually length high (slow); when viewed along *b*, extinction is inclined to cleavage traces and major faces, and grains are generally length high (slow) as $c \wedge \gamma$ is less than 45° for most compositions.

Twinning: {100} simple or polysynthetic is common; when combined with fine polysynthetic twins on {001}, and noted by observation along *b*, it is referred to as herringbone structure (color photo 57).

Hardness: 5–6.

Specific gravity: 3.2–3.6.

Occurrence: common in mafic and ultramafic igneous rocks; also in metamorphic gneisses and granulites.

Alteration: commonly to fibrous amphibole (uralite); may have reaction rims of hornblende; weathering commonly produces chlorite, and less commonly talc.

Characteristic features: typical clinopyroxene shape, cleavage, and types of extinction; moderate 2*V* (+), with high extinction angle; commonly in igneous rocks (color photo 58).

Look-alikes: cannot be distinguished from diopside by ordinary optical methods; aegirine-augite is distinctly green; olivine has poor cleavage; amphibole {110} cleavages intersect at about 56° and 124°, and most varieties are strongly pleochroic.

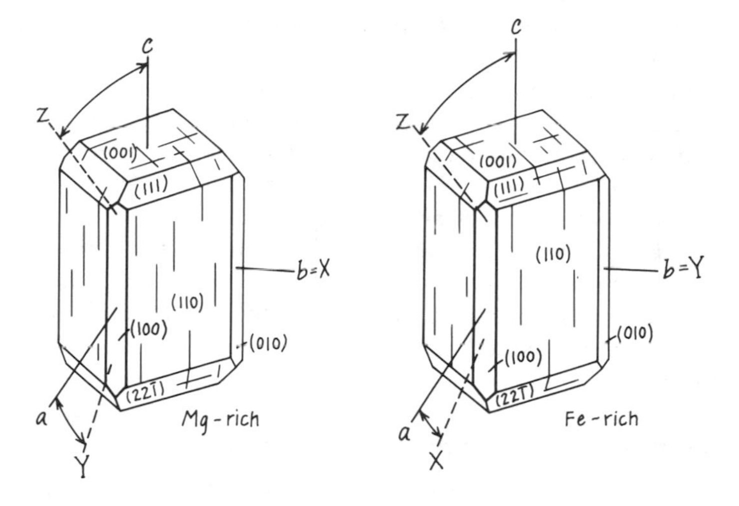

Pigeonite

$Ca(Mg,Fe^{+2},Al)[(Si,Al)O_3]_2$
Monoclinic
$\measuredangle\beta = 108°$
$\alpha = 1.682–1.722$, $\beta = 1.684–1.722$, $\gamma = 1.705–1.751$
$\gamma - \alpha = 0.023–0.029$ (2nd-order yellow)
$2V = 0°–32°$ (+)
$b = X$, $c \wedge Z = 37°–44°$ (more common); $b = Y$, $c \wedge Z = 40°–44°$ (less common)

Pleochroism and color: colorless to pale green or brown: some varieties may be moderately pleochroic, with $Y > X$ or Z.

Dispersion: $r > v$ or $v > r$ weak to moderate.

Habit: euhedral to anhedral; stubby prismatic, elongated on *c* with eight-sided outline when viewed along *c* (due to presence of {100}, {010}, and {110} faces); zoning is common.

Cleavage: {110} good at 87° and 93° (Fig. 172); {001} parting may be present.

Extinction in section: parallel and symmetric when viewed along *c;* inclined when viewed along *b,* and parallel when viewed along *a;* length high (slow) when viewed along *a* or *b.*

Twinning: {100} simple or polysynthetic common.

Hardness: 6.

Specific gravity: 3.30–3.46.

Occurrence: principally with augite or hypersthene as a matrix mineral in volcanic andesite and basalt, or in diabase; uncommon in phenocrysts, and unknown in plutonic rocks.

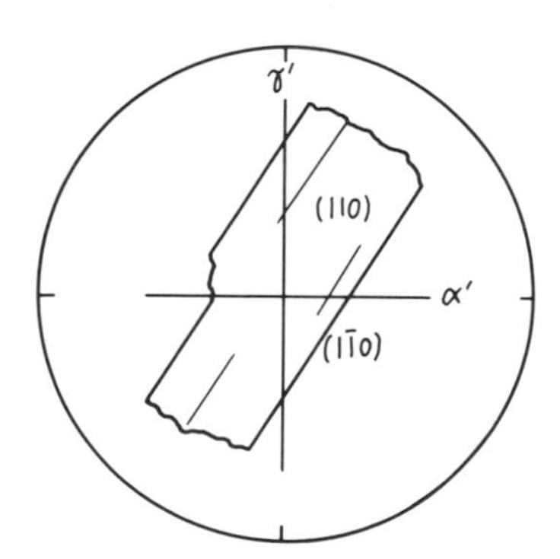

Figure 172. Pigeonite fragments on the {110} cleavage permit estimation of α' and γ' (see Fig. 166A, B). Extinction is inclined (see Fig. 166C); for all compositions $\gamma' \wedge \{110\}$ is between 30° and 35°. Grains are length high (slow) and the figure is nonsymmetric.

Alteration: to fine-grained fibrous amphibole (uralite), or less commonly to chlorite or serpentine; if pigeonite has been inverted to orthopyroxene (as is often the case), its former presence can often be detected by the presence of mosaic grains and lamellae of augite on the relict {001} plane of pigeonite.

Characteristic features: small $2V$; moderate birefringence; typical pyroxene grain shapes and cleavages; occurrence.

Look-alikes: augite, with moderate $2V$, may occur as phenocrysts, and in plutonic igneous rocks; clinoenstatite-clinoferrosilite are extremely uncommon in natural occurrences; olivine has higher birefringence and lacks the good {110} cleavage of pyroxenes.

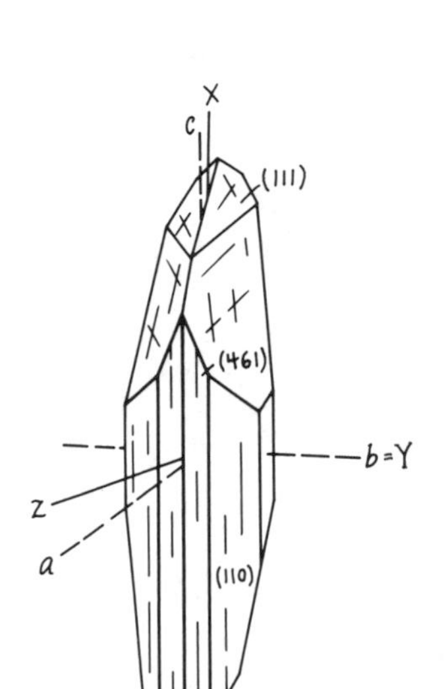

Aegirine—Aegirine-Augite Series

$NaFe^{+3}(SiO_3)_2$ to $(Na,Ca)(Fe^{+3},Fe^{+2},Mg,Al)(SiO_3)_2$. Compositions commonly stated in terms of end-members (for example, Ae_{40}).

Monoclinic

$\measuredangle\beta = 106°$

$\alpha = 1.722–1.776$, $\beta = 1.742–1.820$, $\gamma = 1.758–1.836$ (Ae) (Fig. 173)

$\alpha = 1.700–1.722$, $\beta = 1.710–1.742$, $\gamma = 1.730–1.758$ (AeAug)

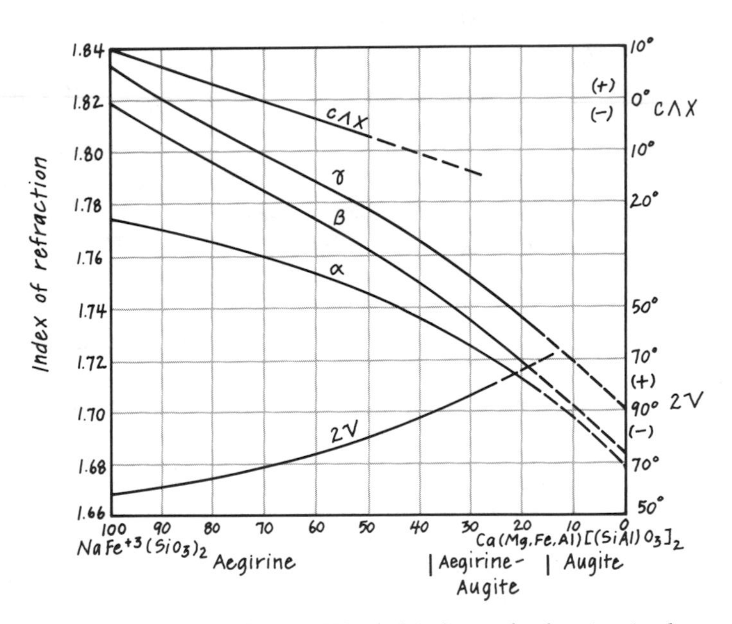

Figure 173. Variation of $c \wedge X$, $2V$, and indices of refraction in the aegirine to aegirine-augite series. (Data from Deer, Howie, and Zussman, 1962)

$\gamma - \alpha = 0.036-0.060$ (up to 4th-order green) (Ae); 0.030–0.036 (up to 2nd-order red) (AeAug)
$2V = 58°-90°$ (−) (Ae); 70°–90° (+) (AeAug) (Fig. 173)
$b = Y$, $c \wedge X = 10°-12°$ (Ae); $b = Y$, $c \wedge X = -12°$ to $-20°$ (AeAug) (Fig. 173)
Pleochroism and color: this series is more highly colored than any of the other common pyroxenes: aegirine is commonly strongly pleochroic in bright green to yellow-green; aegirine-augite is weakly pleochroic in pale green to yellow-green; X = bright green or emerald green, Y = grass green or yellow-green, and Z = pale yellow-green or yellow, with $X > Y > Z$; *acmite* is a variety of aegirine that is pleochroic in brown.
Dispersion: $r > v$ moderate to strong.
Habit: aegirine typically forms euhedral to subhedral elongated or acicular prismatic crystals (parallel to c); aegirine-augite may have the same habit, but tends more towards stubby prismatic crystals (elongated on c), with the typical pyroxene eight-sided cross-section (color photo 59); zoning is common, with brightly colored rims enriched in Fe^{+3}.
Cleavage: {110} good at 87° and 93° (Fig. 174), may have {100} or {001} partings.
Extinction in section: parallel or inclined in prismatic section; grains are length low (fast); when viewed along c, extinction is parallel and symmetric to crystal faces, and symmetric to {110} cleavage traces.
Twinning: {100} simple common.
Hardness: 6.

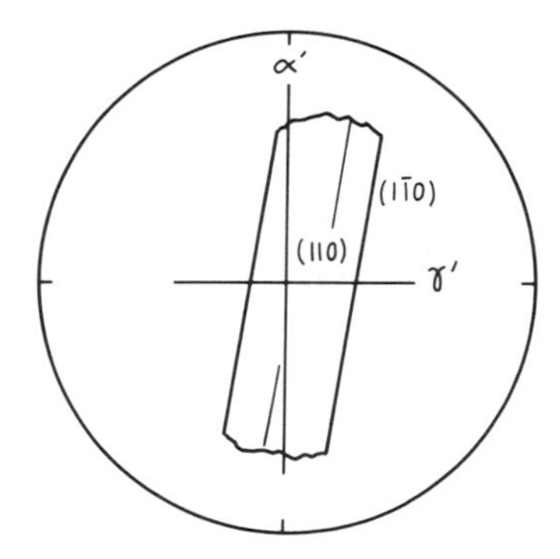

Figure 174. Fragments lying on {110} permit estimation of α' ($\simeq \alpha$) and γ' (1.751–1.829) for aegirine, and α' ($\simeq \alpha$) and γ' (1.721–1.751) for aegirine-augite. Extinction is inclined, with $\alpha' \wedge c = 8°$ to $-11°$ for aegirine, and $\alpha' \wedge c = -11°$ to $-21°$ for aegirine-augite. Cleavage fragments are length low (fast). The figure is nonsymmetric except for Ae_{70}, which is monosymmetric Bxo. (See Figure 173.)

Specific gravity: 3.4–3.6.

Occurrence: typically in alkaline igneous rocks such as syenite, nepheline syenite, trachyte, and phonolite; occasionally in metamorphosed Na-rich rocks, with riebeckite or glaucophane.

Alteration: to chlorite, epidote, and ferric oxide, or to fibrous amphibole (uralite).

Characteristic features: typical pyroxene outline and cleavage; strong coloration and pleochroism; large $2V$ with (−) or (+) sign; length low (fast); occurrence.

Look-alikes: aegirine and aegirine-augite are distinguished from each other by optic sign; other pyroxenes have larger extinction angles and (+) sign, are not as strongly colored or as strongly pleochroic; the {110} cleavages in amphiboles are at about 56° and 124°.

Jadeite

$NaAl(SiO_3)_2$

Monoclinic

$\measuredangle\beta = 107°26'$

$\alpha = 1.654–1.658$, $\beta = 1.657–1.663$, $\gamma = 1.665–1.674$

$\gamma - \alpha = 0.011–0.016$ (up to 1st-order red)

$2V = 68°–72°$ (+)

$b = Y$, $c \wedge Z = 33°–36°$

Pleochroism and color: commonly colorless, but rarely may be weakly pleochroic with X = pale green, Y = colorless, and Z = pale yellow.

Dispersion: $v > r$ moderate.

Habit: fine fibrous aggregates, or anhedral granular.

Cleavage: {110} good at 87° and 93° (Fig. 175), {100} parting uncommon.

Extinction in section: parallel or inclined on prismatic sections; length high (slow); symmetrical extinction against {110} cleavages when viewed along c.

Twinning: {100} fine polysynthetic.

Hardness: 6.

Specific gravity: 3.25–3.40.

Occurrence: in high-pressure, low-temperature metamorphic rocks, in constant association with albite, and occasionally with lawsonite, glaucophane, actinolite, zoisite, analcime, or nepheline.

Alteration: to nepheline and albite at high temperatures; to analcime with weathering.

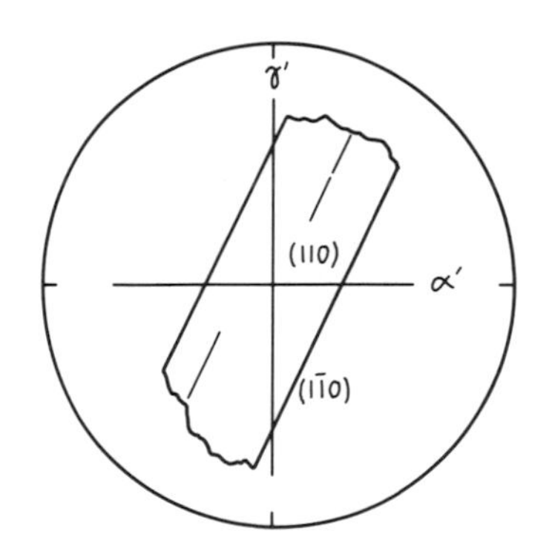

Figure 175. Jadeite fragments lying on {110} permit estimation of α' (1.655–1.660) and γ' (1.664–1.672). Extinction is inclined, with $\gamma' \wedge c$ = 30°–33°, and grains are length high (slow). The figure is nonsymmetric.

Characteristic features: typical pyroxene outline and cleavages; low birefringence and high $2V$ as compared to most other clinopyroxenes; occurrence in high-pressure metamorphic rocks.
Look-alikes: amphiboles are commonly pleochroic, have smaller extinction angles, and intersecting {110} cleavages at about 56° and 124°; omphacite has higher birefringence; diopside and augite have moderate $2V$s; aegirine and aegirine-augite are strongly colored and pleochroic.

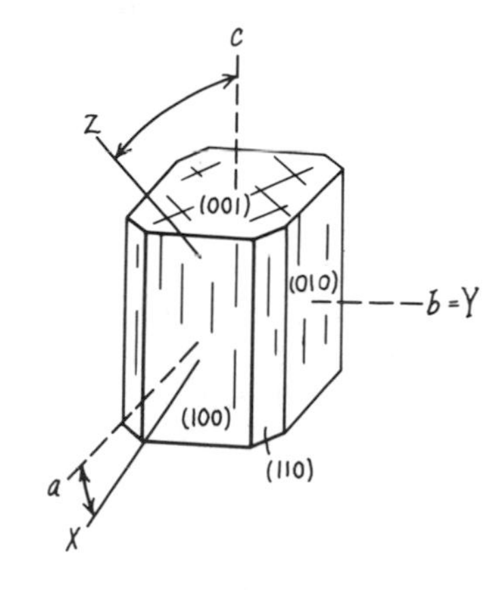

Omphacite

(Ca,Na)(Mg,Fe^{+2},Fe^{+3},Al)$(SiO_3)_2$. Intermediate in composition between diopside and jadeite.
Monoclinic
$\measuredangle\beta = 106°$
$\alpha = 1.667–1.669$, $\beta = 1.674–1.676$, $\gamma = 1.689–1.693$
$\gamma - \alpha = 0.022–0.024$ (2nd-order green)
$2V = 60°–74°$ (+)
$b = Y$, $c \wedge Z = 39°–43°$
Pleochroism and color: colorless to very pale green; weak pleochroism, with $X = Z$ = very pale green, and Y = colorless.
Dispersion: $r > v$ moderate.
Habit: coarse granular.
Cleavage: {110} good at 87° and 93° (Fig. 176).
Extinction in section: symmetric to {110} cleavage when viewed along *c:* in prismatic section extinction is parallel or inclined; length high (slow).
Twinning: {100} polysynthetic.
Hardness: 5–6.

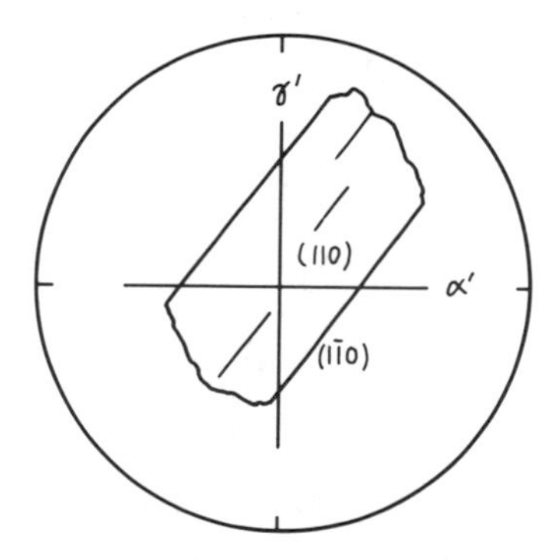

Figure 176. Omphacite fragments lying on {110} permit estimation of α' (1.669–1.671) and γ' (1.686–1.689). Extinction is inclined, with $\gamma' \wedge c = 37°–39°$, and grains are length high (slow). The figure is nonsymmetric.

Specific gravity: 3.29–3.37.

Occurrence: uncommon; in high-grade metamorphic assemblages (particularly eclogite) with garnet, plagioclase, enstatite, kyanite, zoisite, or glaucophane.

Alteration: to fibrous glaucophane or hornblende.

Characteristic features: typical {110} pyroxene cleavage; large $2V$; occurrence.

Look-alikes: jadeite has lower indices and birefringence, and is usually found in the blueschist facies; diopside has a moderate $2V$ and is typically found in contact metamorphic rocks; aegirine is strongly colored.

Spodumene

$LiAl(SiO_3)_2$

Monoclinic

$\measuredangle\beta = 110°20'$

$\alpha = 1.648–1.663$, $\beta = 1.655–1.669$, $\gamma = 1.662–1.679$

$\gamma - \alpha = 0.014–0.027$ (up to 2nd-order yellow)

$2V = 55°–70°$ (+)

$b = Y$, $c \wedge Z = 22°–26°$

Pleochroism and color: colorless.

Dispersion: $v > r$ weak.

Habit: often euhedral prismatic crystals elongated on c, with {110} prominent, with {100} and {010} less so; cross-section viewed along c is eight-sided; size ranges from tiny acicular to gigantic crystals (in pegmatites).

Cleavage: {110} good at 87° and 93° (Fig. 177); {010} parting.

Extinction in section: symmetric to {110} cleavage traces when viewed along c; inclined when viewed along b, and parallel when viewed along a; length high (slow).

Twinning: {100} rare.

Hardness: 6½–7.

Specific gravity: 3.0–3.2.

Occurrence: predominantly in lithium-bearing granitic pegmatites; rarely in aplites and metamorphic gneisses.

Alteration: to eucryptite ($LiAlSiO_4$), albite, and lepidolite.

Characteristic features: crystal size and occurrence; colorless, with small extinction angle.

Look-alikes: aegirine has a very small extinction angle, and is strongly colored; clinoenstatite is a rare mineral that does not occur in pegmatites; diopside has a larger extinction angle.

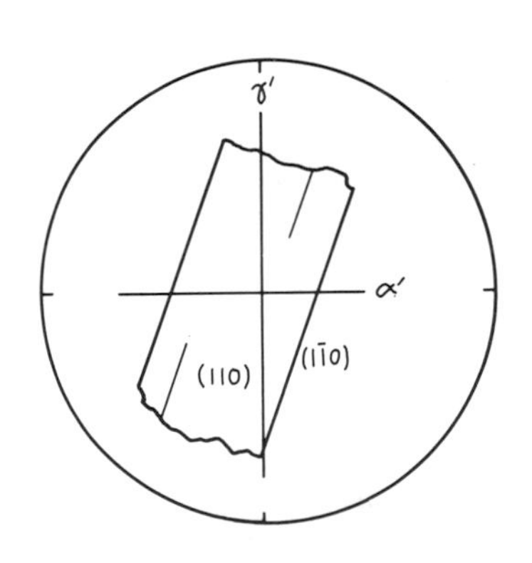

Figure 177. Spodumene fragments lying on {110} permit estimation of α' ($\simeq$1.66) and γ' ($\simeq$1.67). Extinction is inclined, with $\gamma' \wedge c \simeq 21°$, and grains are length high (slow). The figure is nonsymmetric.

Pyrrhotite. See Opaque Minerals (O2).

Quartz. See Silica Group (S4).

Rhodochrosite. See Carbonate Group (C2).

RHODONITE (R1)

$(Mn,Ca,Fe^{+2})SiO_3$
Triclinic
$\measuredangle\alpha = 90°02'$, $\measuredangle\beta = 95°22'$, $\measuredangle\gamma = 103°26'$
$\alpha = 1.711–1.738$, $\beta = 1.714–1.741$, $\gamma = 1.724–1.751$
$\gamma - \alpha = 0.011–0.014$ (up to 1st-order orange)
$2V = 61°–76°$ (+); indices and birefringence decrease and $2V$ increases with increasing Ca content
$a \wedge X \simeq 5°$, $b \wedge Y \simeq 20°$, $c \wedge Z \simeq 25°$
Pleochroism and color: usually colorless; rarely pale pink and weakly pleochroic, with X = orange, Y = rose-pink, and Z = pale yellow-orange.
Dispersion: $v > r$ weak, crossed.
Habit: euhedral to anhedral; commonly tabular parallel to {001}.
Cleavage: {110} and {1$\bar{1}$0} perfect, {001} good. See Figure 178.
Extinction in section: inclined to faces and cleavage traces; if grain is tabular on {001} it is length low (fast).
Twinning: {010} polysynthetic uncommon.
Hardness: 5½–6½.
Specific gravity: 3.55–3.76.
Occurrence: hydrothermal ore deposits with rhodochrosite and Mn-oxides; contact metamorphic skarns and marble. See Figure 179.
Alteration: to rhodochrosite and pyrolusite.
Characteristic features. occurrence; consistently inclined extinction; high indices and low birefringence.
Look-alikes: clinopyroxenes have higher birefringence.

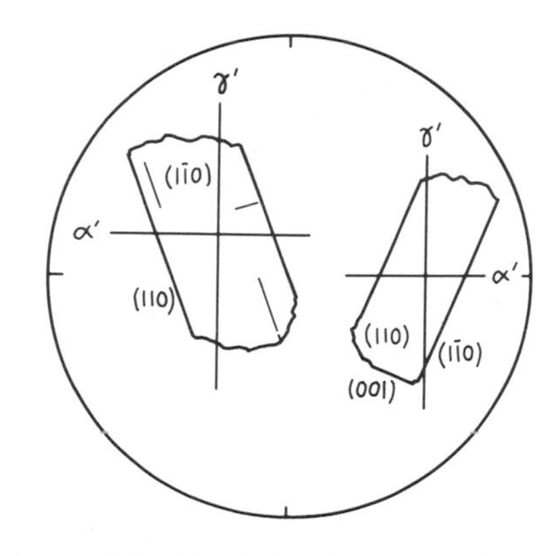

Figure 178. Optical orientation data for rhodonite are not sufficiently precise to permit calculation of extinction angles and indices on cleavage surfaces. Fragments lying on {110} and {1$\bar{1}$0} permit estimation of γ' (fairly close to γ in value) and α' (unknown). Extinctions are inclined, with γ' at an acute angle to the length of the grain. The good {001} cleavage on some fragments is almost perpendicular to the principal cleavage edge. The figures are nonsymmetric.

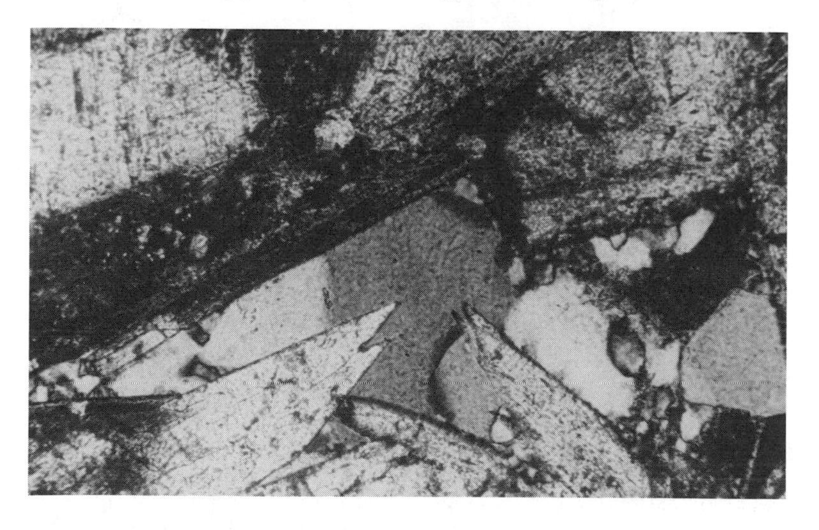

Figure 179. Rhodonite (high relief) surrounding several grains of quartz (center). Crossed nicols. Photo length is 0.55 mm.

Riebeckite. See Amphibole Group (A2).

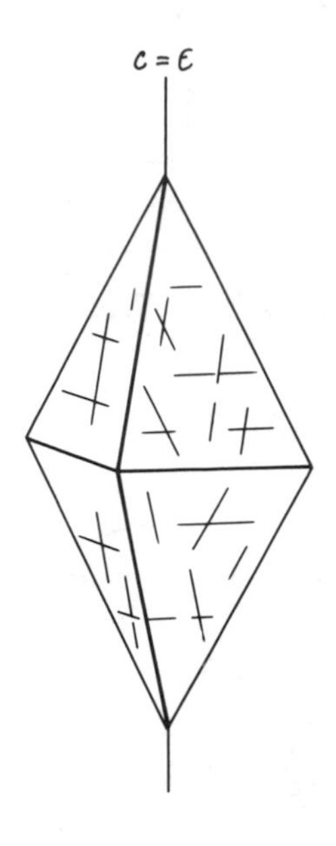

RUTILE GROUP (R2)

Anatase (Octahedrite)

TiO_2
Tetragonal
$\omega = 2.561$, $\varepsilon = 2.488$
$\omega - \varepsilon = 0.073$ (white of the higher order)
Uniaxial (−); rarely biaxial with a small $2V$
Pleochroism and color: color irregularly or zonally arranged; may be pale to bright yellow, brown, blue, green, or black; pleochroism is weak, with abs. $E > O$, rarely $O > E$.
Habit: anhedral to pyramidal, tabular or prismatic.
Cleavage: {001} and {111} perfect. See Figure 180.
Extinction in section: parallel or symmetric to principal faces and cleavage traces; length high (slow) or length low (fast) when extinction is parallel to principal faces.
Twinning: {112} rare.
Hardness: 5½–6.
Specific gravity: 3.82–3.97.
Occurrence: rare; anatase, the low-temperature TiO_2 polymorph, is found within veins and cavities in hydrothermally altered igneous and metamorphic rocks; most common as an accessory detrital mineral in sediments.
Alteration: to rutile.
Characteristic features: coloration; very high relief and birefringence; uniaxial (−); restricted occurrence.
Look-alikes: rutile is uniaxial (+), and brookite is biaxial (+).

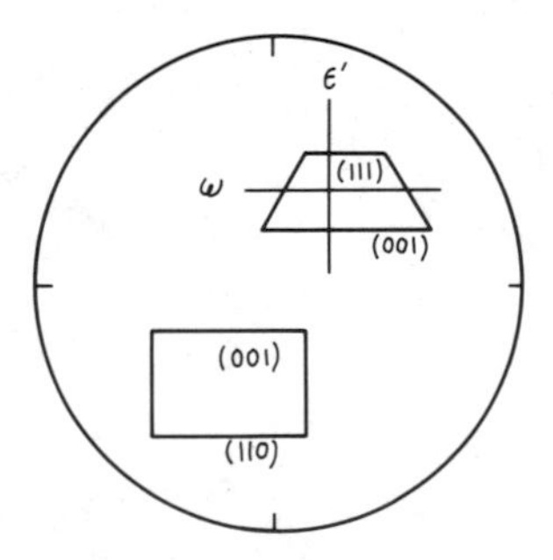

Figure 180. Anatase fragments lying on {001} permit estimation of ω, remain in extinction during stage rotation, and yield an O.A. figure. Fragments lying on {111} permit estimation of ω and ε' (2.493), have parallel and symmetric extinction, and yield an inclined O.A. figure.

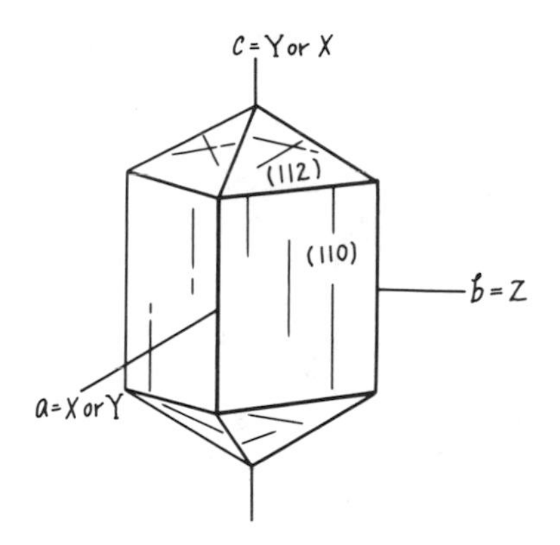

Brookite

TiO_2
Orthorhombic
$\alpha = 2.583$, $\beta = 2.584$, $\gamma = 2.700$
$\gamma - \alpha = 0.117$ (white of the higher order)
$2V = 0°–30°$ (+)
$a = X$, $b = Z$, $c = Y$ for red to yellow light; $a = Y$, $b = Z$, $c = X$ for green to blue light
Pleochroism and color: yellow-brown, red-brown, or deep brown; weak pleochroism in shades of yellow, orange, and brown, with $Z > Y > X$.
Dispersion: very strong crossed axial plane dispersion, resulting in non-black isogyres and dominant dispersion fringes in interference figures.
Habit: euhedral, with {010} dominant, or elongated on *c*.
Cleavage: {120} poor, {001} very poor (Fig. 181).
Extinction in section: commonly parallel to major faces; length high (slow) or low (fast) depending upon grain orientation.
Twinning: {120} rare.
Hardness: 5½–6.
Specific gravity: 4.12.
Occurrence: accessory mineral in pegmatites, schists, gneisses, and hydrothermal veins; commonly with anatase and rutile; a detrital mineral in sediments.
Alteration: to rutile.
Characteristic features: extremely high indices and birefringence; euhedral grains; unusual dispersion effects seen in interference figures.
Look-alikes: rutile is uniaxial (+) and has good cleavage; anatase is uniaxial (−) and has good cleavage.

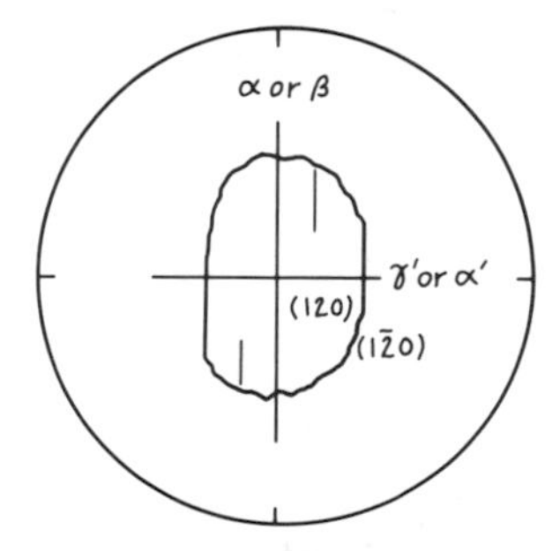

Figure 181. Most brookite fragments are in random orientation. Rare fragments lying on the poor {120} cleavage permit estimation of α or β (as a function of wavelength) parallel to direction of elongation. Grains are length high (slow) or length low (fast). The figure is either monosymmetric Bxa or monosymmetric Bxo.

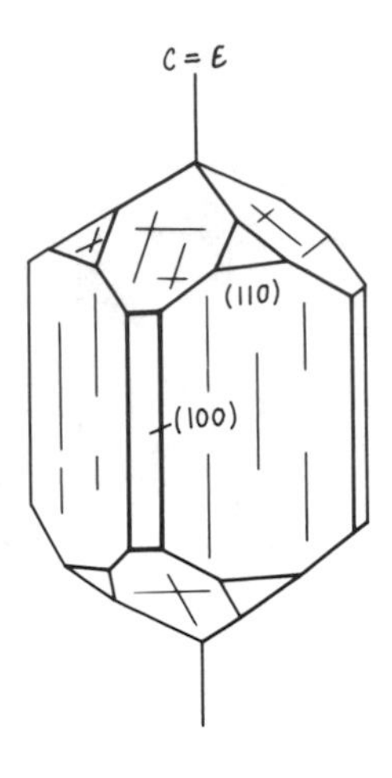

Rutile

TiO_2
Tetragonal
$\omega = 2.605–2.616$, $\varepsilon = 2.890–2.903$
$\varepsilon - \omega = 0.285–0.296$ (white of the higher order)
Uniaxial (+)
Pleochroism and color: deep red-brown, yellow-brown, or almost opaque; pleochroism usually weak, with abs. $E > O$.
Habit: euhedral, with {100} and {110} dominant; acicular; as inclusions in quartz, phlogopite, or corundum; a major constituent of the fine-grained aggregate known as *leucoxene.*
Cleavage: {110} and {100} good (Fig. 182), {111} poor, {902} or {101} parting may be well developed.
Extinction in section: parallel to major faces and cleavage traces; length high (slow).
Twinning: {011} simple or cyclic common, yielding a variety of shapes; {902} glide twinning uncommon; {031} simple uncommon.
Hardness: 6–6½.
Specific gravity: 4.2–5.6.
Occurrence: common as an accessory mineral in many igneous and metamorphic rocks; in sedimentary rocks as detrital grains.
Alteration: stable.
Characteristic features: red-brown color; extreme relief and birefringence. See Figure 183.
Look-alikes: anatase and hematite are optically (−); brookite is biaxial.

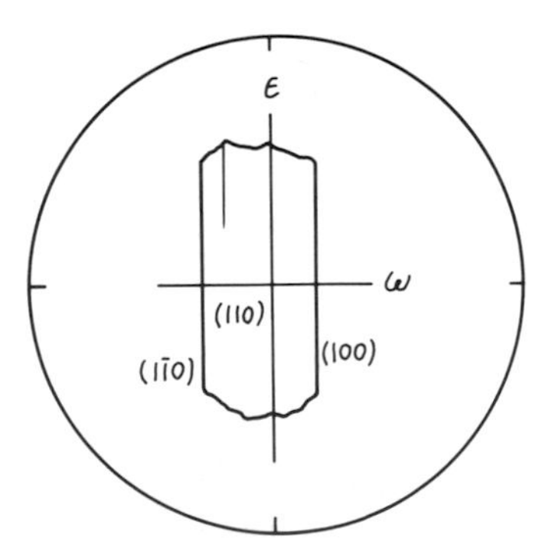

Figure 182. Rutile fragments lying on {110} or {100} permit estimation of ε and ω, have parallel extinction, and are length high (slow). The figure is O.N.

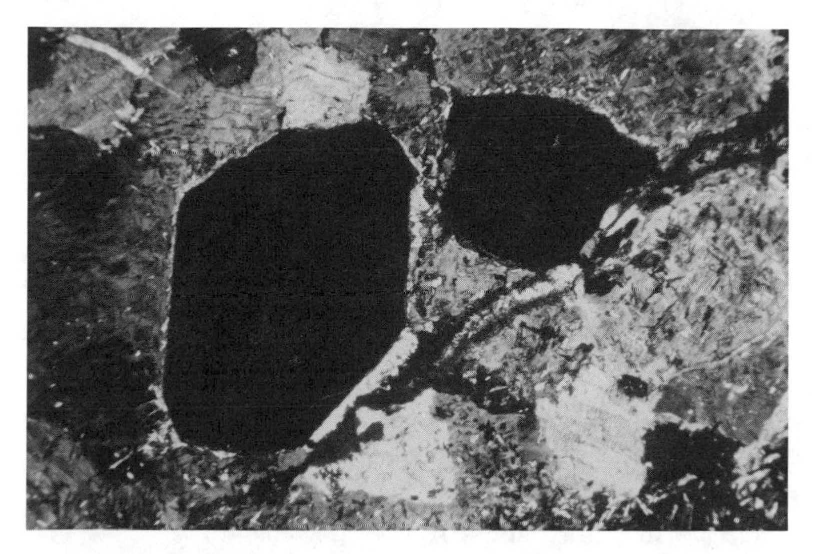

Figure 183. The two twinned grains of rutile (center) are almost opaque. The surrounding material is altered oligoclase. Crossed nicols. Photo length is 2.22 mm.

Salite. See Pyroxene Group (P8).

Sanidine. See Feldspar Group (F1).

SAPPHIRINE (S1)

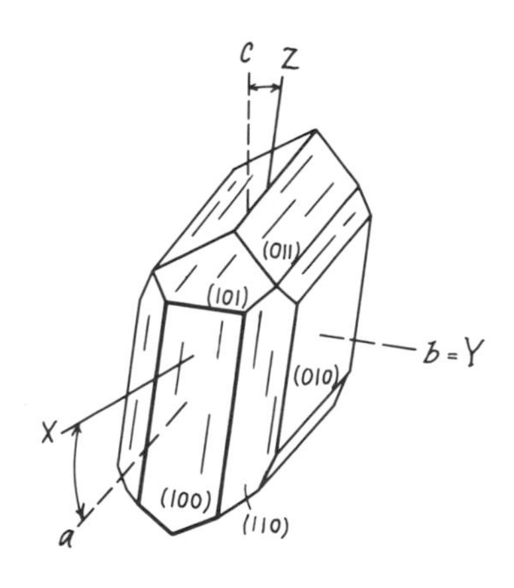

$(Mg,Fe)_2Al_4O_6(SiO_4)$
Monoclinic
$\measuredangle\beta = 125°20'$
$\alpha = 1.701–1.729$, $\beta = 1.703–1.732$, $\gamma = 1.705–1.734$
$\gamma - \alpha = 0.004–0.006$ (1st-order gray, or anomalous grayish purple)
$2V = 51°–69° (-)$
$b = Y$, $c \wedge Z = -6°$ to $-15°$
Pleochroism and color: pale blue or gray; weak pleochroism, with X = colorless, buff, greenish blue, pale brown, or yellow, Y = greenish blue, blue, or dark blue, and Z = deep blue, yellow-green, or green; abs. $Z > Y > X$.
Dispersion: $v > r$ strong, inclined.
Habit: euhedral to anhedral, tabular {010} and somewhat elongate on c.

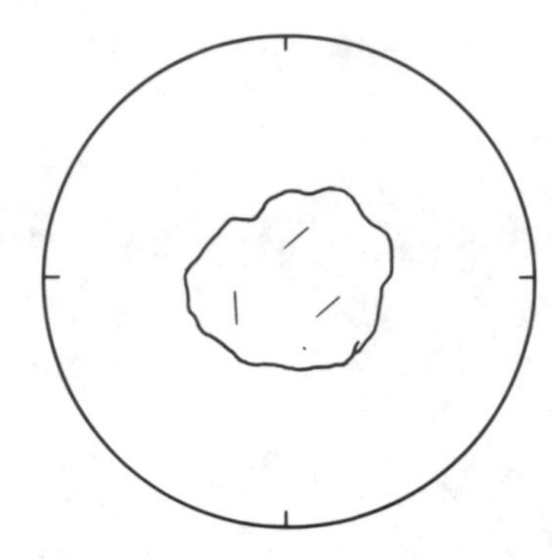

Figure 184. With sapphirine, lack of good cleavage results in random orientation of fragments. With proper choice of grains, any index can be determined, and any type of figure obtained.

Cleavage: {100}, {010}, and {001} poor (Fig. 184).
Extinction in section: parallel to major faces when viewed along *a* or *c*; length high (slow) or low (fast) as a function of orientation.
Twinning: {010} and {100} polysynthetic uncommon.
Hardness: 7½.
Specific gravity: 3.40–3.58.
Occurrence: a high-temperature mineral of regional or contact metamorphism; found in rocks with low SiO_2 and high MgO and Al_2O_3; in schists, gneisses, granulites, amphibolites, hornfelses, and contact emery deposits; may be associated with anthophyllite, cordierite, biotite, hornblende, hypersthene, sillimanite, anorthite, corundum, or diaspore. See Figure 185.

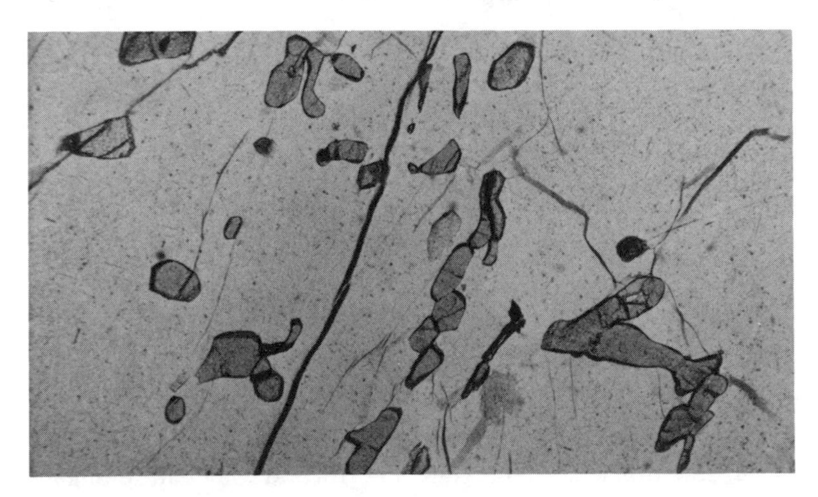

Figure 185. Anhedral grains of sapphirine within feldspar. Crossed nicols. Photo length is 2.22 mm.

Alteration: easily to corundum and biotite.
Characteristic features: pale blue color; low birefringence; dispersion; occurrence.
Look-alikes: kyanite has good cleavages; corundum is uniaxial; cordierite has lower indices.

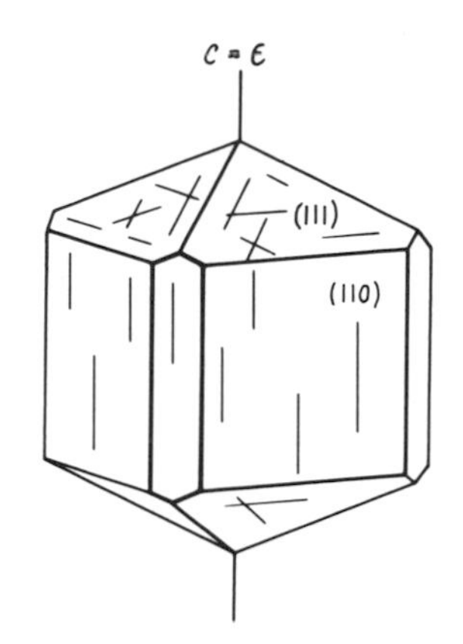

SCAPOLITE GROUP (S2)

Marialite-Meionite Series

$3Na(AlSi_3)O_8 \cdot NaCl$ to $3Ca(Al_2Si_2)O_8 \cdot CaCO_3$
Tetragonal
$\omega = 1.540 - 1.600$, $\varepsilon = 1.535 - 1.565$ (Fig. 186)
$\omega - \varepsilon = 0.004 - 0.037$ (up to 2nd-order red)
Uniaxial (−)
Pleochroism and color: colorless.
Habit: granular to prismatic crystals.
Cleavage: {100} and {110} good (Fig. 187).
Extinction in section: parallel to major faces and cleavage traces; length low (fast).
Twinning: none.
Hardness: 5½–6.
Specific gravity: 2.54–2.81.

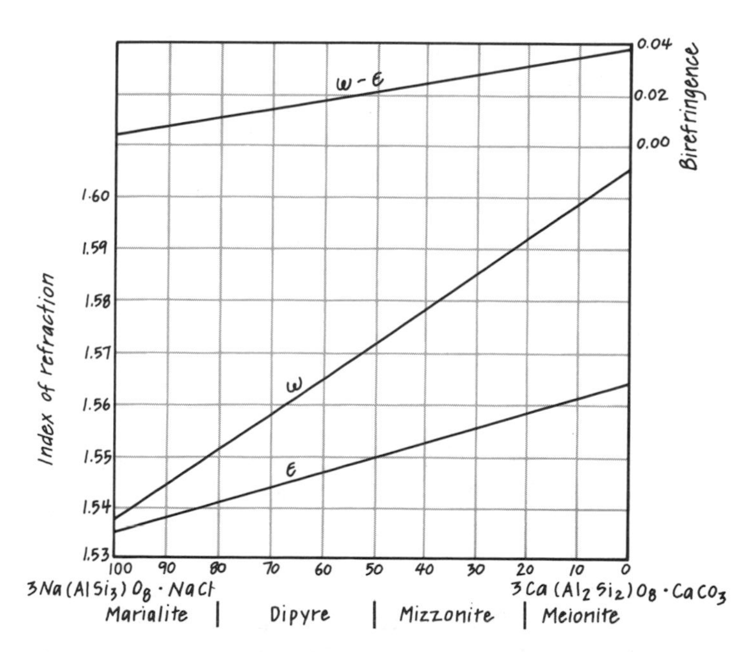

Figure 186. Variation in refractive indices and birefringence in the scapolite minerals (from data of Ulbrich, 1973).

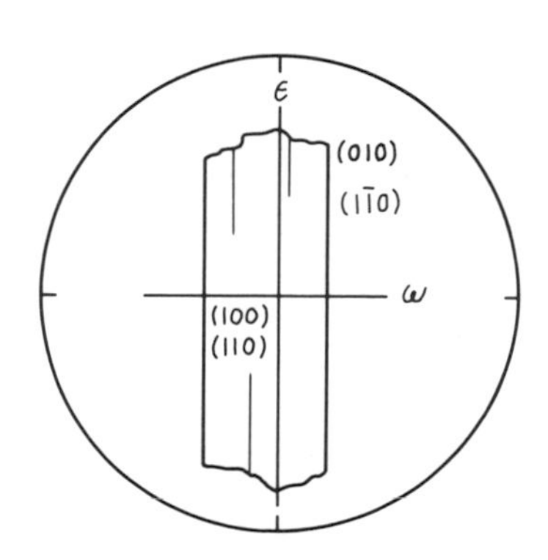

Figure 187. Scapolite fragments lying on {100} or {110} permit estimation of ω and ε; extinction is parallel and grains are length low (fast). The figure is O.N.

Occurrence: metamorphic and metasomatic environments in which abundant brine (NaCl) or CO_2 is available; except for uncommon pegmatite occurrences, scapolite is not a primary igneous mineral.

Alteration: scapolite may form from or alter to plagioclase; also to a fine-grained aggregate that may contain calcite, chlorite, sericite, epidote, plagioclase, idocrase, zeolites, or clays.

Characteristic features: uniaxial (−); occurrence; marialite has low relief and birefringence; meionite has moderate relief and birefringence; both have prismatic cleavage.

Look-alikes: quartz lacks good cleavage and is uniaxial (+); feldspars and cordierite are biaxial and are commonly twinned; wollastonite and andalusite have higher relief and are biaxial (−).

Scolecite. See Zeolite Group (Z1).

Sericite. See Clays and Clay Minerals (C6) and Mica Group (M2).

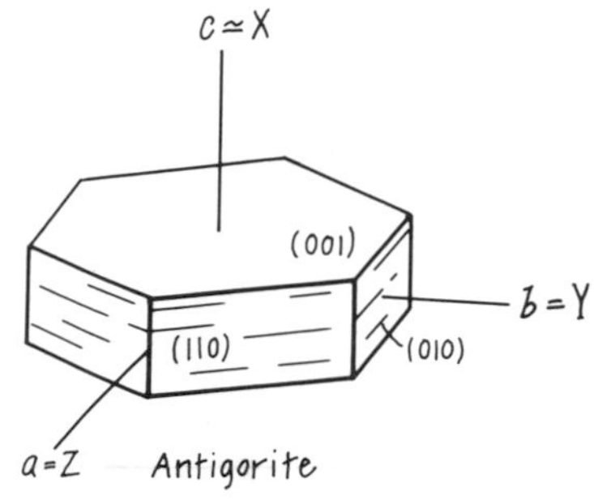

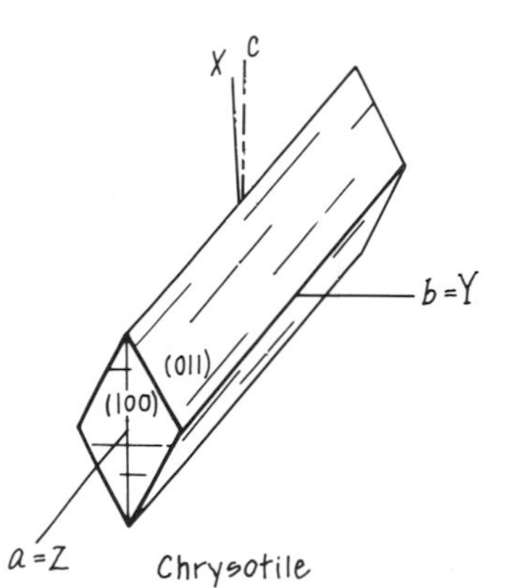

SERPENTINE GROUP (S3)

Chrysotile, Lizardite, Antigorite

Chrysotile (asbestos) is usually fibrous, lizardite is usually platy, and antigorite is either fibrous or platy.

$Mg_3Si_2O_5(OH)_4$. Small differences in composition are common (minor Fe and Al, as well as differences in Fe^{+3}/Fe^{+2} ratios).

Monoclinic. The three species are best characterized by X-ray diffraction; optical properties, similar for all three, are combined below.

$\measuredangle\beta = 90°–93°$

$\alpha = 1.529–1.567$, $\beta = 1.530–1.573$, $\gamma = 1.537–1.574$

$\gamma - \alpha = 0.004–0.009$ (up to 1st-order white, but often gray or essentially isotropic)

$2V = 20°–60°$ (−)

$b = Y, c \wedge X = 0°–7°$

Pleochroism and color: colorless to pale green; may be weakly pleochroic, with $Z > Y = X$; X = colorless or greenish yellow, Y = colorless or yellow-green, and Z = yellow or green.

Dispersion: $r > v$ weak (antigorite).

Habit: chrysotile is commonly asbestiform and elongate on a; veins with cross fibers are common; antigorite and lizardite form netlike and leaflike intergrowths with undulatory or mottled extinction.

Cleavage: {001} perfect (Fig. 188).
Extinction in section: platy grains have parallel or near-parallel extinction and are length high (slow); fibers have parallel extinction and are length high (slow).
Twinning: probably common on a submicroscopic level.
Hardness: 2½–3½.
Specific gravity: 2.5–2.6.
Occurrence: always secondary, and commonly forms from alteration of Mg-rich olivine, pyroxene, and amphibole in ultramafic rocks such as dunite, peridotite, and pyroxenite (color photo 60); the major constituent of serpentinite (Fig. 189); also may form in metamorphic contact zones by alteration of forsterite in marble (color photo 48).
Alteration: serpentine itself is an alteration product of Mg-rich olivine and pyroxene: it rarely alters to chlorite.
Characteristic features: low birefringence; fine-grained fibrous or platy character; association with ultramafic rocks or forsteritic marble.
Look-alikes: many chlorites have stronger color, higher birefringence, or anomalous interference colors; brucite commonly has anomalous interference colors and is uniaxial.

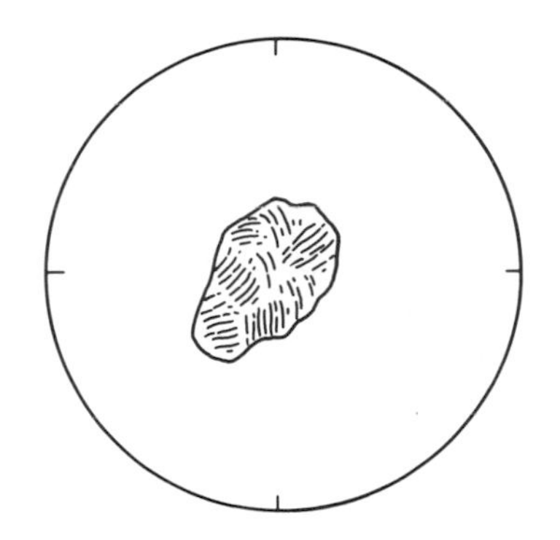

Figure 188. As a result of fine grain size, it is normally impossible to measure specific indices for serpentine minerals (except in the case of a serpentine pseudomorph after olivine or pyroxene). Extinction is parallel or near-parallel to platelets and fibers, and grains are length high (slow). Figures are normally not obtainable.

Figure 189. Serpentine, consisting of randomly oriented fibers, flakes, and semi-isotropic material. Crossed nicols. Photo length is 2.22 mm.

Sheridanite. See Chlorite Group (C4).

Siderite. See Carbonate Group (C2).

SILICA GROUP (S4)

Chalcedony

SiO_2 (with up to 10 percent submicroscopic inclusions of H_2O)
Hexagonal
$\omega = 1.526-1.544$, $\varepsilon = 1.532-1.553$ (chalcedony with no H_2O has indices equal to those of quartz)
$\varepsilon - \omega = 0.005-0.009$ (up to 1st-order white)
Uniaxial (+); some aggregates may be biaxial (+)
Pleochroism and color: nonpleochroic and usually colorless; some highly colored varieties may show weak color in section.
Habit: often fibrous, with elongation normal to *c* axis, and less commonly parallel to *c;* also spherulitic, banded, interstitial, colloform, vuggy, or pseudomorphic after other minerals or fossils; interference figures are obtainable from interlayered granular silica, but usually not from chalcedony (due to fine grain size).
Cleavage: none (Fig. 190).
Extinction in section: parallel to fibers; commonly length low (fast) and less commonly length high (slow).
Twinning: not obvious due to fine grain size.
Hardness: 6½–7.
Specific gravity: 2.57–2.64.
Occurrence: usually forms at or near surface conditions in cavities, veins, or as a replacement of preexisting minerals or fossils; may form from devitrification of gelatinous silica (derived from earlier weathering of silicate minerals) or from late hydrothermal solutions; common as nodules and as a cementing agent in sediments (color photo 61) and associated with terrestrial and submarine volcanic rocks; also common as detrital grains in sediments.
Alteration: generally stable, but may tend to recrystallize to quartz with time, or be partially dissolved by alkaline solutions.
Characteristic features: fine grain size; low birefringence; low relief; parallel extinction.
Look-alikes: fibrous serpentine is often pale green; fibrous quartz or cristobalite are often indistinguishable optically.

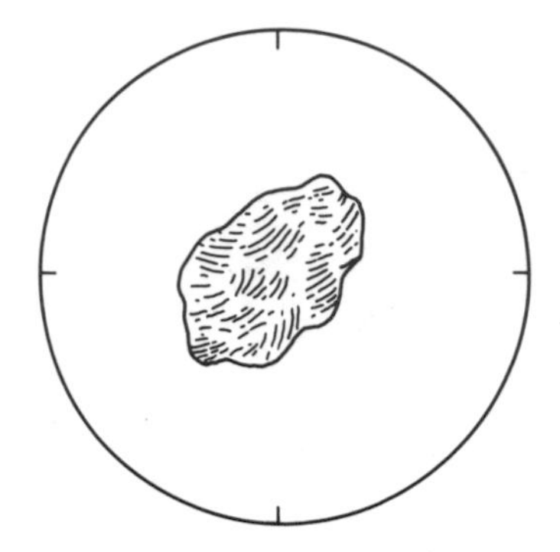

Figure 190. With chalcedony, fine grain size permits estimation of only an average index of refraction. If fibers are visible, extinction is parallel and the length is high (slow) or low (fast). Figures are not obtainable generally, but somewhat coarser quartz grains may yield uniaxial figures in various orientations.

α-Cristobalite

SiO_2
Tetragonal (the isometric form, β-cristobalite, inverts to α-cristobalite at temperatures below 268°C)
$\omega = 1.487$, $\varepsilon = 1.484$
$\omega - \varepsilon = 0.003$ (1st-order gray)

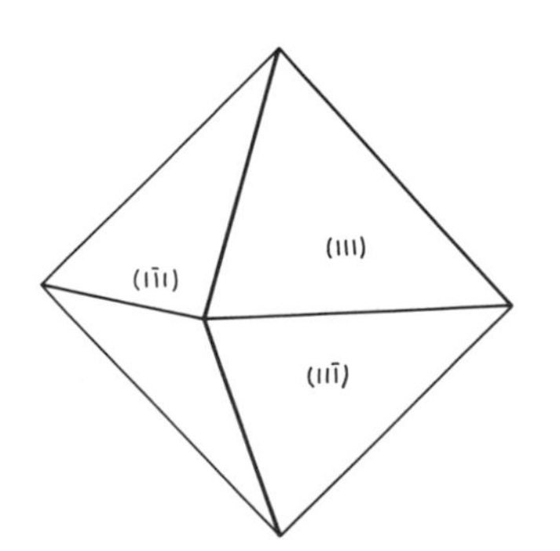

Uniaxial (−), but may be anomalously biaxial (−) with $2V = 25°$

Pleochroism and color: colorless.

Habit: commonly as octahedral or cubo-octahedral pseudomorphs after β-cristobalite; crystals tend to have octahedral faces indented; optic axis of α-cristobalite is parallel to an octahedral axis; also common as spherulites, and intergrown with sanidine and tridymite; the variety *lussatite* is an intergrowth of fibrous α- and β-cristobalite with fibrous cryptocrystalline quartz; a vitreous submicroscopic form of cristobalite constitutes the material called *common opal.*

Cleavage: none; curved fractures common (Fig. 191).

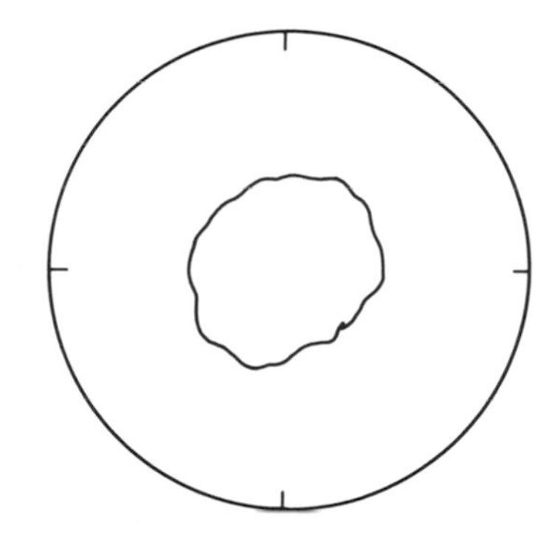

Figure 191. Cristobalite fragments have no cleavage and will therefore have irregular shapes and random orientation.

Extinction in section: often appears essentially isotropic; fibrous varieties have parallel extinction and may be length high (slow) or length low (fast).

Twinning: {111} polysynthetic common in one or two directions.

Hardness: 6½.

Specific gravity: 2.33.

Occurrence: volatile deposition in vugs and fractures of siliceous volcanic rocks; often associated with tridymite, quartz, sanidine, or anorthoclase; less commonly in metamorphic contact zones and meteorites; may form at low temperatures in chalcedony, opal, bentonite, siliceous sinter, and biochemical deposits.

Alteration: may invert to fine-grained quartz.

Characteristic features: very low birefringence; negative relief; equant crystals; association with silicic volcanic rocks.

Look-alikes: may be indistinguishable from tridymite, which is biaxial (+), has lower indices, and is more common; quartz and sanidine have higher indices than cristobalite; zeolites are more common in basalt.

Opal

SiO_2 (with 5–10 percent H_2O)

Amorphous (submicroscopic spherulites of tridymite and/or cristobalite)

$n = 1.40–1.46$ (indices decrease with increasing H_2O content; usually about 1.435 to 1.455)

Isotropic

Pleochroism and color: colorless, rarely may be pale brown or yellow.

Habit: amorphous, colloform, botryoidal, or reniform; may contain patches of chalcedony, tests of microfossils, or woody structure.

Cleavage: none.

Extinction in section: essentially isotropic.
Hardness: 5½–6½.
Specific gravity: 1.99–2.25.
Occurrence: a secondary mineral precipitated from circulating meteoric water in cavities and veins in igneous rocks; also may form opaline sedimentary deposits from the accumulation of silica-secreting organisms, or surface deposits of siliceous sinter and geyserite in active geothermal areas.
Alteration: to chalcedony, or may dissolve in hot alkaline solutions.
Characteristic features: high negative relief; isotropic; occurrence.
Look-alikes: volcanic glass has higher indices and is limited to rapidly cooled volcanic rocks; analcime has complex polysynthetic twinning and weak birefringence; fluorite has perfect octahedral cleavage.

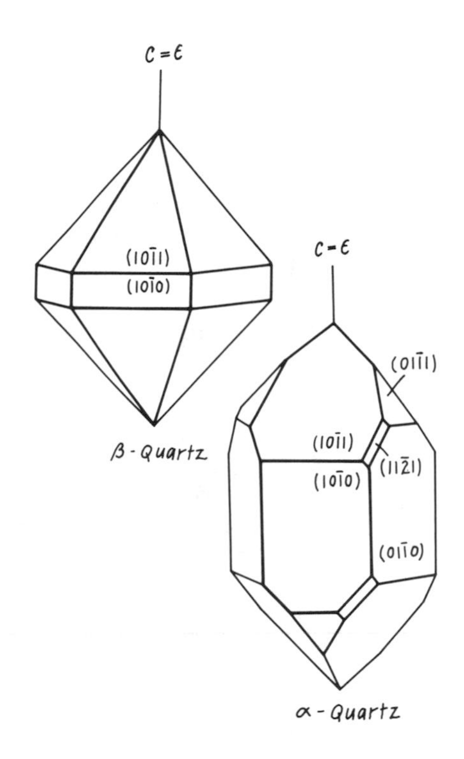

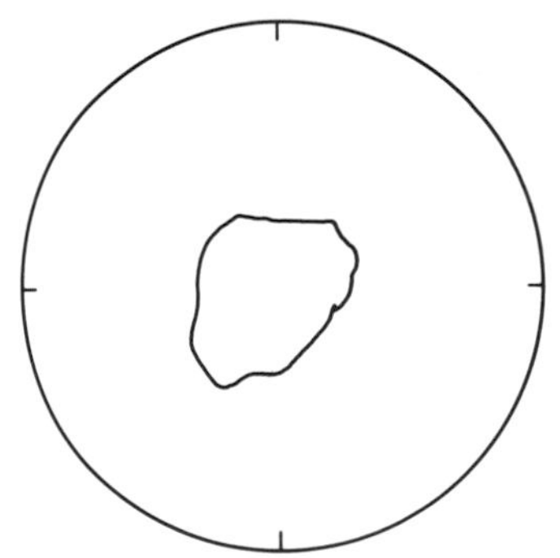

Figure 192. Quartz fragments have no cleavage, and will therefore have irregular shapes and random orientation.

α-Quartz

SiO_2
Hexagonal (the higher temperature polymorph, β-quartz, is also hexagonal)
$\omega = 1.544$, $\varepsilon = 1.553$
$\varepsilon - \omega = 0.009$ (1st-order white)
Uniaxial (+) (rarely anomalously biaxial)
Pleochroism and color: colorless.
Habit: in volcanic rocks, quartz phenocrysts commonly form the stubby hexagonal dipyramids that are characteristic of β-quartz; during cooling, β-quartz inverts to α-quartz at 573°C, with retention of the high-temperature forms; in plutonic and metamorphic rocks quartz is commonly anhedral; quartz in low-temperature veins often forms euhedral singly terminated prisms.
Cleavage: none (Fig. 192).
Extinction in section: when viewed ⊥ to *c*, pseudomorphs after β-quartz commonly show symmetrical extinction to dipyramid faces; prismatic quartz crystals have parallel extinction and are length high (slow); deformation of quartz (especially in metamorphic rocks) often results in undulatory extinction, or deformation bands or lamellae (subplanar bands with a slightly different index and optical orientation than the host crystal). Unusually thick quartz (≥1 mm) does not show complete extinction when viewed along the *c* axis, due to the *optical activity* of quartz, which results in a rotation of the plane of polarization of transmitted light; this is seen in an inter-

Figure 193. Interlocking quartz grains in quartzite. Inclusions are abundant within the quartz. The smaller euhedral grains are muscovite. Crossed nicols. Photo length is 0.55 mm.

ference figure as either a rotation or elimination of the isogyres at and adjacent to the melatope.

Twinning: common, but rarely observed in sections or fragments.

Hardness: 7.

Specific gravity: 2.65.

Occurrence: extremely common; may be in all igneous rocks, except those which are silica-undersaturated; quartz is stable throughout the entire range of regional and contact metamorphism; it is extremely common in detrital sedimentary deposits (Fig. 193), as well as hydrothermal veins; quartz is *not* found in association with silica-deficient minerals such as the feldspathoids, Mg-rich olivines, or corundum. See color photos 61 and 62.

Alteration: stable and unaltered.

Characteristic features: low relief and birefringence; uniaxial (+); lack of cleavage and alteration; extremely common.

Look-alikes: feldspar is biaxial, often twinned and altered, and has two directions of cleavage; cordierite is commonly twinned and/or altered, and may have a subtle color and contain pleochroic haloes; staining techniques may be necessary to determine the precise amount of quartz (versus feldspar or cordierite) in a section; nepheline, beryl, and scapolite are uniaxial (−).

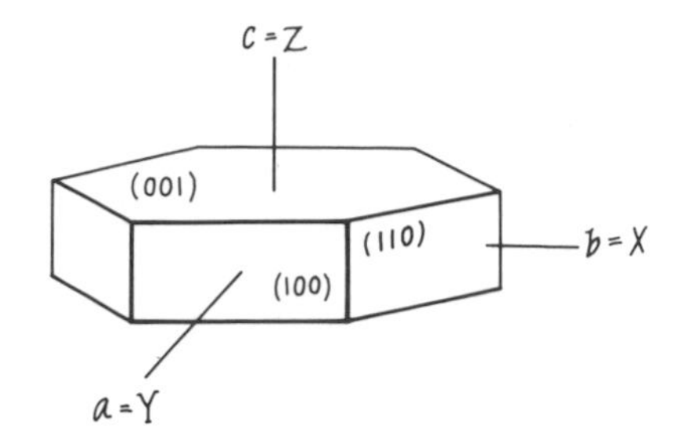

α-Tridymite

SiO_2

Orthorhombic (inverted from a hexagonal high-temperature form)

$\alpha = 1.469–1.479$, $\beta = 1.469–1.480$, $\gamma = 1.473–1.483$

$\gamma - \alpha = 0.002–0.004$ (1st-order gray)

$2V = 35°–90° (+)$

$a = Y, b = X, c = Z$

Pleochroism and color: colorless.

Habit: very small (1–2 mm) tabular euhedral crystals with hexagonal prismatic $\{10\bar{1}0\}$ outline; fan-shaped and spherical groups.

Cleavage: none (Fig. 194).

Extinction in section: parallel to major faces; length low (fast).

Twinning: common on former $\{10\bar{1}0\}$ surfaces, forming wedge- or pie-shaped aggregates (mainly composed of three wedges). See Figure 195.

Hardness: 7.

Specific gravity: 2.27.

Occurrence: typically in vugs and fractures in silicic to intermediate volcanic rocks; may also be in groundmass, and less commonly as phenocrysts; in lithophysae in rhyolite and obsidian with quartz, cristobalite, sanidine, or fayalite; rarely present in vugs within mafic volcanic and contact metamorphic rocks; common in meteorites; synthetic tridymite is a major constituent of silica bricks.

Alteration: to fine-grained α-quartz.

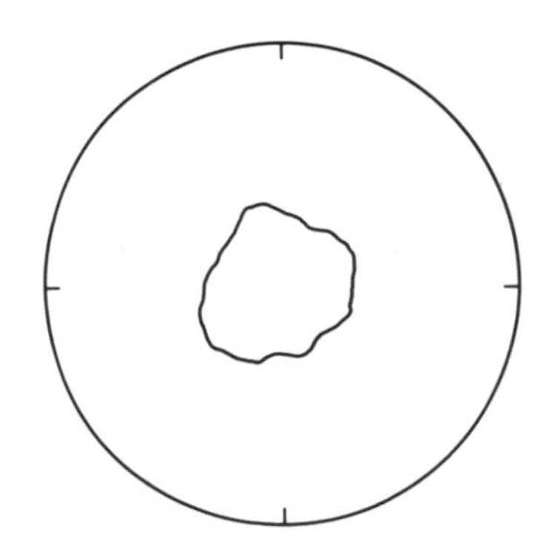

Figure 194. Tridymite fragments have no cleavage and will therefore have irregular shapes and random orientations.

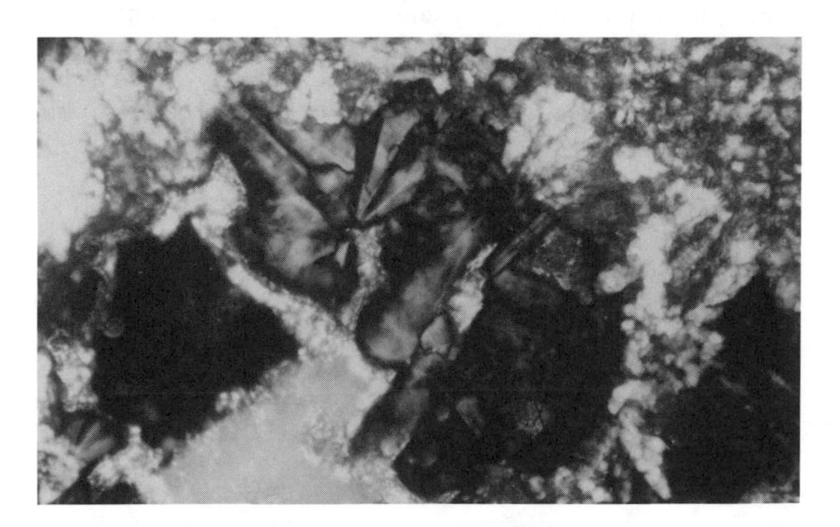

Figure 195. Wedge-shaped grains of tridymite in a silicic lava. Crossed nicols. Photo length is 0.55 mm.

Characteristic features: wedge-shaped twins; low birefringence and indices; occurrence.
Look-alikes: may be indistinguishable from cristobalite if figure is unavailable; zeolites tend to occur in mafic volcanic rocks, where tridymite is uncommon.

Sillimanite. See Aluminosilicate Group (A1).

Smithsonite. See Carbonate Group (C2).

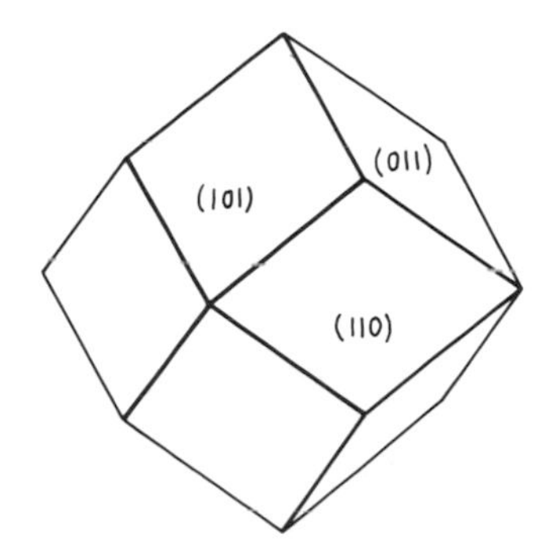

SODALITE GROUP (S5)

Hauyne

$(Na,Ca)_{8-4}Al_6Si_6O_{24}(SO_4)_{1-2}$
Isometric
$n = 1.496-1.505$

Noselite

$Na_8Al_6Si_6O_{24}SO_4$
Isometric
$n = 1.485-1.495$

Sodalite

$Na_8Al_6Si_6O_{24}Cl_2$
Isometric
$n = 1.483-1.487$

Isotropic
Pleochroism and color: hauyne is usually blue, and less commonly blue-green; noselite and sodalite are colorless to pale gray-blue; the blue color in all varieties may be zonally or irregularly distributed.
Habit: commonly dodecahedral crystals with a hexagonal outline, but may be anhedral.
Cleavage: {110} poor (Fig. 196).
Twinning: {111} rare.
Hardness: 5½–6.
Specific gravity: 2.27–2.40.
Occurrence: hauyne and noselite typically occur in silica-poor volcanic rocks; sodalite occurs in both volcanic and plutonic silica-poor rocks (Fig. 197) and is common in nepheline syenite; sodalite minerals may occur in metasomatized carbonate contact rocks near igneous contacts; *lazurite* is the sodalite group

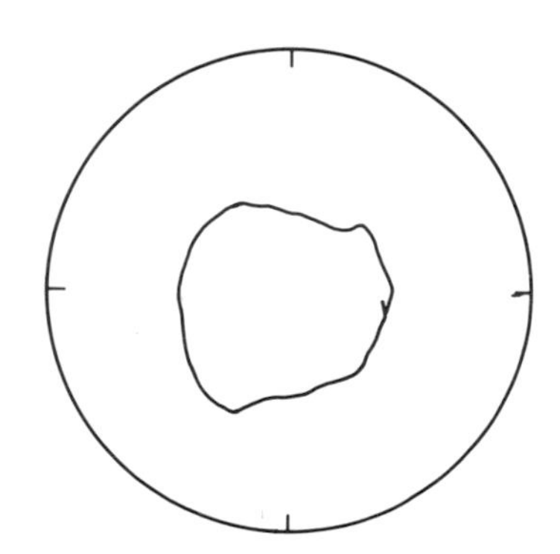

Figure 196. Sodalite fragments lack a good cleavage, are isotropic, and permit estimation of a single index.

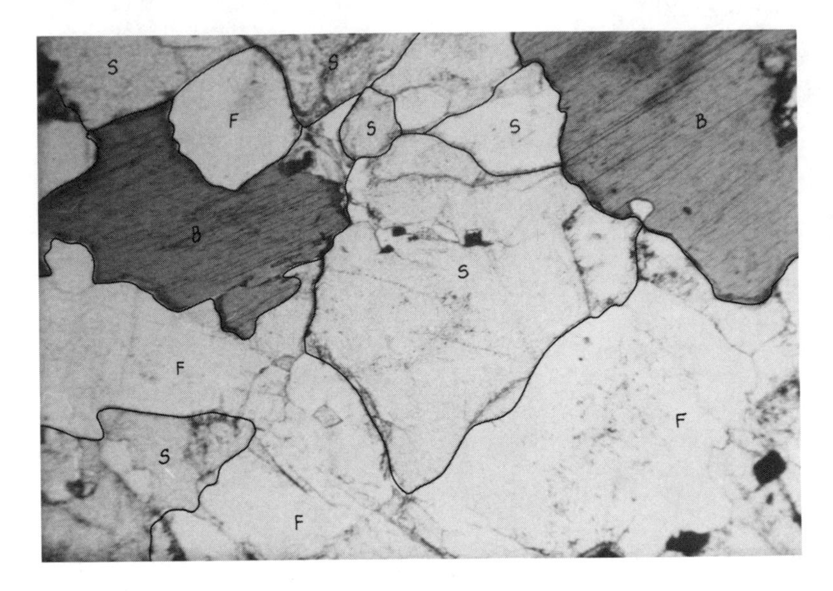

Figure 197. Anhedral sodalite (S), associated with biotite (B) and feldspar (F). Plane-polarized light. Photo length is 2.22 mm.

mineral (a sulfide-bearing hauyne) that is dominant in *lapislazuli* (a mixture of lazurite, diopside, muscovite, calcite, pyrite, and amphibole).

Alteration: to fibrous zeolites, clays, or cancrinite.

Characteristic features: isotropic, low index; often blue; six-sided outline; limited to silica-poor rocks.

Look-alikes: analcime and leucite, when euhedral, are eight-sided; leucite is weakly birefringent and twinned, and analcime has a poor cubic cleavage.

Spessartine. See Garnet Group (G1).

SPHALERITE (S6)

(Zn,Fe)S

Isometric

n = 2.37–2.50 (increases with Fe content)

Pleochroism and color: colorless, pale yellow, pale brown; color may be nonuniform.

Habit: tetrahedral or dodecahedral, coarse or fine granular, concretionary.

Cleavage: {110} perfect (Fig. 198).

Twinning: {111} simple or polysynthetic common.

Hardness: 3½–4.

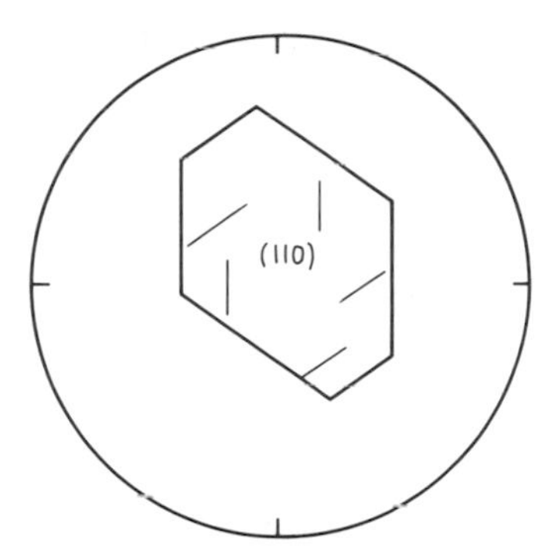

Figure 198. Isotropic sphalerite fragments lie on the {110} cleavage and present an outline in which up to six cleavage edges can be observed. Assuming the presence of six cleavage edges, the angle of intersection of any two adjacent edges is either 110° or 125°.

Specific gravity: 3.9–4.1.
Occurrence: common in hydrothermal veins (almost always with galena, and often chalcopyrite), as well as contact metamorphic deposits and Mississippi Valley type ore deposits; rarely in lignite and coal deposits. See Figure 199.
Alteration: to the zinc sulfate *goslarite.*
Characteristic features: extreme relief; isotropic; cleavage; occurrence.
Look-alikes: none.

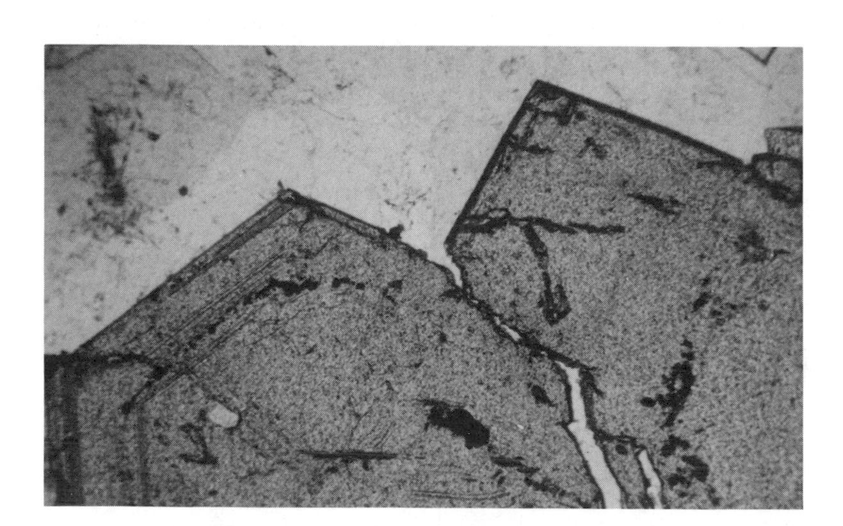

Figure 199. Euhedral high-relief grains of sphalerite in a quartz matrix. Plane-polarized light. Photo length is 2.25 mm.

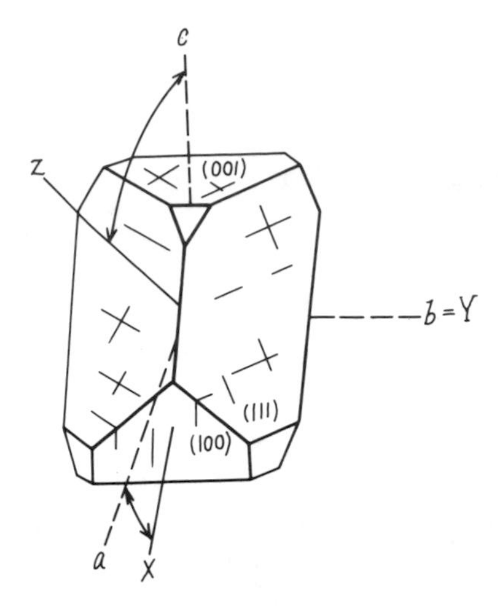

SPHENE (TITANITE) (S7)

$CaTiSiO_5$
Monoclinic
$\measuredangle\beta = 120°$
$\alpha = 1.843–1.950$, $\beta = 1.870–2.034$, $\gamma = 1.943–2.110$
$\gamma - \alpha = 0.100–0.192$ (white of the higher order)
$2V = 17°–40°\ (+)$
$b = Y, c \wedge Z \simeq 51°$
Pleochroism and color: colorless, pale brown, or yellow-brown (color photo 63); weak pleochroism, with abs. $Z > Y > X$; X = colorless or pale yellow, Y = pale yellow-brown, pale yellow-green, or pink, and Z = orange-brown, green, or pink.
Dispersion: $r > v$ strong, inclined.
Habit: commonly euhedral with rhombic cross-section; also as anhedral lensoid grains; secondary sphene may be present along cleavages within earlier minerals; detrital sphene is irregular to rounded.
Cleavage: {110} good, {221} parting; see Fig. 200.
Extinction in section: inclined to principal faces and cleavage traces in most orientations; grains are length low (fast) when viewed along *b*.
Twinning: {100} simple common, dividing rhombic sections near the long diagonal.
Hardness: 5–5½.
Specific gravity: 3.45–3.56.
Occurrence: a very common accessory in most igneous, metamorphic, and detrital sedimentary rocks.
Alteration: to the fine-grained nearly opaque (anatase- or rutile-rich) aggregate known as *leucoxene*.

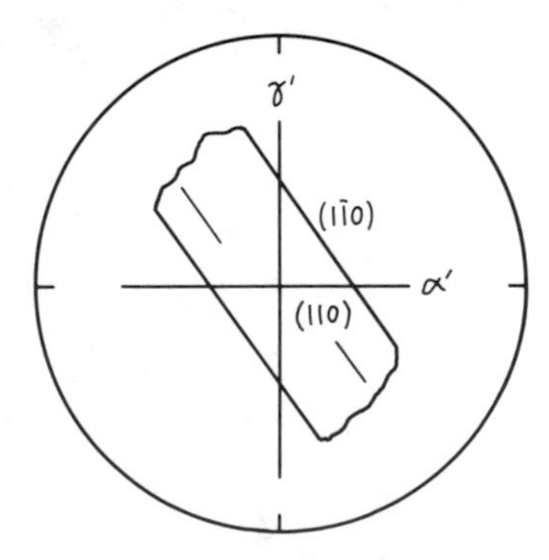

Figure 200. Sphene fragments on the {110} cleavage surface permit estimation of α' (1.85–1.99) and γ' (1.91–2.06). The extinction is inclined, with $\gamma' \wedge \{110\} = 34°–42°$. The figure is nonsymmetric.

Characteristic features: high indices, very high birefringence; characteristic rhombic outline with inclined extinction (color photo 64).

Look-alikes: monazite has lower refractive indices and birefringence; the hexagonal carbonates are colorless, have symmetrical extinction, and rhombohedral cleavage; cassiterite, rutile, and xenotime (YPO_4) are uniaxial.

SPINEL GROUP (S8)

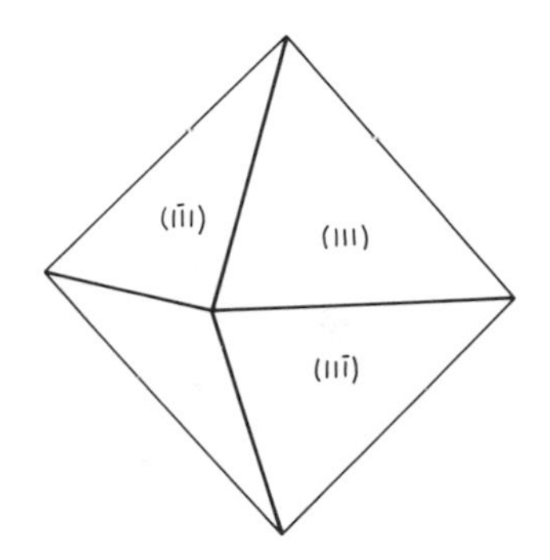

Spinel

$MgAl_2O_4$
Isometric
$n = 1.72–1.74$
Pleochroism and color: colorless.[1]

Pleonaste

$(Mg,Fe)Al_2O_4$
Isometric
$n = 1.75–1.79$
Pleochroism and color: green to blue-green.

Hercynite

$(Fe,Mg)Al_2O_4$
Isometric
$n = 1.78–1.80$
Pleochroism and color: dark green.

Picotite

$(Fe,Mg)(Al,Cr)_2O_4$
Isometric
$n = 2.00$
Pleochroism and color: olive brown to brown.

(Chromite and magnetite, listed under Opaque Minerals, are members of this group.)

Habit: spinels are euhedral to anhedral; octahedra and cubes are common; detrital spinel is commonly rounded and often displays a conchoidal fracture.

Cleavage: none (Fig. 201); {111} parting rare.

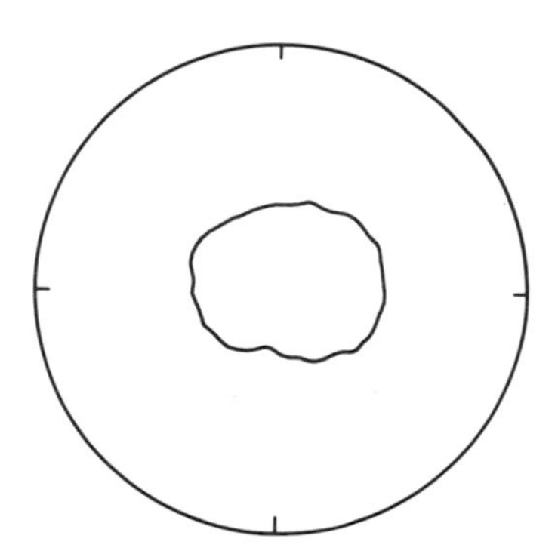

Figure 201. Spinel fragments lack cleavage, are isotropic, and permit estimation of a single index.

---

[1] The colors of each member of the group are highly variable; anomalous pleochroism is uncommon.

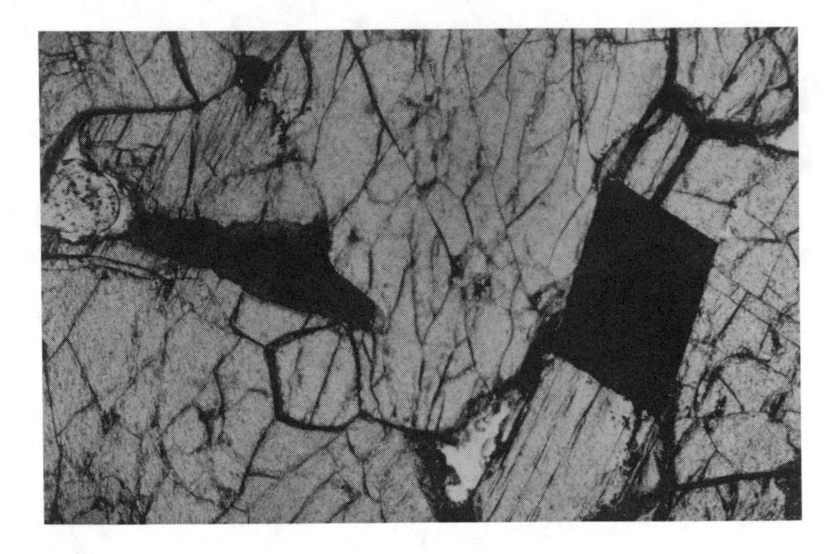

Figure 202. Spinel (black) in olivine (left) and pyroxene (right). Crossed nicols. Photo length is 2.22 mm.

Twinning: {111} (spinel law) is common; usually simple.
Hardness: 7½–8.
Specific gravity: 3.55–4.04.
Occurrence: typically found in high-grade aluminum-rich metamorphic rocks; also is an accessory in mafic and silica-poor igneous rocks (Fig. 202); common as a detrital mineral in sediments.
Alteration: generally stable.
Characteristic features: octahedral shape and lack of inclusions; isotropic; generally deep color; high indices. See color photos 65 and 66.
Look-alikes: garnets do not show octahedral cross-sections and are generally not strongly colored; it may be necessary to determine the index of refraction, density, and length of the cell edge to obtain the approximate composition; chromite and magnetite are semiopaque to opaque respectively; periclase has perfect cubic cleavage.

Spodumene. See Pyroxene Group (P8).

SPURRITE (S9)

$Ca_5(SiO_4)_2CO_3$
Monoclinic
$\measuredangle\beta = 123°$
$\alpha = 1.637–1.641$, $\beta = 1.672–1.676$, $\gamma = 1.676–1.681$
$\gamma - \alpha = 0.039–0.040$ (3rd-order blue-green)
$2V = 35°–41° (-)$
$b = X$, $c \wedge Y \simeq 33°$, $a \simeq Z$

Pleochroism and color: colorless.
Dispersion: $r > v$ weak, crossed.
Habit: usually anhedral, granular.
Cleavage: {001} good, {100} poor (Fig. 203).
Extinction in section: parallel to {001} and {010} pinacoids (when present) and {001} cleavage traces; length high (slow) or length low (fast) as a function of grain orientation.
Twinning: {20$\bar{5}$} simple and {001} polysynthetic at 57°.
Hardness: 5.
Specific gravity: 3.01.
Occurrence: uncommon; high-grade contact metamorphic carbonate rocks, with *larnite* ($Ca_4(SiO_4)_2$) and *merwinite* ($Ca_3Mg(SiO_4)_2$).
Alteration: to *afwillite* ($Ca_3Si_2O_4(OH)_6$).
Characteristic features: occurrence; biaxial (−); polysynthetic twins.
Look-alikes: larnite is biaxial (+), has birefringence of 0.023, and generally inclined extinction against good cleavage traces; merwinite is biaxial (+), has birefringence of 0.008–0.023, perfect cleavage, and two sets of polysynthetic twins.

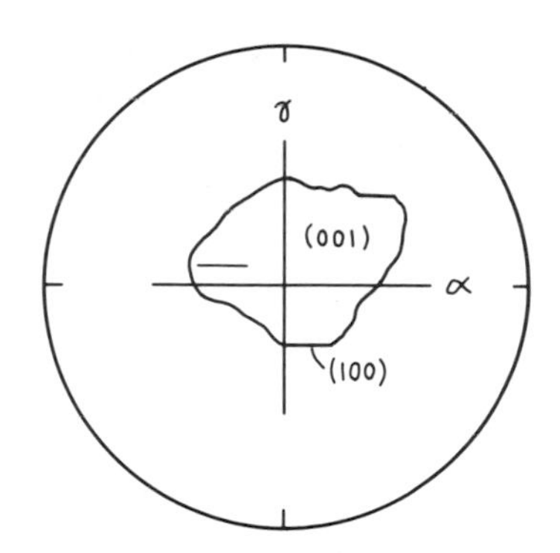

Figure 203. Spurrite fragments on the good {001} cleavage permit estimation of α and a rough estimation of γ. Extinction is parallel to traces of the poor {100} cleavage (if present) and grains tend to be length low (fast). The figure is essentially bisymmetric O.N.

STAUROLITE (S10)

$Fe_2^{+2}Al_9Si_4O_{22}(O,OH)_2$
Monoclinic (pseudo-orthorhombic with $\measuredangle\beta \simeq 90°$)
$\alpha = 1.736–1.747$, $\beta = 1.740–1.754$, $\gamma = 1.745–1.762$
$\gamma - \alpha = 0.009–0.015$ (1st-order orange)
$2V = 80°–90°$ (+)
$a = Y, b = X, c = Z$
Pleochroism and color: golden yellow; moderately pleochroic from deep yellow to colorless; abs. $Z > Y > X$.
Dispersion: $r > v$ weak to moderate.
Habit: commonly as large euhedral crystals elongated on *c;* prominant {110} and {010} faces result in a six-sided outline when viewed along *c*; crystals are four-sided when viewed in the zone normal to *c;* porphyroblasts are commonly poikiloblastic (sieve structure—see color photos 67 and 68).
Cleavage: {010} poor (Fig. 204).
Extinction in section: symmetric and parallel when viewed along *c:* usually length low (fast); parallel extinction and length high (slow) when viewed normal to *c*.
Twinning: penetration twins common on {023} or {232}, forming cruciform twins at about 90° and 60° (Fig. 205).
Hardness: 7–7½.
Specific gravity: 3.74–3.83.

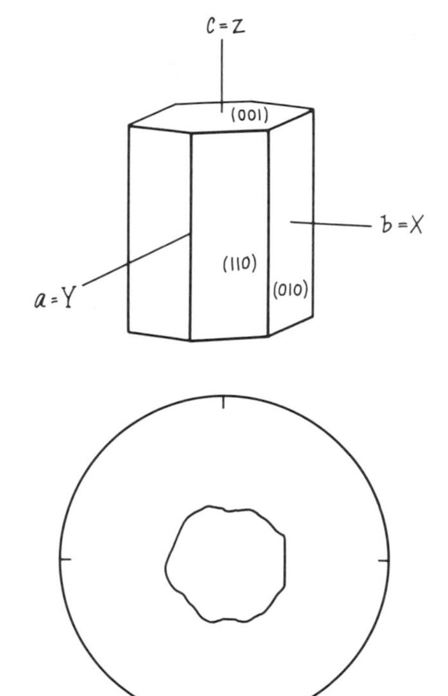

Figure 204. Most staurolite fragments are in random orientation. Those few that lie on the poor {010} cleavage permit estimation of β and γ and have a bisymmetric Bxo figure.

Figure 205. Cruciform staurolite twin within a quartz-mica schist. The fine-grained acicular crystals at the upper left are sillimanite. Crossed nicols. Photo length is 5.24 mm. (From J. Hinsch, E. Leitz, Inc.)

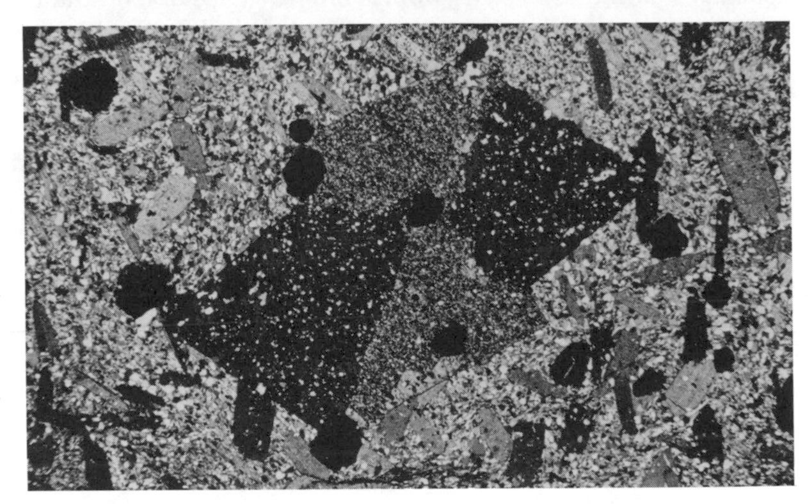

Figure 206. Staurolite twin containing abundant inclusions of quartz. The groundmass consists of quartz, mica, and garnet. Crossed nicols. Photo length is 8.25 mm. (From J. Hinsch, E. Leitz, Inc.)

Occurrence: common as porphyroblasts in medium-grade metapelitic schists, with kyanite, almandine, muscovite, and biotite (Fig. 206); less common in low-grade schist with chloritoid and chlorite; common as a detrital mineral in sediments.

Alteration: resistent to weathering, but may alter to sericite and chlorite; with prograde metamorphism, may be converted to almandine and either kyanite or sillimanite; with retrograde metamorphism, may be converted to chlorite or sericite.

Characteristic features: yellow pleochroism; low birefringence; porphyroblastic habit; euhedral; lack of cleavage. See color photo 69.

Look-alikes: tourmaline is uniaxial (−) and length low (fast).

Stilbite. See Zeolite Group (Z1).

STILPNOMELANE (S11)

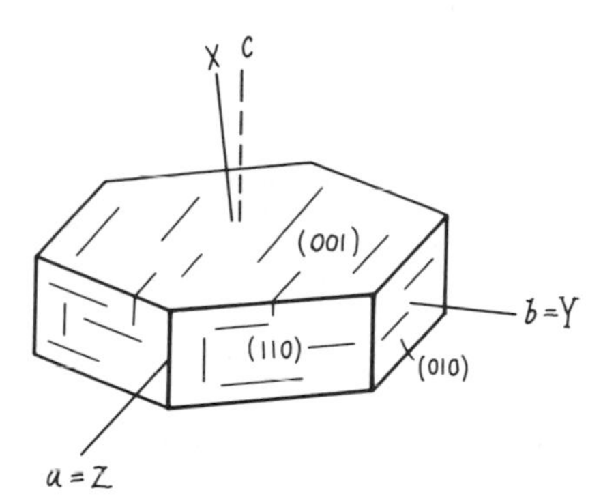

$(K,Na,Ca)_{0-1.4}(Fe^{+3},Fe^{+2},Mg,Al,Mn)_{5.9-8.2}[Si_8O_{20}]$
$(OH)_4(O,OH,H_2O)_{3.6-8.5}$
Monoclinic (pseudohexagonal)
$\measuredangle\beta = 96°$
$\alpha = 1.543-1.634$, $\beta = \gamma = 1.576-1.745$
$\gamma - \alpha = 0.030-0.110$ (up to white of the higher order)
$2V \simeq 0°$, but rarely may be up to 40° (−)
$b = Y$, $c \wedge X = 7°$, $a \simeq Z$

Pleochroism and color: green, brown, or yellow; strongly pleochroic, with X = pale yellow, golden yellow, pale brown, colorless, and $Y = Z$ = deep brown, deep red-brown, nearly black, deep yellow brown, light greenish yellow, with abs. $Z = Y \gg X$; ferroan varieties are generally yellow to green, and ferrian varieties are brown, dark brown, or black.

Habit: thin micaceous {001} plates that may form radial, plumose, or subparallel aggregates; the outline is pseudohexagonal on the basal section; zoning is fairly common with rims more Fe^{+3}-rich than cores.

Cleavage: {001} perfect, {010} poor. See Figure 207.

Extinction in section: parallel to cleavage traces and basal surfaces; length high (slow); sensibly isotropic when viewed normal to {001}; the bird's-eye maple structure shown by micas is not present.

Twinning: not reported.

Hardness: 3–4.

Specific gravity: 2.59–2.96.

Occurrence: relatively common in low-grade schists that are rich in Fe and Mn, and associated with chlorite, garnet, epidote,

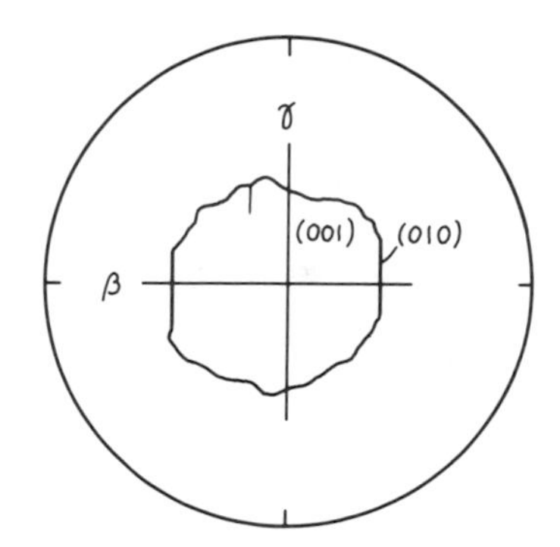

Figure 207. All stilpnomelane fragments lie on the perfect {001} cleavage, and permit estimation of β and γ (which are equal). The poor {010} cleavage may be present. Fragments are in total extinction, and the figure commonly appears to be of the uniaxial O.A. type.

and actinolite; also found in high-pressure metamorphic rocks with glaucophane and lawsonite.

Alteration: to clays and iron oxides.

Characteristic features: strongly pleochroic; pseudohexagonal plates; 2*V* usually 0°; lacks bird's-eye maple structure; perfect {001} cleavage with parallel extinction.

Look-alikes: stilpnomelane is easily mistaken for biotite: biotite lacks poor {010} cleavage, has bird's-eye maple structure, and has a lower birefringence than Fe^{+3}-rich stilpnomelane (color photo 70).

Strontianite. See Carbonate Group (C2).

SULFATE GROUP (S12)

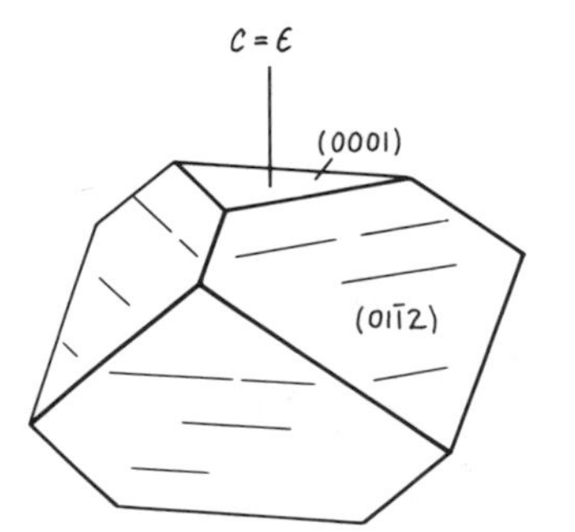

Alunite

$KAl_3(SO_4)_2(OH)_6$
Hexagonal
$\omega = 1.569-1.578$, $\varepsilon = 1.590-1.601$
$\varepsilon - \omega = 0.021-0.023$ (up to 2nd-order blue-green)
Uniaxial (+)
Pleochroism and color: colorless.
Habit: cube-like rhombohedral or tabular {0001}.
Cleavage: {0001} good, {01$\bar{1}$2} poor (Fig. 208).
Extinction in section: parallel or symmetrical to principal faces, and parallel to cleavage traces.
Twinning: none.
Hardness: 3½–4.0.

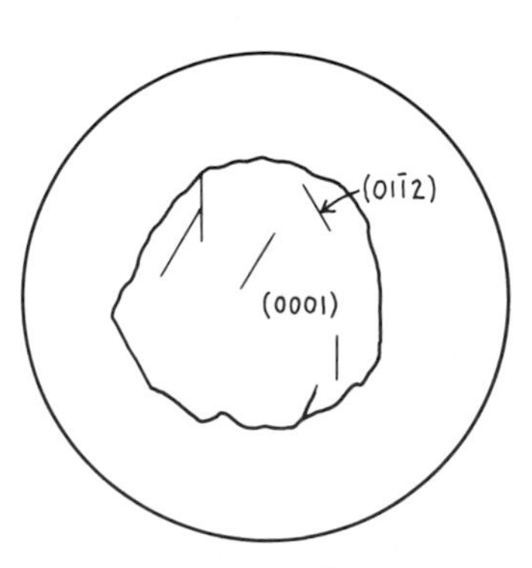

Figure 208. Alunite fragments on the good {0001} cleavage are in total extinction; ω may be measured; the figure is centered O.A.

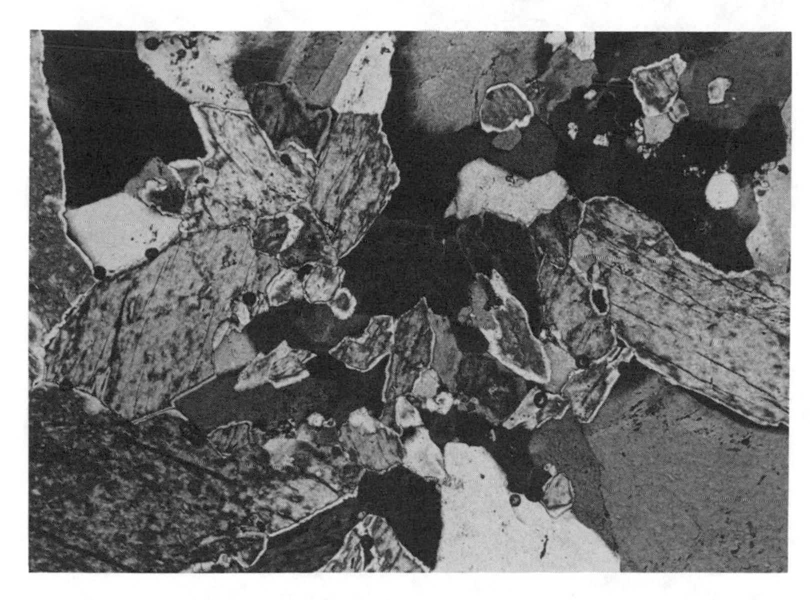

Figure 209. Aggregate of alunite grains. Crossed nicols. Photo length is 2.06 mm. (From J. Hinsch, E. Leitz, Inc.)

Specific gravity: ≃2.71.

Occurrence: hydrothermal alteration product of feldspars in volcanic flows and pyroclastic rocks; from pyrite in Al-rich rocks; with quartz, clays, and pyrite. See Figure 209.

Alteration: not described.

Characteristic features: moderate birefringence, moderate indices; occurrence.

Look-alikes: muscovite is biaxial (−); scapolite is uniaxial (−); brucite occurs in metamorphic rocks and commonly exhibits anomalous interference colors.

Anhydrite

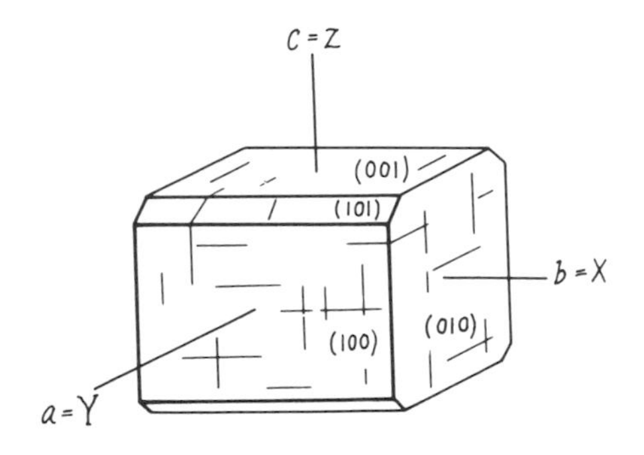

$CaSO_4$

Orthorhombic

$\alpha = 1.570$, $\beta = 1.576$, $\gamma = 1.614$

$\gamma - \alpha = 0.044$ (3rd-order green)

$2V = 44°$ (+)

$a = Y$, $b = X$, $c = Z$

Pleochroism and color: colorless.

Dispersion: $v > r$ strong.

Habit: subhedral fine to coarse granular, with {100}, {010}, and {001} common; rarely fibrous or radiating aggregates.

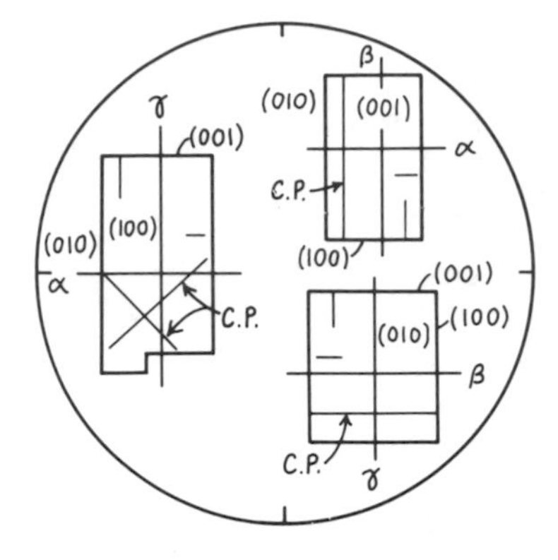

Figure 210. Most anhydrite grains lie on {010}, have parallel extinction, yield β and γ indices, and show a bisymmetric Bxo figure. Grains on {001} have parallel extinction, yield α and β indices, and show a bisymmetric Bxa figure. Grains on {100} have parallel extinction, yield α and γ indices, and show a bisymmetric O.N. figure. Twinning in two directions (C.P. = composition plane) may be seen on {100}, and a single direction on {010} and {001}.

Figure 211. Rectangular anhydrite, showing pinacoidal cleavage. The irregular line (top) is a stylolite. Crossed nicols. Photo length is 0.55 mm.

Cleavage: perfect {010}, {001} and {100} good. See Figure 210.
Extinction in section: generally parallel to pinacoidal grain edges and cleavages.
Twinning: simple or polysynthetic common on {011}; twins on (011) and (01$\bar{1}$) intersect at 83.5° and 96.5°.
Hardness: 3–3½.
Specific gravity: ≈2.96.
Occurrence: sedimentary evaporite deposits with gypsum, halite, dolomite, or calcite; less commonly in oxidized zones of ore deposits, fumarolic deposits, or within amygdules in mafic volcanic rocks with zeolites, prehnite, or quartz; rarely in contact metamorphic rocks. See Figure 211.
Alteration: commonly to gypsum.
Characteristic features: pseudocubic cleavages; generally parallel extinction; moderate birefringence.
Look-alikes: polyhalite has lower indices, larger $2V$, and a single perfect cleavage; gypsum has lower relief and birefringence, and commonly inclined extinction to cleavage traces; barite and celestite have higher indices and low birefringence.

Barite

$BaSO_4$
Orthorhombic
$\alpha = 1.636–1.637$, $\beta = 1.637–1.639$, $\gamma = 1.647–1.649$
$\gamma - \alpha = 0.011–0.012$ (1st-order yellow)

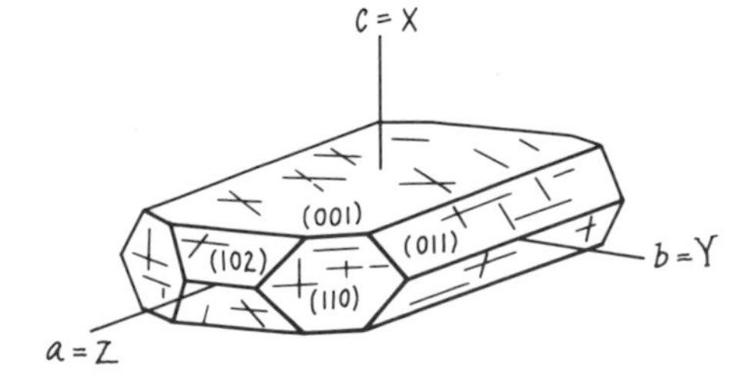

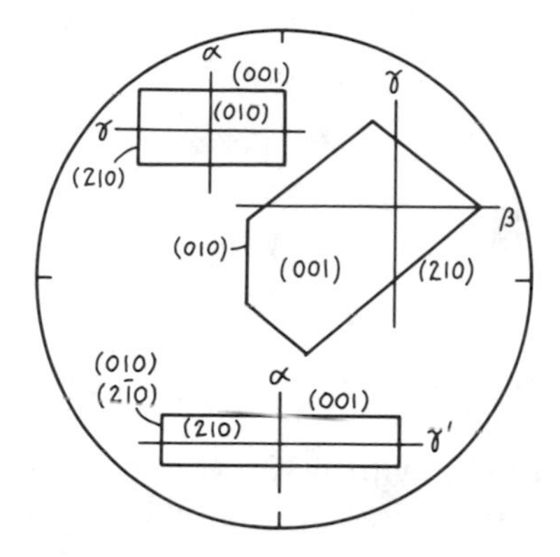

Figure 212. Most barite grains lie on {001}, have symmetrical and parallel extinctions, yield β and γ indices, and have a bisymmetric Bxo figure. Grains on {210} have parallel extinction, yield α and γ' (1.642) indices, and have a monosymmetric Bxa figure. Grains on {010} have parallel extinction, yield α and γ indices, and have a bisymmetric O.N. figure.

$2V = 36°–37°$ (+)
$a = Z, b = Y, c = X$
Pleochroism and color: typically colorless.
Dispersion: $v > r$ weak.
Habit: commonly tabular {001}, with {101} and {011}.
Cleavage: {001} perfect, {210} perfect to good at 78° and 102°, {010} good to poor. See Figure 212.
Extinction in section: on {001}, parallel to well-developed faces and symmetric to prismatic cleavages; when viewed ⊥ to *c*, extinction is parallel to the longer {001} grain edge and {001} cleavage; length high (slow).
Twinning: {110} polysynthetic may be frequent.
Hardness: 2½–3½.
Specific gravity: 4.5.
Occurrence: in limestone and dolomite; as a gangue mineral in metalliferous hydrothermal veins; associated minerals are fluorite, sphalerite, galena, calcite, or dolomite.
Alteration: to witherite; may be replaced by a large variety of other minerals.
Characteristic features: occurrence; generally parallel or symmetric extinction; moderate 2*V* (+); low birefringence.
Look-alikes: greatly resembles celestite, which has lower indices and birefringence; anhydrite has higher birefringence and lower indices, and extinction is always parallel to cleavage traces; gypsum has inclined extinction.

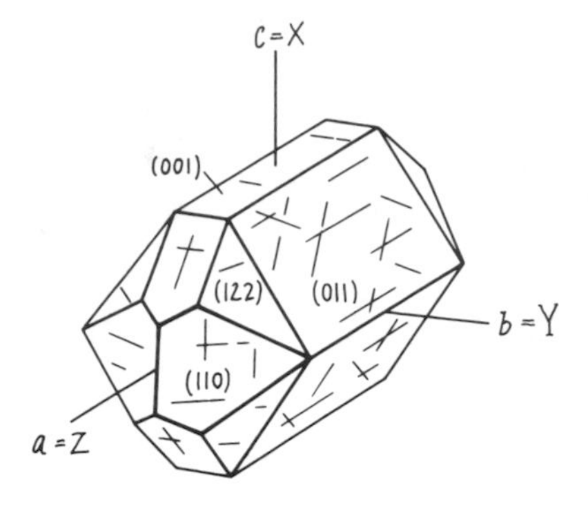

Celestite

$SrSO_4$
Orthorhombic
$\alpha = 1.621-1.622$, $\beta = 1.623-1.624$, $\gamma = 1.630-1.632$
$\gamma - \alpha = 0.009-0.010$ (1st-order whitish yellow)
$2V \simeq 50°$ (+)
$a = Z, b = Y, c = X$
Pleochroism and color: colorless.
Dispersion: $v > r$ moderate.
Habit: tabular on {001} with {110}, or elongated prismatic on *a* with {011} dominant; may be fibrous or granular; less commonly elongate on *b* or *c*.
Cleavage: {001} perfect, {210} good, {010} poor. See Figure 213.
Extinction in section: usually parallel or symmetric to major faces and cleavage traces; length high (slow) in most orientations.
Twinning: rare.
Hardness: 3–3½.
Specific gravity: ≃3.97.
Occurrence: evaporite deposits and carbonate sediments, with calcite, dolomite, anhydrite, gypsum, and fluorite. See Figure 214.
Alteration: to strontianite.
Characteristic features: occurrence; low birefringence; moderate 2*V* (+); parallel extinction common.
Look-alikes: barite has a smaller 2*V* and higher indices; gypsum has inclined extinction; anhydrite has three mutually perpendicular cleavages.

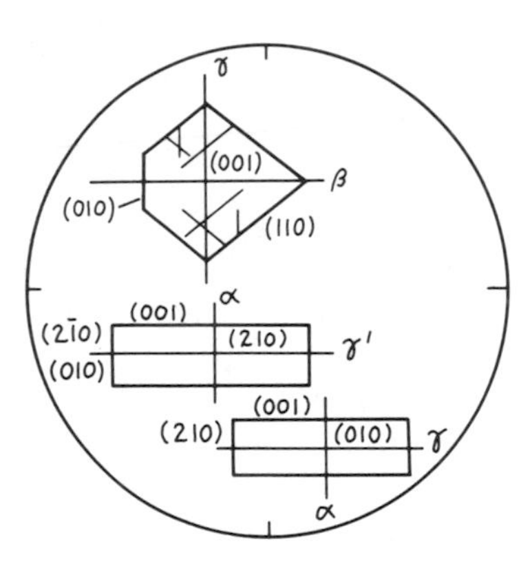

Figure 213. Most celestite grains lie on {001}, yield β and γ indices, have mainly symmetric extinction, and have bisymmetric Bxo figures. A lesser number of grains lie on {210}, yield α and γ' (1.626), have parallel extinction, and have a monosymmetric Bxa figure. An uncommon type of grain lies on {010}, yields α and γ indices, has parallel extinction, and has a bisymmetric O.N. figure.

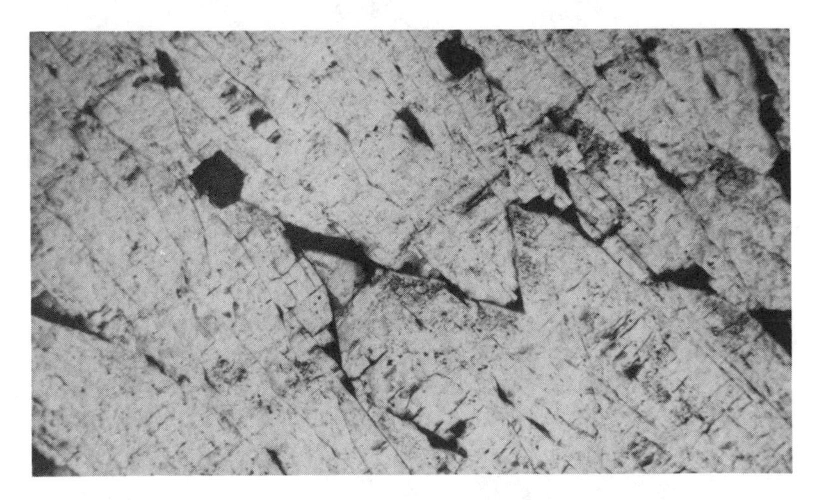

Figure 214. Celestite intergrowth. Grains are elongate on *c* and show {001} and {210} cleavages. Crossed nicols. Photo length is 2.22 mm.

Gypsum

$CaSO_4 \cdot 2H_2O$
Monoclinic
$\measuredangle\beta \simeq 127°$
$\alpha = 1.519–1.521$, $\beta = 1.522–1.526$, $\gamma = 1.529–1.531$
$\gamma - \alpha = 0.010$ (1st-order pale yellow)
$2V = 58°$ (+)
$b = Y, c \wedge Z \simeq 52°$
Pleochroism and color: colorless.
Dispersion: $r > v$ strong, inclined.
Habit: commonly as euhedral crystals with {010} dominant, and {110} and {011} less so; stubby prismatic or acicular (parallel to *c*) common; massive or granular.
Cleavage: {010} perfect, {100} and {$\bar{1}$11} good. See Figure 215.
Extinction in section: commonly parallel to {010} faces and cleavage traces, length high (slow) or low (fast) depending upon orientation; extinction inclined against other surfaces and cleavages.
Twinning: simple {100} common, but may be polysynthetic; rare {$\bar{1}$01}.
Hardness: 2.
Specific gravity: 2.31.
Occurrence: in evaporite deposits with anhydrite, calcite, dolomite, and halite; in caprock of salt domes; in some limestones

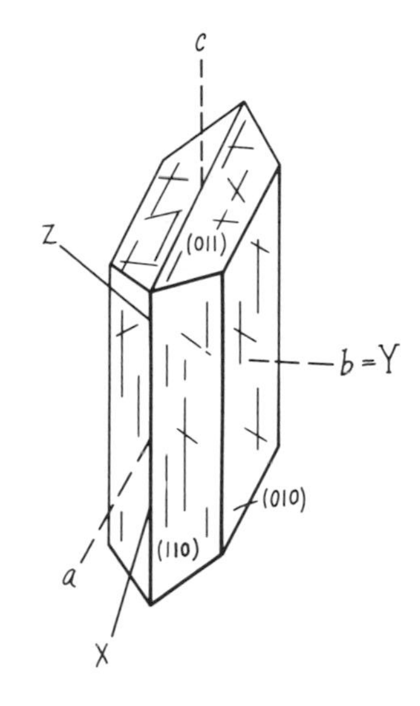

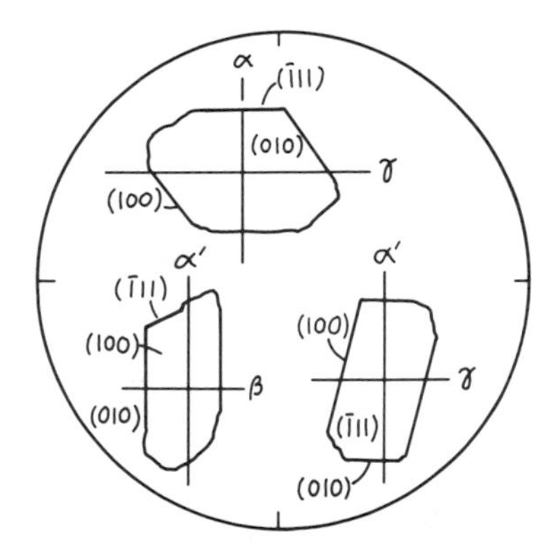

Figure 215. Most gypsum grains lie on {010} and permit estimation of α and γ. The extinction of $\gamma \wedge \{\bar{1}11\} = 1°$, $\gamma \wedge \{100\} = 52°$, and the figure is bisymmetric O.N. Less common grains on {100} permit estimation of effectively identical indices α' (1.524) and β. Extinction against {010} is parallel, $\beta \wedge \{\bar{1}11\} = 23°$, and the figure is close to O.A. Less common grains on {$\bar{1}$11} permit estimation of α' (~1.524) and γ. Extinction of $\gamma \wedge \{010\}$ is parallel, $\gamma \wedge \{100\} = 77°$, and the figure is monosymmetric Bxo.

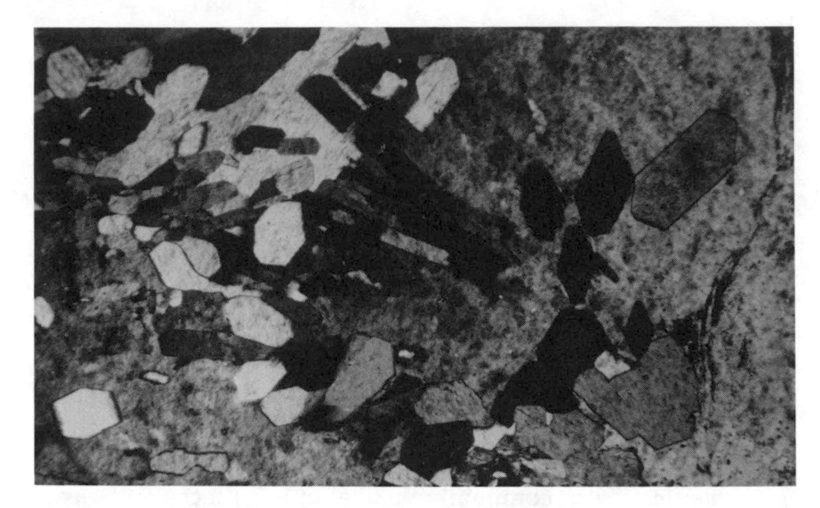

Figure 216. Small grains of gypsum enclosed by gypsum. Crossed nicols. Photo length is 2.22 mm.

and dolomites with anhydrite, barite, celestite, and fluorite; from oxidation of sulfide minerals in ore deposits.

Alteration: may be replaced by carbonates, chalcedony, opal, anhydrite, or celestite.

Characteristic features: low birefringence; negative relief; occurrence; cleavages. See Figure 216.

Look-alikes: anhydrite has positive relief and pseudocubic cleavage; feldspars are not found in evaporite deposits.

Sylvite. See Halite Group (H1).

Tachylite. See Glass (G2).

TALC (T1)

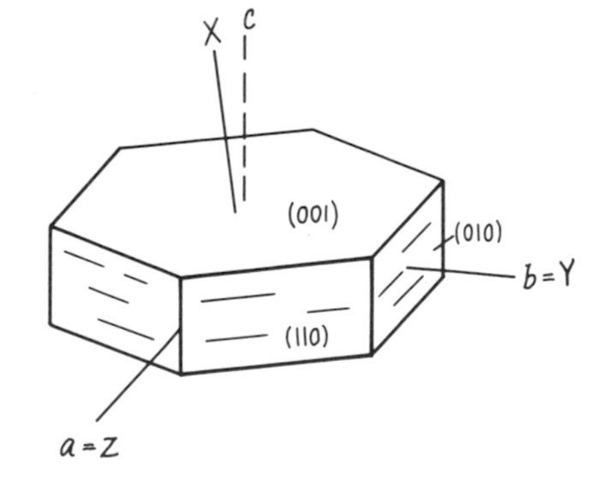

$Mg_3Si_4O_{10}(OH)_2$
Monoclinic
$\measuredangle\beta = 98°55'$
$\alpha = 1.539–1.550$, $\beta = 1.589–1.594$, $\gamma = 1.589–1.600$
$\gamma - \alpha \simeq 0.05$ (3rd-order yellow)
$2V = 0°–30° (-)$
$b = Y, c \wedge X \simeq 10°, a \simeq Z$
Pleochroism and color: colorless.
Dispersion: $r > v$ moderate.
Habit: massive aggregates of {001} plates in random, subparallel, radial, or concentric arrangement; curved plates are common.

Cleavage: {001} perfect (Fig. 217).
Extinction in section: parallel to cleavage traces; length high (slow).
Twinning: rare.
Hardness: 1.
Specific gravity: 2.6–2.8.
Occurrence: forms during metamorphism of impure dolomite (Fig. 218) and is associated with tremolite, calcite, serpentine, chlorite, and anthophyllite; secondary alteration of ultramafic rocks may yield massive impure talc rock (steatite, soapstone) or talc-magnesite rocks.
Alteration: stable, but may convert to magnesite and quartz if sufficient CO_2 is available.
Characteristic features: closely resembles muscovite and pyrophyllite; occurrence and associated minerals (color photo 71).
Look-alikes: easily mistaken for muscovite or pyrophyllite (which have larger 2*V*s); talc has a somewhat smeared or diffuse appearance when viewed with crossed nicols as compared to muscovite and pyrophyllite; brucite often shows anomalous interference colors and is uniaxial (+); may be difficult to distinguish optically from sericite.

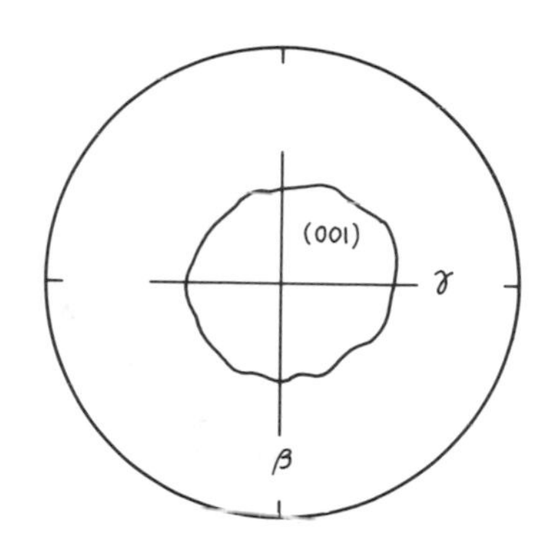

Figure 217. All talc fragments lie on the perfect {001} cleavage and permit estimation of β and γ. The figure is essentially bisymmetric Bxa.

Figure 218. Fine-grained talc matrix enclosing larger blocky and elongate grains of tremolite. Crossed nicols. Photo length is 5.24 mm. (From J. Hinsch, E. Leitz, Inc.)

Thomsonite. See Zeolite Group (Z1).

Thulite. See Epidote Group (E1).

Titanite. See Sphene (S7).

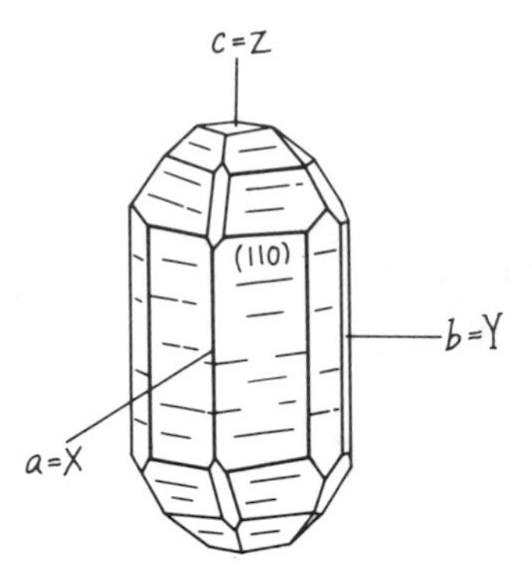

TOPAZ (T2)

$Al_2(SiO_4)(F,OH)_2$
Orthorhombic
$\alpha = 1.609–1.630$, $\beta = 1.612–1.632$, $\gamma = 1.618–1.640$ (Fig. 219A)
$\gamma - \alpha = 0.009–0.010$ (1st-order yellowish white)
$2V = 48°–69°$ (+) (see Fig. 219B)
$a = X, b = Y, c = Z$
Pleochroism and color: colorless; in detrital grains, $X = Y$ = yellow and Z = pink.
Dispersion: $r > v$ moderate.

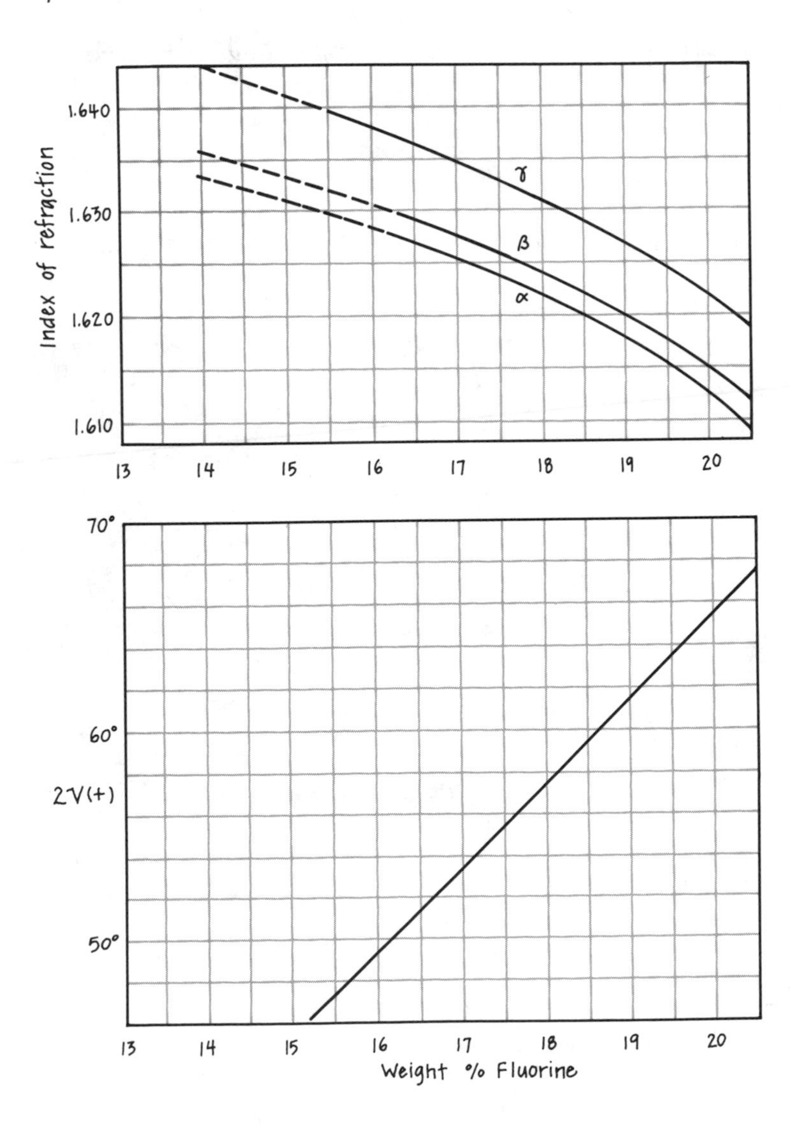

Figure 219. (A) Variation in indices of refraction of topaz as a function of fluorine content (after Ribbe and Rosenberg, 1971). (B) Variation in $2V$ as a function of fluorine content (after Ribbe and Rosenberg, 1971).

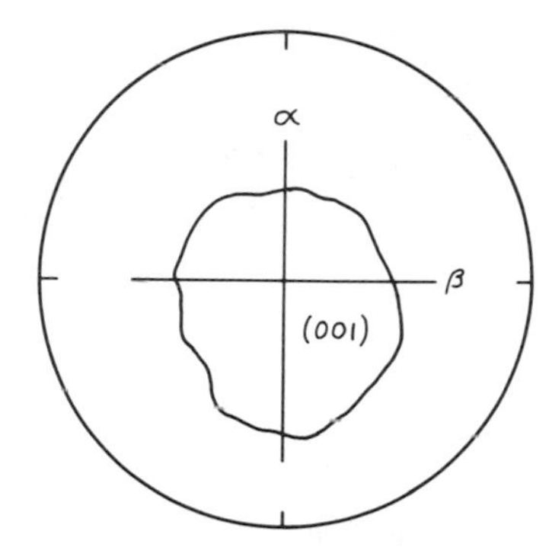

Figure 220. Most topaz fragments lie on the {001} cleavage and permit estimation of α and β. The figure is bisymmetric Bxa.

Habit: commonly euhedral, elongate on *c* with {110} and {010} or {110} and {120} predominant; often anhedral.
Cleavage: {001} perfect (Fig. 220).
Extinction in section: elongated grains commonly have parallel extinction and are length high (slow); when viewed along the *c* axis, the extinction is symmetrical.
Twinning: rare.
Hardness: 8.
Specific gravity: 3.49–3.57.
Occurrence: in vugs and fissures in siliceous igneous rocks, usually as a result of late-stage pneumatolytic action; also in granitic pegmatites, greisens, and some metamorphic rocks; as a heavy mineral in detrital sediments.
Alteration: very stable, but may be altered by hydrothermal solutions to sericite or clay minerals.
Characteristic features: occurrence; low birefringence, moderate relief, one cleavage; moderate $2V$ (+).
Look-alikes: apatite is uniaxial (−) and length low (fast); quartz and feldspar have lower relief.

TOURMALINE GROUP (T3)

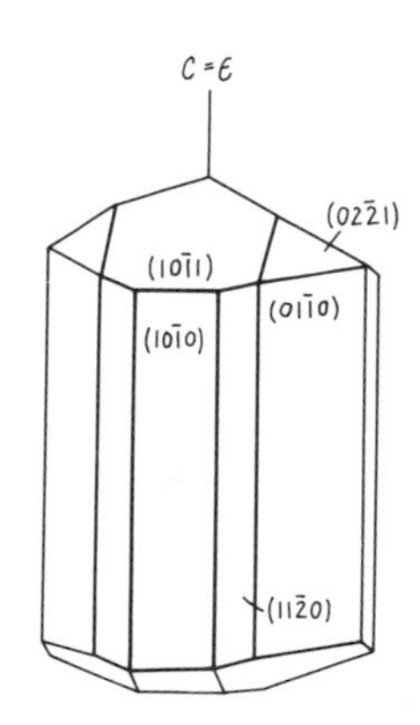

$Na(Mg,Fe,Li,Al)_3Al_6(Si_6O_{18})(BO_3)_3(OH,F)_4$
Hexagonal
$\omega = 1.635–1.675$, $\varepsilon = 1.610–1.650$
$\omega - \varepsilon = 0.017–0.035$ (up to 2nd-order red)
Uniaxial (−)
Pleochroism and color: very variable and often zoned within a single grain; black, brown, green, blue, yellow, pink, or colorless; strongly pleochroic with abs. $O > E$. See color photo 72.

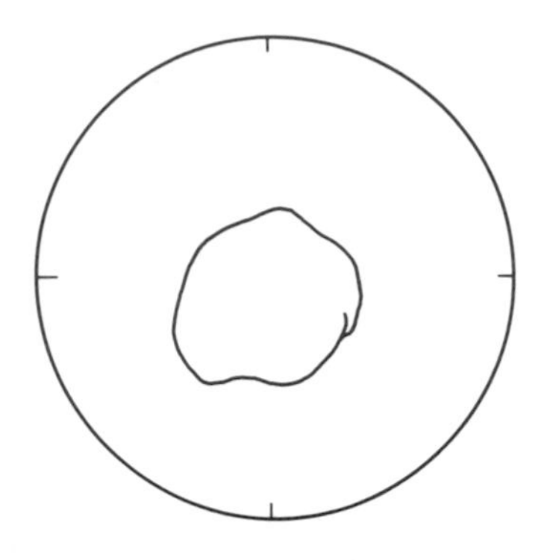

Figure 221. Tourmaline fragments are in random orientation, have random outline, and have a variety of interference figures.

Habit: usually euhedral, ranging from stubby prisms to acicular aggregates; cross-sections may be hexagonal or triangular (with convex curvature).

Cleavage: $\{11\bar{2}0\}$ and $\{10\bar{1}1\}$ poor, {0001} parting often well-developed (Fig. 221).

Extinction in section: parallel to predominant prism faces, and length low (fast); cross-sections are in continuous extinction.

Twinning: rare.

Hardness: 7–7½.

Specific gravity: 3.0–3.25.

Occurrence: a late-stage pneumatolytic accessory mineral in many granites, pegmatites, schists, and gneisses (Fig. 222); also in metamorphosed and metasomatized carbonate rocks; common as a detrital mineral in sediments.

Alteration: to illite, sericite, chlorite, or biotite.

Characteristic features: strong pleochroism, moderate birefringence; uniaxial (−); parallel extinction; rounded triangular cross-section.

Look-alikes: the maximum absorption direction of biotite and stilpnomelane is parallel to the direction of elongation, whereas in tourmaline it is normal to the direction of elongation; biotite has one perfect cleavage and bird's-eye maple structure; hornblende has two good cleavages, maximum absorption at or near the direction of elongation, and is biaxial; topaz is biaxial; apatite has low birefringence.

Figure 222. Poikilitic grain of tourmaline (center) viewed normal to the *c* axis. The surrounding grains are quartz, feldspar, and mica. Crossed nicols. Photo length is 2.22 mm.

Tremolite. See Amphibole Group (A2).

Tridymite. See Silica Group (S4).

Tschermakite. See Amphibole Group (A2).

Uvarovite. See Garnet Group (G1).

Vermiculite. See Clays and Clay Minerals (C6).

VESUVIANITE (IDOCRASE) (V1)

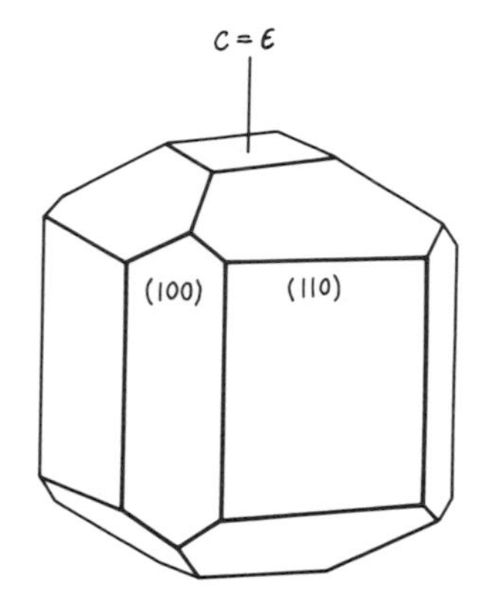

$Ca_{10}(Mg,Fe^{+2})_2Al_4(SiO_4)_5(Si_2O_7)_2(OH)_4$
Tetragonal
ω = 1.703–1.752, ε = 1.700–1.746
$\omega - \varepsilon$ = 0.001–0.008 (up to 1st-order white, but usually dark gray)
Uniaxial (−), rarely (+); some varieties are biaxial ($2V$ = 5°–65°), generally (−) and rarely (+)
Pleochroism and color: usually colorless, but may be pale shades of green, brown, or yellow, with weak pleochroism, abs. $O > E$; color may be zoned or irregularly distributed; anomalous deep blue, brown, green, or purple interference colors common (color photo 73).
Habit: euhedral to anhedral; commonly crystals are stubby prismatic, with prisms and dipyramids of the same order.
Cleavage: {110} and {001} poor (Fig. 223).
Extinction in section: parallel to prismatic faces; length low (fast); continuous extinction when viewed along *c*.
Twinning: uncommon, but basal sections of biaxial varieties may be divided into four sectors.
Hardness: 6–7.
Specific gravity: 3.33–3.45.
Occurrence: typically in contact metamorphic rocks such as marble, hornfels, and calc-silicate schist, with calcite, diopside, grossular, sphene, wollastonite, and epidote; uncommonly in nepheline syenite, or in veins cutting ultramafic rocks.
Alteration: stable.
Characteristic features: occurrence; high relief; uniaxial (−); low birefringence; anomalous interference colors.
Look-alikes: zoisite and clinozoisite are biaxial (+) and have a good cleavage parallel to their direction of elongation; andalusite is biaxial (−) and has a higher birefringence; melilite has lower indices and is commonly in undersaturated lavas.

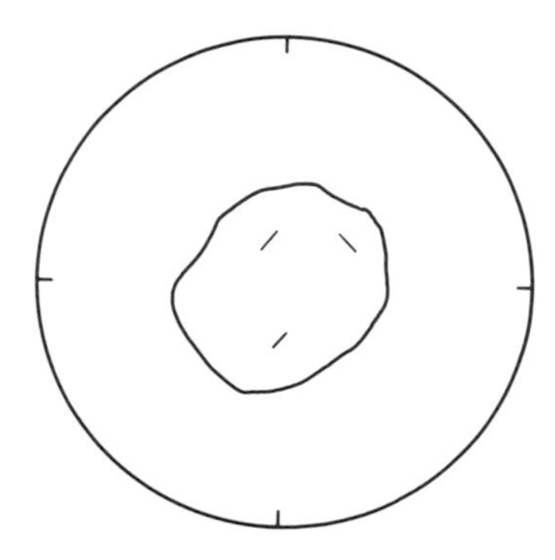

Figure 223. With vesuvianite, the lack of good cleavage results in random orientation and shape of fragments.

Vishnevite. See Cancrinite (C1).

Witherite. See Carbonate Group (C2).

WOLLASTONITE (W1)

$CaSiO_3$ (most natural wollastonite is close to $CaSiO_3$ in composition, but extensive $FeSiO_3$ solid solution is possible).
Triclinic
$\measuredangle\alpha = 90°02'$, $\measuredangle\beta = 95°22'$, $\measuredangle\gamma = 103°26'$
$\alpha = 1.616–1.640$, $\beta = 1.628–1.650$, $\gamma = 1.631–1.653$ (lower index values are typical)
$\gamma - \alpha = 0.013–0.014$ (1st-order orange)
$2V = 38°–60°\ (-)$
$b \wedge Y = 0°–5°$, $c \wedge X = -33°$ to $-44°$, $a \wedge Z = -37°$ to $-50°$, optic plane is almost parallel to {010}.
Pleochroism and color: colorless.
Dispersion: $r > v$ weak, inclined.
Habit: commonly columnar, bladed, or fibrous; elongated on b, with either {100} or {001} dominant.
Cleavage: {100} perfect, {001} and {$\bar{1}$02} good; {100} $\wedge$ {001} = 84.5° and {001} $\wedge$ {$\bar{1}$02} = 45° as seen on {010} sections. See Figure 224.

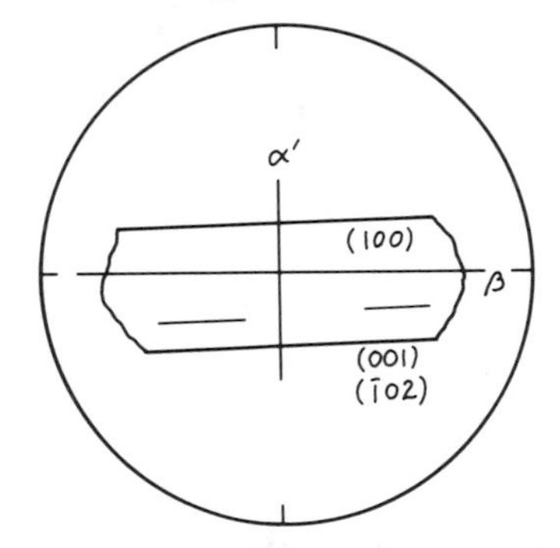

Figure 224. Most wollastonite fragments lie on the perfect {100} cleavage and permit estimation of α' (1.621) and β. Extinction is essentially parallel to (001) or ($\bar{1}$02) cleavage edges and the elongation direction is length high (slow). The figure is essentially monosymmetric Bxo. Fragments on the good {001} cleavage permit estimation of α' (1.624) and β, have essentially parallel extinction, and are length high (low). The figure is essentially monosymmetric Bxo. As the {$\bar{1}$02} cleavage is almost normal to X, fragments on {$\bar{1}$02} permit estimation of β and γ, have essentially parallel extinction, and are length low (fast). The figure is bisymmetric Bxa.

Figure 225. Aggregate of wollastonite in various orientations. Crossed nicols. Photo length is 2.22 mm.

Extinction in section: elongate grains have parallel to near-parallel extinction, and are length low (fast) when viewed along *c*, and length high (slow) when viewed along *a;* when viewed along *b*, extinction is highly inclined, with $X \wedge \{100\}$ cleavage trace $\simeq$ 30°–44°.

Twinning: {100} polysynthetic common.

Hardness: 4½–5.

Specific gravity: 2.9–3.1.

Occurrence: typically in contact metamorphic and metasomatized carbonate rocks, with calcite, tremolite, diopside, grossular, vesuvianite, epidote, and anorthite; rarely in alkaline undersaturated igneous rocks. See Figure 225.

Alteration: weathers to calcite; with high temperatures it may react to form other calc-silicate minerals, such as larnite ($Ca_4(SiO_4)_2$) and monticellite.

Characteristic features: occurrence; three cleavages; low birefringence (color photo 74); elongate grains with essentially parallel extinction; moderate $2V$ (−).

Look-alikes: tremolite, diopside, and pectolite have higher birefringence and a different optic orientation; zoisite and clinozoisite have higher indices and $2V$; the rare high-temperature polymorph, *pseudowollastonite,* has a birefringence of 0.034–0.041 and is biaxial (+) with a small $2V$.

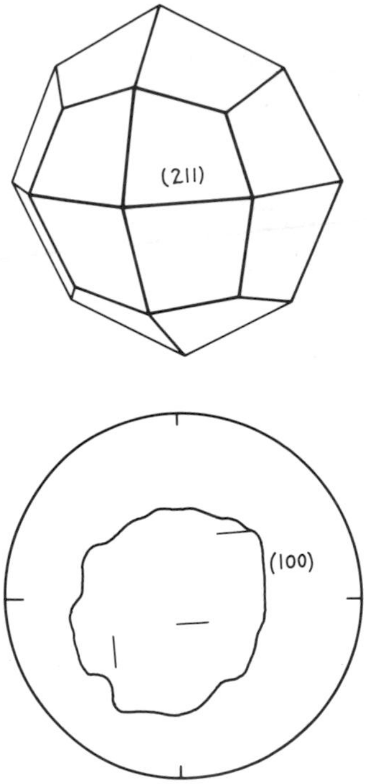

THE ZEOLITE GROUP (Z1)

The zeolite group has the general formula (Na_2,K_2,Ca,Sr,Ba) [(Al,Si)O_2]$_n$·xH_2O, and is characterized by the atomic ratios O:(Al+Si) = 2 and Al_2O_3:(Na_2,K_2,Ca,Sr,Ba)O = 1. The zeolites have a very open framework of (Si,Al)O_4 tetrahedra; cavities are able to accommodate large cations and neutral water molecules. As a consequence, dehydration is continuous with increasing temperature (and partially reversible with decreasing temperature).

Zeolites are of three structural types, based on the nature of silicate tetrahedra linkages. Some are fibrous (laumontite, natrolite, scolecite, and thomsonite); some are foliated (heulandite and stilbite); and others are blocky (analcime and chabazite).

Because of their open framework structure, zeolites have low refractive indices, low birefringence, and low density. Zeolites are stable under surface and hydrothermal conditions, as well as in the lower grades of metamorphism. Thus, zeolites are formed in preexisting igneous rocks by hydrothermal processes, or in sedimentary rocks that have undergone diagenesis or incipient metamorphism. In the latter case, recrystallization is often incomplete, and the validity of an optical identification of a particular zeolite is often questionable; the identification should be verified by supplementary techniques (such as X-ray diffraction).

(211)

(100)

Figure 226. Isotropic analcime fragments of random shape and orientation permit estimation of a single index of refraction. Traces of the poor {100} cleavage may be present.

Analcime (analcite)

$NaAlSi_2O_6 \cdot H_2O$
Isometric
$n = 1.479–1.493$
Normally isotropic, but birefringence may reach 0.002
Pleochroism and color: colorless.
Habit: trapezohedral {211}.
Cleavage: {100} poor (Fig. 226).
Twinning: birefringent sections may show fine polysynthetic twins on {100} and {110}.
Hardness: 5½.
Specific gravity: 2.25.
Occurrence: commonly in vein and cavity fillings, or may be primary in silica-deficient igneous rocks; authigenic mineral in sediments; may form from alteration of Na-feldspars; associated minerals are calcite, other zeolite minerals, fluorite, quartz, pyrite, and K-feldspar.
Alteration: stable at low pressure and temperature.
Characteristic features: isotropic, low index, occurrence.
Look-alikes: leucite has a higher index, and is restricted to volcanic rocks; sodalite is a primary mineral, and has poor {110} cleavage; opal has $n \simeq 1.45$ and forms in crusts and veinlets.

Chabazite

$Ca(Al_2Si_4)O_{12}\cdot6H_2O$
Hexagonal
$\omega = 1.472-1.494$, $\varepsilon = 1.470-1.485$
$\omega - \varepsilon = 0.002-0.010$ (1st-order yellow)
Uniaxial (−); may be (+), with an anomalous $2V$ of 0°–30°
Pleochroism and color: colorless.
Habit: euhedral rhombohedral $\{10\bar{1}1\}$ crystals that approximate cubes.
Cleavage: $\{10\bar{1}1\}$ good (Fig. 227).
Extinction in section: symmetric to cleavage traces and dominant faces.
Twinning: {0001} simple penetration common, $\{10\bar{1}1\}$ simple less common.
Hardness: 4½.
Specific gravity: 2.1.
Occurrence: common in cavities and joints within basalts, andesites, and related rocks; associated with other zeolites; also in hydrothermal and thermal spring deposits, and in some low-grade metamorphic rocks.
Alteration: relatively stable, but may alter to montmorillonite.
Characteristic features: occurrence; uniaxial (−); negative relief; rhombohedral cleavage.
Look-alikes: optically indistinguishable from the uncommon zeolites gmelinite and levynite; analcime has poor cleavage and lower birefringence.

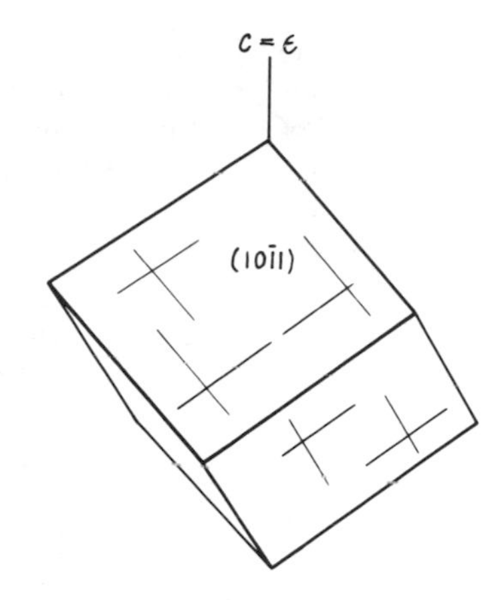

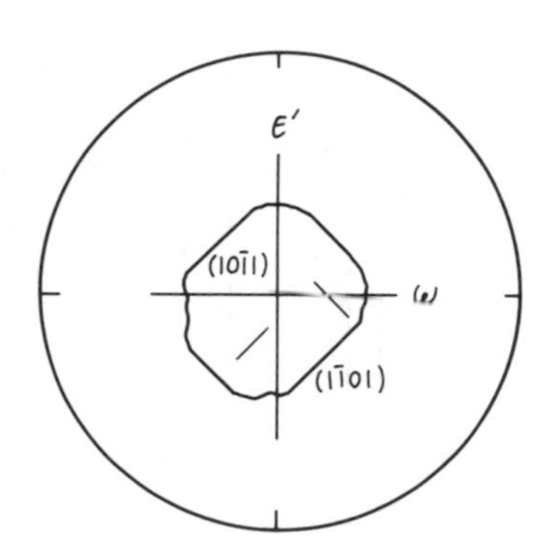

Figure 227. Most chabazite fragments lie on the good $\{10\bar{1}1\}$ cleavage and permit estimation of ω and ε' (1.471–1.488). Extinction is symmetric and the figure is inclined O.A.

Heulandite

$(Ca,Na_2)(Al_2Si_7)O_{18}\cdot6H_2O$
Monoclinic
$\measuredangle\beta = 116°25'$
$\alpha = 1.487-1.499$, $\beta = 1.487-1.501$, $\gamma = 1.488-1.505$
$\gamma - \alpha = 0.001-0.007$ (up to 1st-order grayish white)
$2V \simeq 30°$ (variable, 0°–50°) (+)
$b = Z$, $c \wedge Y = -8°$ to $-32°$
Pleochroism and color: colorless.
Dispersion: $r > v$ moderate, crossed.
Habit: tabular {010}, equant, anhedral.
Cleavage: {010} perfect (Fig. 228).
Extinction in section: tabular {010} grains when viewed in the zone normal to b have parallel extinction and are length low (fast); most other orientations yield inclined extinction.
Twinning: none.
Hardness: 3½–4.
Specific gravity: 2.2.

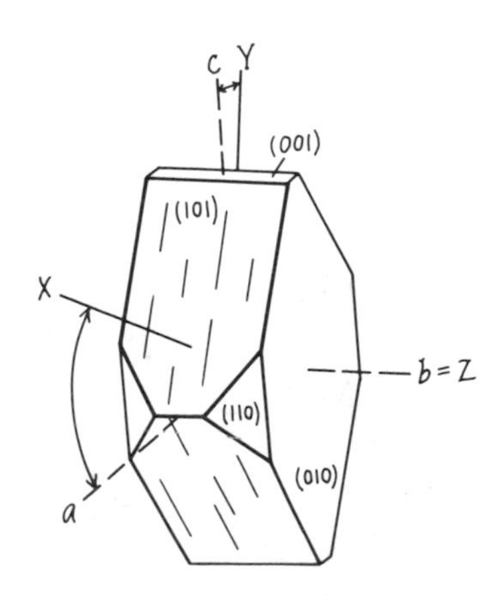

Figure 228. Most heulandite fragments lie on the perfect {010} cleavage and permit estimation of α and β. The figure is bisymmetric Bxa, with distinct crossed dispersion.

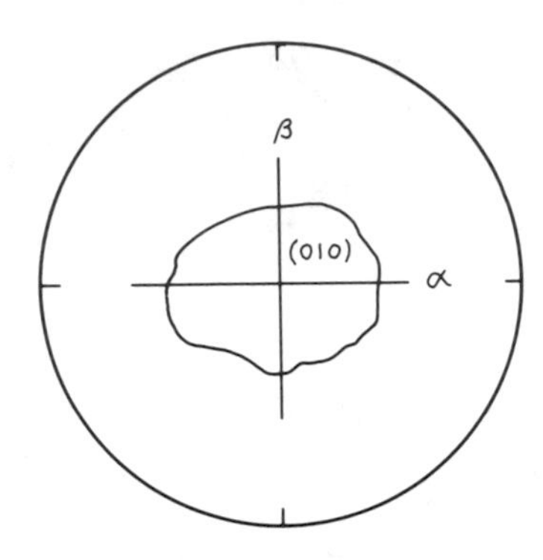

Occurrence: in cavities and veins within mafic volcanic rocks as a result of hydrothermal activity; with other zeolites; also in low-grade reactive metamorphic rocks such as vitric tuffs.
Alteration: relatively stable, but may alter to montmorillonite.
Characteristic features: occurrence; negative relief; small $2V$ (+); distinct dispersion; one perfect cleavage.
Look-alikes: stilbite is (−) and is often twinned.

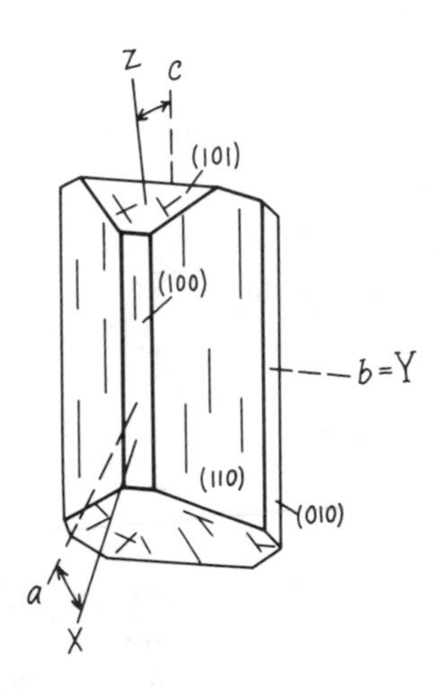

Laumontite

$Ca(Al_2Si_4)O_{12}\cdot 4H_2O$
Monoclinic
$\measuredangle\beta = 111°30'$
$\alpha = 1.502–1.514$, $\beta = 1.512–1.522$, $\gamma = 1.514–1.525$
$\gamma - \alpha = 0.008–0.016$ (up to 1st-order orange)
$2V = 25°–47°$ (−)
$b = Y$, $c \wedge Z = 8°$ to $40°$
Pleochroism and color: colorless.
Dispersion: $v > r$ strong, inclined.
Habit: columnar to prismatic on *c*, with parallel or radial arrangement.
Cleavage: {010} and {110} perfect. See Figure 229.
Extinction in section: parallel and symmetric to cleavage traces when viewed along *c*; parallel to cleavage traces when viewed along *a*, and length high (slow); inclined to cleavage traces when viewed along *b*, and length high (slow).
Twinning: {100} simple common.
Hardness: 3–4.

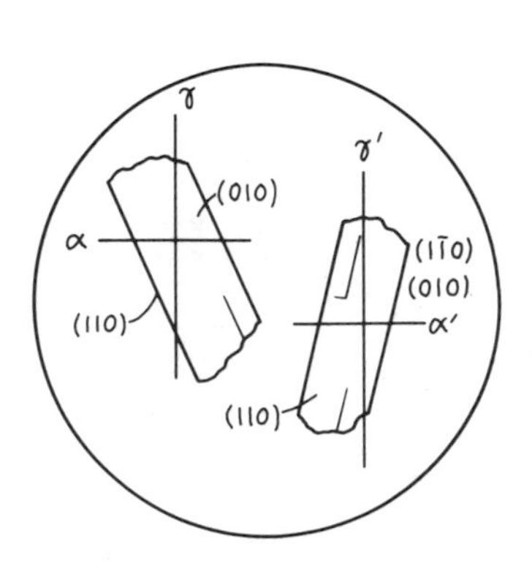

Figure 229. Laumontite fragments on {010} permit estimation of α and γ, have an inclined extinction of 8°–40° ($c \wedge Z$), and are length high (slow); the figure is bisymmetric O.N. Fragments on {110} permit estimation of α' (1.505–1.518) and γ' (1.514–1.525) and have a variable extinction angle as a function of composition; fragments are length high (slow), and the figure is generally nonsymmetric.

Specific gravity: 2.23–2.41.
Occurrence: common in cavities and veins in a wide variety of igneous and metamorphic rocks; common in low-grade metamorphic rocks with prehnite, pumpellyite, epidote, and calcite.
Alteration: stable, or with dehydration converts to leonhardite, a zeolite with lower indices, higher extinction angle, and length low (fast) on {100}.
Characteristic features: biaxial (−); inclined extinction; columnar to prismatic; negative relief; strong dispersion.
Look-alikes: heulandite and stilbite have a single perfect cleavage; scolecite is length low (fast); the zeolite phillipsite is biaxial (+).

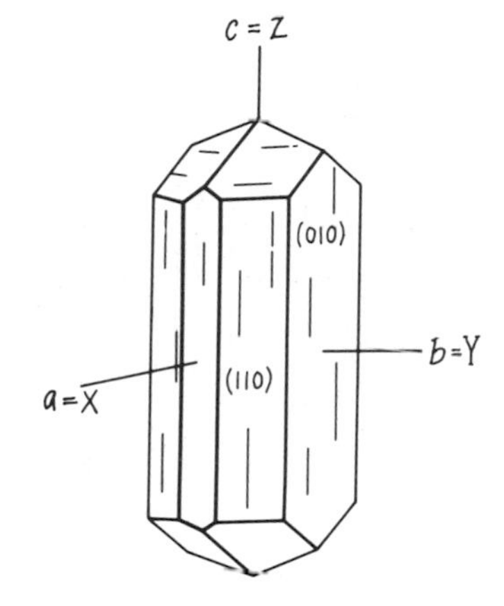

Natrolite

$Na_2(Al_2Si_3)O_{10}{\cdot}2H_2O$
Orthorhombic (pseudotetragonal)
$\alpha = 1.473–1.483$, $\beta = 1.476–1.486$, $\gamma = 1.485–1.496$
$\gamma - \alpha = 0.012–0.013$ (1st-order yellow-orange)
$2V = 58°–64°$ (+)
$a = X, b = Y, c = Z$
Pleochroism and color: colorless.
Dispersion: $v > r$ weak.
Habit: prismatic to acicular, with elongation on *c;* parallel or radial arrangement.
Cleavage: {110} perfect, {010} parting. See Figure 230.
Extinction in section: parallel to cleavage traces and dominant faces; length high (slow).
Twinning: {110}, {011}, {031} rare.
Hardness: 5.
Specific gravity: 2.20–2.26.
Occurrence: common in cavities and veins in mafic volcanic rocks and serpentinite; with other zeolites; an alteration product of

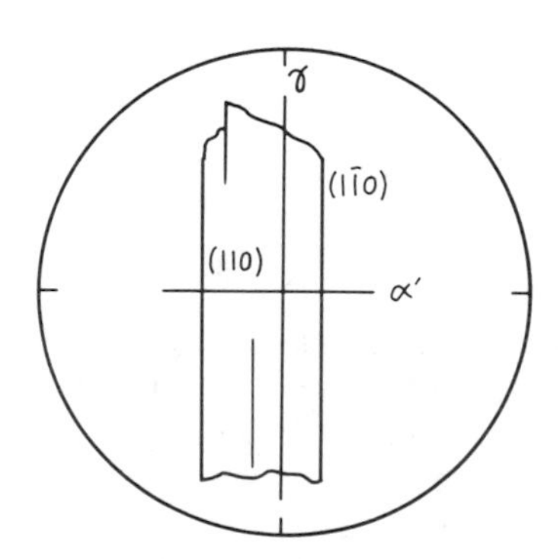

Figure 230. Natrolite fragments lie on the perfect {110} cleavage and permit estimation of α' (≃1.48) and γ. Extinction is parallel and grains are length high (slow). The figure is monosymmetric Bxo.

feldspathoids, analcime, and plagioclase in Si-undersaturated igneous rocks.

Alteration: relatively stable, but may alter to montmorillonite.

Characteristic features: two perfect cleavages with parallel extinction; negative relief; length high (slow); low birefringence; moderate $2V$ (+).

Look-alikes: scolecite has inclined extinction; thomsonite may be length low (fast).

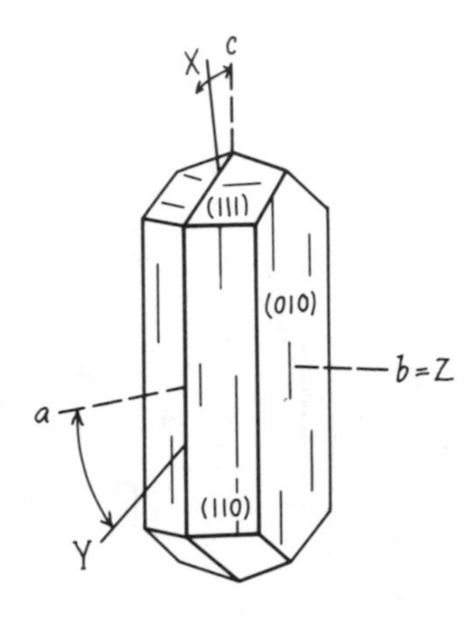

Scolecite

$Ca(Al_2Si_3)O_{10}{\cdot}3H_2O$
Monoclinic (pseudotetragonal)
$\measuredangle\beta = 90°39'$
$\alpha = 1.507–1.513$, $\beta = 1.516–1.520$, $\gamma = 1.517–1.521$
$\gamma - \alpha = 0.008–0.010$ (up to 1st-order yellow-white)
$2V = 36°–56°$ (−)
$b = Z$, $c \wedge X \simeq 18°$

Pleochroism and color: colorless.

Dispersion: $v > r$ strong, parallel.

Habit: acicular elongation on *c;* parallel or radial arrangement.

Cleavage: {110} perfect (Fig. 231).

Extinction in section: parallel or inclined to principal faces and cleavage traces when viewed normal to elongation direction; length low (fast); symmetric extinction against cleavage traces when viewed along *c*.

Twinning: {100} polysynthetic common, {110} and {001} rare.

Hardness: 5.

Specific gravity: 2.25–2.29.

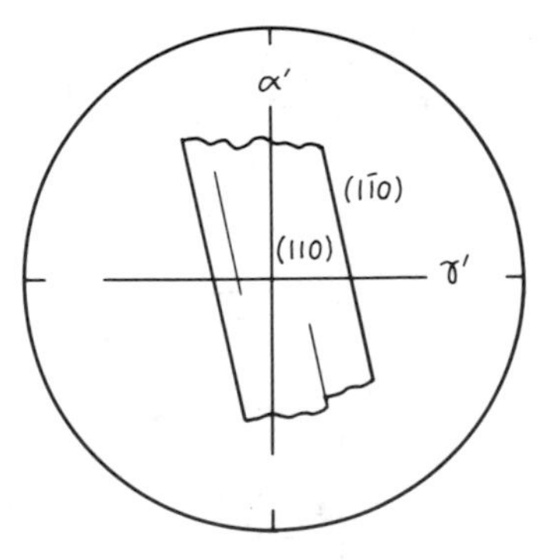

Figure 231. Scolecite fragments lie on the perfect {110} cleavage and permit estimation of α' (1.51) and γ' (1.52). Extinction is inclined, with $\alpha' \wedge c \simeq 12°$, and the grains are length low (fast). The figure is nonsymmetric.

Occurrence: typically in cavities and veins in mafic volcanic rocks with other zeolites; in some low-grade metamorphic rocks.
Alteration: relatively stable, but may alter to montmorillonite.
Characteristic features: needle-like crystals with typically inclined extinction; occurrence; negative relief; length low (fast); low birefringence.
Look-alikes: stilbite has lower indices, does not show strong parallel dispersion, and has a single perfect cleavage.

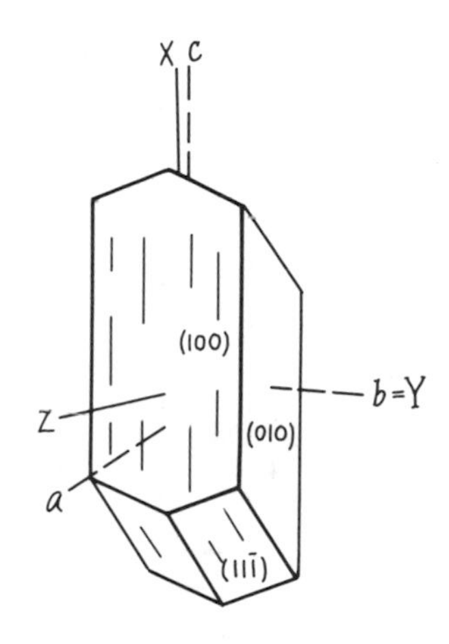

Stilbite

$(Ca,Na_2,K_2)(Al_2Si_7)O_{18}\cdot 7H_2O$
Monoclinic
$\measuredangle\beta = 129°10'$
$\alpha = 1.482-1.500$, $\beta = 1.491-1.507$, $\gamma = 1.493-1.513$
$\gamma - \alpha \simeq 0.010$ (1st-order yellow)
$2V = 30°-49°$ (−)
$b = Y, c \wedge X = 0°-7°$
Pleochroism and color: colorless.
Dispersion: $v > r$, inclined.
Habit: elongate on c and somewhat platy on {010}; often in radial aggregates.
Cleavage: {010} perfect, {100} poor. See Figure 232.
Extinction in section: slightly inclined (up to 7°) or parallel to {110} faces when viewed on {010} section; length low (fast); parallel to cleavage traces when viewed along a, and length low (fast).
Twinning: {001} cruciform penetration, very common.
Hardness: 3½–4.
Specific gravity: 2.1–2.2.

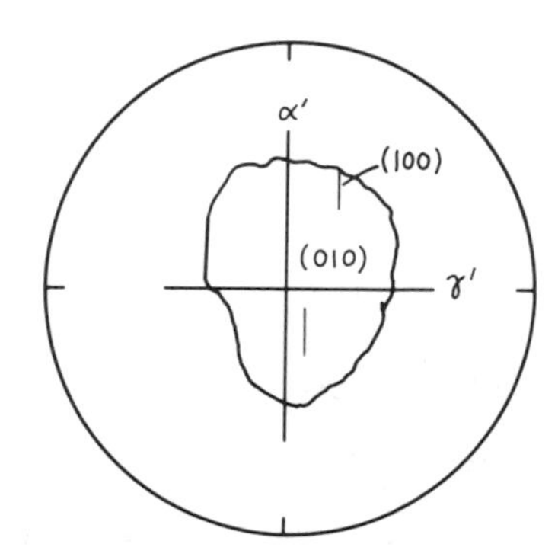

Figure 232. Stilbite fragments lie on the perfect {010} cleavage and permit estimation of α and γ. Extinction of α against traces of the poor {100} cleavage is inclined at 0°–7°, and the grains are length low (fast). The figure is bisymmetric O.N.

Occurrence: in cavities and veins in igneous rocks (particularly volcanic types) with calcite, heulandite, and other zeolites; also in thermal springs and hydrothermal veins.
Alteration: relatively stable, but may alter to montmorillonite.
Characteristic features: cruciform twins; extinction parallel or near-parallel to cleavage; negative relief; occurrence.
Look-alikes: scolecite has strong parallel dispersion, two directions of cleavage, and higher indices; heulandite is (+).

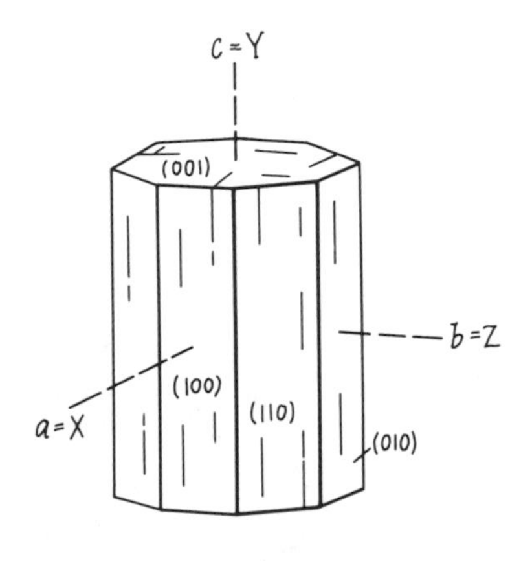

Thomsonite

$NaCa_2[(Al,Si)_5O_{10}]_2 \cdot 6H_2O$
Orthorhombic (pseudotetragonal)
$\alpha = 1.497–1.530$, $\beta = 1.513–1.533$, $\gamma = 1.518–1.544$
$\gamma - \alpha = 0.006–0.021$ (up to 2nd-order violet-blue)
$2V = 42°–75°$ (+)
$a = X$, $b = Z$, $c = Y$
Pleochroism and color: colorless.
Dispersion: $r > v$ strong.
Habit: columnar to acicular, with elongation parallel to *c*.
Cleavage: {010} perfect, {100} good. See Figure 233.
Extinction in section: parallel to cleavage traces and dominant faces; as grains are elongate on *Y*, they may be either length high (slow) or low (fast), as a function of orientation.
Twinning: {110} rare.
Hardness: 5–5½.

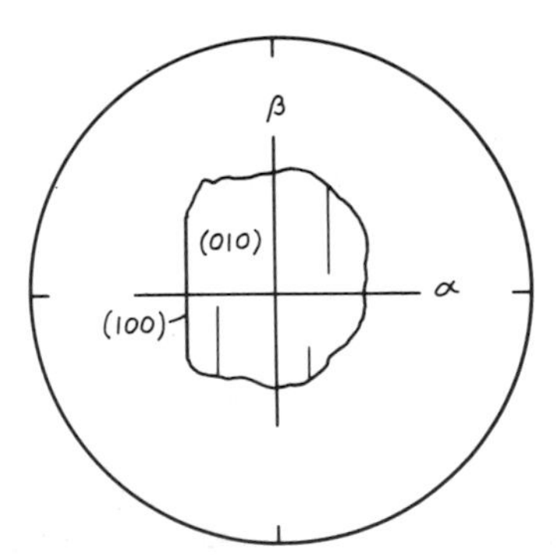

Figure 233. Most thomsonite fragments lie on the perfect {010} cleavage and permit estimation of α and β. Extinction is parallel to traces of the good {100} cleavage, and grains are length high (slow). The figure is bisymmetric Bxa. Fragments on {100} (not shown) permit estimation of β and γ. Extinction is parallel, and grains are length low (fast). The figure is bisymmetric Bxo.

Specific gravity: 2.10–2.39.
Occurrence: commonly in cavities and fractures within mafic volcanic rocks; with other zeolites.
Alteration: relatively stable, but may alter to montmorillonite.
Characteristic features: elongate crystals with parallel extinction that are length high (slow) or length low (fast) as a function of grain orientation; negative relief, low to moderate birefringence; occurrence.
Look-alikes: the zeolite *mesolite* can be similarly length high and low, but its birefringence is 0.001.

ZIRCON (Z2)

$ZrSiO_4$
Tetragonal
$\omega = 1.923–1.960$, $\varepsilon = 1.968–2.015$
$\varepsilon - \omega = 0.042–0.065$ (up to 4th-order)
Uniaxial (+); rarely biaxial (+) with $2V < 10°$
Pleochroism and color: normally colorless, but may be pale brown with weak pleochroism, abs. $E > O$.
Habit: tiny euhedral crystals with {110} dominant; may be somewhat rounded or embayed.
Cleavage: {110} poor (Fig. 234).
Extinction in section: parallel to dominant faces; length high (slow).
Twinning: {111} uncommon.
Hardness: 7½.
Specific gravity: 4.6–4.7.
Occurrence: a common accessory mineral in plutonic igneous rocks, schists, and gneisses; large crystals may be found in pegmatites; common as a detrital mineral in sediments.
Alteration: resistant to weathering; radioactive elements in zircon may partially or completely destroy its structure, producing metamict isotropic pseudomorphs, as well as pleochroic haloes within surrounding minerals.
Characteristic features: small euhedral crystals; very high relief and birefringence; length high (slow); often surrounded by pleochroic halo (color photos 23, 26, and 75).
Look-alikes: sphene is biaxial (+) and has characteristic diamond-shaped cross-section; monazite is biaxial (+); rutile and cassiterite are usually strongly colored.

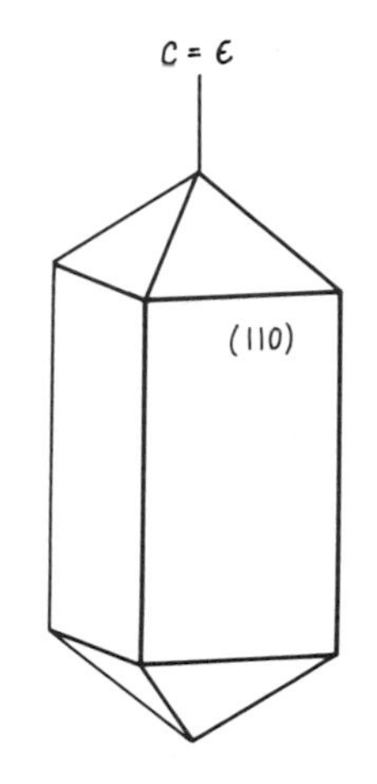

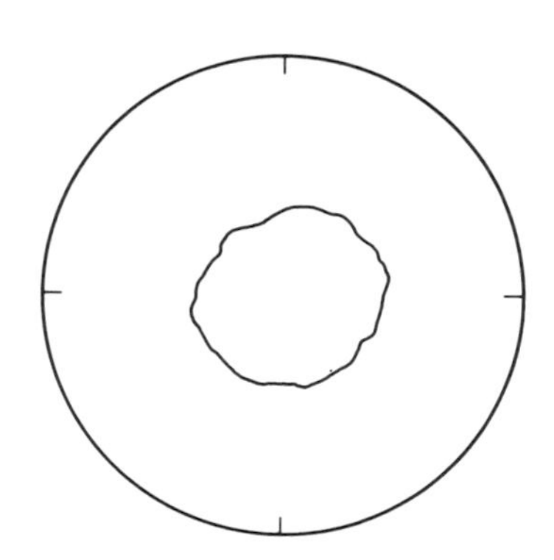

Figure 234. Zircon grains are in random orientation, permit estimation of ω and ε, and yield a variety of interference figures.

Zoisite. See Epidote Group (E1).

Appendix

DESCRIPTION OF VOLUME 1

Optical Mineralogy provides in two volumes the necessary information to identify the common rock-forming minerals in both thin sections and grain mounts by use of the petrographic (polarizing) microscope.

Volume 1 describes optical theory and techniques. The use of grain mounts is emphasized in the explanations as their use facilitates an understanding of optical theory. The described optical techniques are useful, however, for both grain mounts and thin sections.

CONTENTS FOR VOLUME 1

Bibliography

Albee, A. L. 1962. Relationships between the mineral association, chemical composition and physical properties of the chlorite series. *American Mineralogist 47,* 851–870.

Anderson, A. T., and T. L. Wright. 1972. Phenocrysts and glass inclusions and their bearing on oxidation and mixing of basaltic magmas, Kilauea volcano, Hawaii. *Journal of Petrology 57,* 188–216.

Atkins, F. G. 1969. Pyroxenes of the Bushveld intrusion, South Africa. *Journal of Petrology 10,* 222–249.

Bayliss, P. 1975. Nomenclature of the trioctahedral chlorites. *Canadian Mineralogist 13,* 178–180.

Bloss, F. D. 1978. The spindle stage: A turning point for optical crystallography. *American Mineralogist 63,* 433–477.

Bloss, F. D. 1981. *The Spindle Stage: Principles and Practice.* Cambridge, U.K.: Cambridge University Press, 340 pp.

Bohlen, S. R., and E. J. Essene. 1978. Igneous pyroxenes from metamorphosed anorthosite massifs. *Contributions to Mineralogy and Petrology 65,* 433–442.

Bonnichsen, B. 1969. Metamorphic pyroxenes and amphiboles in the Biwabik Iron formation, Dunka River area, Minnesota. *Mineralogical Society of America Special Paper 2,* 217–239.

Borg, I. Y. 1967. Optical properties and cell parameters in the glaucophane-riebeckite series. *Contributions to Mineralogy and Petrology 15,* 67–92.

Brown, G. M., and E. A. Vincent. 1963. Pyroxenes from the late stages of fractionation of the Skaergaard Intrusion, East Greenland. *Journal of Petrology 4,* 175–197.

Buchanan, D. L. 1978. A combined transmission electron microscope and electron microprobe study of Bushveld pyroxenes from the Bethal Area. *Journal of Petrology 20,* 327–354.

Burri, C., R. L. Parker, and E. Wenk. 1967. *Die optische Orientierung der Plagioklase.* Basel: Birkhauser Verlag, 334 pp.

Campbell, I. H., and G. D. Borley. 1974. The geochemistry of pyroxenes from the lower layered series of the Jimberlana Intrusion, Western Australia. *Contributions to Mineralogy and Petrology 47*, 281–297.

Davidson, L. R. 1968. Variation in ferrous iron-magnesium distribution coefficients of metamorphic pyroxenes from Quairading, Western Australia. *Contributions to Mineralogy and Petrology 19*, 239–259.

Deer, W. A., R. A. Howie, and J. Zussman. *Rock-Forming Minerals.* New York: John Wiley & Sons.
Vol. 1, 1963: *Ortho- and Ring Silicates,* 333 pp.
Vol. 1A, 1982: *Orthosilicates,* 919 pp.
Vol. 2, 1962: *Chain Silicates,* 379 pp.
Vol. 2A, 1978: *Single-Chain Silicates,* 668 pp.
Vol. 3, 1963: *Sheet Silicates,* 270 pp.
Vol. 4, 1963: *Framework Silicates,* 435 pp.
Vol. 5, 1962: *Nonsilicates,* 371 pp.

Deer, W. A., R. A. Howie, and J. Zussman. 1966. *An Introduction to the Rock-Forming Minerals.* New York: John Wiley & Sons, 528 pp.

Deer, W. A., and L. R. Wager. 1938. Two new pyroxenes included in the system clinoenstatite, clinoferrosilite, diopside, and hedenbergite. *Mineralogical Magazine 25,* 15–22.

Ehlers, E. G., and H. Blatt. 1982. *Petrology: Igneous, Sedimentary, and Metamorphic.* San Francisco, W. H. Freeman, 732 pp.

Fleischer, M., R. E. Wilcox, and J. J. Matzko. 1984. *Microscopic Determination of the Nonopaque Minerals.* U.S. Geological Survey Bulletin 1627, 453 pp.

Fyfe, W. S., and A. R. McBirney. 1975. Subduction and the structure of andesitic volcanic belts. *American Journal of Science 275-A,* 285–297.

von Gruenewaldt, G., and K. Weber-Diefenbach. 1977. Coexisting Ca-poor pyroxenes in the Main Zone of the Bushveld Complex. *Contributions to Mineralogy and Petrology 65,* 11–18.

Hartshorne, N. H., and A. Stuart. 1960. *Crystals and the Polarising Microscope.* 3d ed. London: Edward Arnold, 473 pp.

Hartshorne, N. H., and A. Stuart. 1969. *Practical Optical Crystallography,* 2d ed. New York: American Elsevier, 326 pp.

Hess, H. H. 1949. Chemical composition and optical properties of common clinopyroxenes. *American Mineralogist 34,* 621–666.

Hey, M. H. 1954. A new review of the chlorites. *Mineralogical Magazine 30,* 277–292.

Huber, N. K., and C. D. Rinehart. 1966. Some relationships between the refractive index of fused glass beads and the petrologic affinity of volcanic rock suites. *Geological Society of America Bulletin 77,* 101–110.

Huebner, J. S., 1980. Pyroxene phase equilibria at low pressure, in *Reviews in Mineralogy, Pyroxenes,* Vol. 7, ed. C. T. Prewitt, Mineralogical Society of America, 213–288.

Huebner, J. S., and A. C. Turnock. 1980. The melting relations at 1 bar of pyroxenes composed largely of Ca-, Mg-, and Fe-bearing components. *American Mineralogist 65,* 225–271.

Immega, I. P. and C. Klein, Jr. 1976. Mineralogy and petrology of some metamorphic Precambrian iron formations in southwestern Montana. *American Mineralogist 61,* 117–144.

Jaffe, H. W., P. Robinson, R. J. Tracy, and M. Ross. 1975. Orientation of pigeonite exsolution lamellae in metamorphic augite: Correlation with composition and calculated optimal phase boundaries. *American Mineralogist 60,* 9–28.

Johannsen, A. 1918. *Manual of Petrographic Methods,* 2d ed. New York: McGraw-Hill. (Reprinted by Dover Press, New York.)

Klein, C., Jr. 1964. Cummingtonite-grunerite series: A chemical, optical and X-ray study. *American Mineralogist 40,* 963–982.

Kuno, H. 1966. Review of pyroxene relations in terrestrial rocks in the light of recent experimental works. *Mineralogical Journal* (Japan) *5,* 21–43.

Larsen, E. S., and H. Bermen. 1934. *The Microscopic Determination of the Nonopaque Minerals.* U.S. Geological Survey Bulletin 848, 266 pp.

Leake, B. E. 1968. Optical properties and composition in the orthopyroxene series. *Mineralogical Magazine 36,* 745–747.

Leake, B. E. 1978. Nomenclature of amphiboles. *American Mineralogist 63,* 1023–1052.

MacKenzie, W. S., and J. V. Smith. 1956. The alkali feldspars, III. An optical and X-ray study of the high-temperature feldspars. *American Mineralogist 41,* 405–427.

Mertie, J. B., Jr. 1942. Nomonograms of optic angle formulae. *American Mineralogist 27,* 538–551.

Morse, S. A. 1968. Revised dispersion method for low plagioclase. *American Mineralogist 53,* 105–115.

Muir, I. D. 1951. Clinopyroxenes of the Skaergaard intrusion, eastern Greenland. *Mineralogical Magazine 29,* 690–714.

Nakamura, Y., and I. Kushiro. 1970. Equilibrium relations of hypersthene, pigeonite and augite in crystallizing magmas: Microprobe study of a pigeonite andesite from Weiselberg, Germany. *American Mineralogist 55,* 1999–2015.

Nobugai, K., M. Tokonami, and N. Morimoto. 1978. A study of subsolidus relations of the Skaergaard pyroxenes by analytical electron microscopy. *Contributions to Mineralogy and Petrology 67,* 111–117.

Nurse, R. W., and H. G. Midgley. 1953. Studies in the melilite solid solutions. *Journal of the Iron and Steel Institute 174,* 121–131.

Phillips, W. R. 1964. A numerical system of classification for chlorites and septechlorites. *Mineralogical Magazine 33,* 1114–1124.

Phillips, W. R., and D. T. Griffin. 1981. *Optical Mineralogy, the Nonopaque Minerals.* San Francisco: W. H. Freeman, 677 pp.

Philpotts, A. R. 1966. Origin of the anorthosite-mangerite rocks in southern Quebec, *Journal of Petrology 7,* 1–64.

Poldervaart, A. 1950. Correlation of physical properties and chemical composition in the plagioclase, olivine, and orthopyroxene series. *American Mineralogist 35,* 1067–1079.

Poldervaart, A., and H. H. Hess. 1951. Pyroxenes in the crystallization of basaltic magma. *Journal of Geology 59,* 472–489.

Ramdohr, P. 1969. *The Ore Minerals and Their Intergrowths.* New York: Pergamon Press, 1174 pp.

Ribbe, P. H., and P. E. Rosenberg. 1971. Optical and X-ray determinative methods for fluorine in topaz. *American Mineralogist 56,* 1812–1821.

Robinson, P., H. W. Jaffe, M. Ross, and C. Klein. 1971. Orientation of exsolution lamellae in clinopyroxenes clinoamphiboles: Consideration of optimal phase boundaries. *American Mineralogist 56,* 909–939.

Rosenfeld, J. L. 1950. Determination of all principal indices of refraction on difficulty oriented minerals by direct measurement. *American Mineralogist 35,* 902–905.

Shelley, D. 1985. *Manual of Optical Mineralogy,* 2d ed. Amsterdam: Elsevier, 321 pp.

Smith, J. R. 1958. The optical properties of heated plagioclases. *American Mineralogist 43,* 1179–1194.

Starkey, J. 1967. On the relationship of periclase and albite twinning to the composition and structural state of plagioclase feldspars. *Schweizerische Mineralogische und Petrographische Mitteilungen 47,* 257–268.

Tobi, A. C. 1956. A chart for the measurement of optic axial angles. *American Mineralogist 41,* 516–519.

Tobi, A. C., and H. Kroll. 1975. Optical determination of the An-content of plagioclases twinned by Carlsbad-law: A revised chart. *American Journal of Science 275,* 731–736.

Tomita, T. 1934. Variations in optical properties according to chemical composition in the pyroxenes. *Journal of the Shanghai Science Institute, Sec. 2, Geology 1,* 41–58.

Troeger, W. E. 1979. *Optical Determination of the Rock-forming Minerals,* 4th ed. English edition by H. U. Bambauer, F. Taborszky, and H. D. Trochim. Stuttgart: Schweizerbart, 188 pp.

Tsuboi, S. 1924. On the discrimination of mineral species in rocks. *Japanese Journal of Geology and Geography III,* no. 1, 19–26.

Tuttle, O. F. 1952. Optical studies on alkali-feldspars. *American Journal of Science,* Bowen Volume, 553–567.

Tuttle, O. F., and N. L. Bowen. 1958. Origin of granite in the light of experimental studies in the system $NaAlSi_3O_8$–$KAlSi_3O_8$–SiO_2–H_2O. *Geological Society of America Memoir 74*, 153 pp.
Ulbrich, H. H. 1973. Crystallographic data and refractive indices of scapolites. *American Mineralogist 58*, 81–92.
Wahlstrom, E. E. 1979. *Optical Crystallography*, 5th ed. New York: John Wiley & Sons, 488 pp.
Walker, K. R., N. G. Ware, and J. F. Lovering. 1973. Compositional variations in the pyroxenes of the differentiated Palisades Sill, New Jersey. *Geological Society of America Bulletin 84*, 89–110.
Wheeler, E. P., 2nd. 1965. Fayalite olivine in Northern Newfoundland-Labrador. *Canadian Mineralogist 8*, 339–346.
Wilson, A. F. 1976. Aluminium in coexisting pyroxenes as a sensitive indicator of changes in metamorphic grade within the mafic granulite terrane of the Fraser Range, Western Australia. *Contributions to Mineralogy and Petrology 56*, 255–277.
Winchell, A. N., and H. Winchell. 1951. *Elements of Optical Mineralogy, An Introduction to Microscopic Petrography*. 4th ed., Part II. *Descriptions of Minerals*. New York: John Wiley & Sons, 551 pp.
Winchell, H. 1958. The composition and physical properties of garnet. *American Mineralogist 43*, 595–600.

Index

Definitions and major descriptions are indicated by **bold face** page numbers.